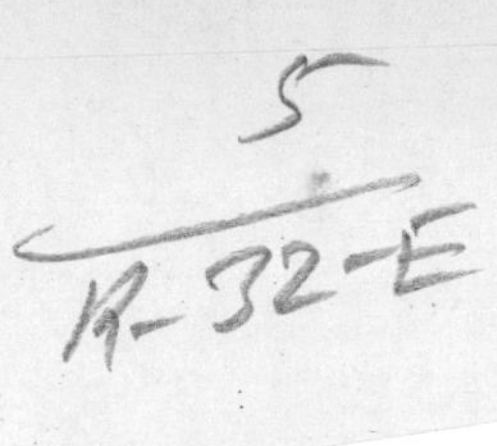

MOLECULAR PHYSICS OF PLANT PHYSIOLOGY

ADVANCES IN PLANT PHYSIOLOGY SERIES—4

Molecular Physics of Plant Physiology

D. K. Arora
Seema Gupta

Anmol Publications Pvt Ltd
NEW DELHI-110002 (INDIA)

Anmol Publications Pvt. Ltd.
4374/4B, Ansari Road, Daryaganj,
New Delhi-110002

Molecular Physics of Plant Physiology

First Edition 1996

ISBN 81-7488-238-3

PRINTED IN INDIA

Published by J. L. Kumar for Anmol Publications, Pvt. Ltd. New Delhi and printed at Chetna Offset Press, Delhi.

Preface

'Molecular Physics of Plant Physiology' offers a comprehensive account of the fundamental principles of plant physiology. It incorporates and organizes the information concerning Molecular Physics of Plant Physiology from relevant authentic sources. It is an humble attempt to present a correct and precise account of the subject matter of this important branch of Botany which forms an integral part of the studies undertaken by the undergraduate and post-graduate students of this subject. The principles have been outlined in considerable detail alongwith this fundamental facts and theories that explain the life processes of plants. The main object of the present book is to provide a comprehensive and well illustrated account of the prescribed subject matter in a very simple and lucid language and concise manner yet—still maintaining its desired academic standard.

Most of illustrations have been taken from professional papers, standard books and journals. Subject matter has also been obtained from a large number of standard books, reviews and research papers by internationally renowned workers and established authors. That is why there can be no claim of originality except in the manner of treatment.

Though the authors have taken special care to present a current account, yet they are fully aware of their limitations, and the readers may come across the mistakes of various types. For all types of mistakes they extend their due apology.

The authors express their thanks to their friends and colleagues whose continuous inspirations have invited them to bring out this text.

Contents include : functional Unit of Plants, Soil Water, Photophysiology, Solutes in Plants, Water relations of Plant Cells.

Authors

Contents

Contents

1

Functional Unit of Plants

The typical cell of the higher plants is a tiny compartment enclosed by a tough elastic wall. The wall of cells consists of two major parts : (1) the *middle lamella* and (2) the *primary wall.* In walls of many plant cells a third structural component, the *secondary wall*, is also present. Although some plant cells are known which do not have not well defined cell wall, this structure is so generally present in plant cells as to be considered one of their characteristic features.

Lining the interior of the wall and occupying more or less of the cell cavity is the *protoplasm*. The protoplasm of active cells is a transparent, slightly viscous, granular material that lacks any conspicuous structural background. It is not homogeneous, however, and contains a number of definite structures. One of these, the *nucleus*, is a denser body which is more or less spheroidal in shape and is separated from the remaining protoplasm by a definite membrane, the *nuclear membrane*. Within the membrane surrounding the nucleus are : (1) a clear liquid known as the *nuclear sap*, (2) a delicate network of denser material, the *reticulum*, and (3) one or more small spherical masses of material known as the *nucleolus* or nucleoli.

All of the protoplasm outside of the nucleus of the cell constitutes the *cytoplasm*. In a typical mature plant cell the cytoplasm is present as a thin layer lining the inner surface of the cell wall. The two boundary surfaces of the cytoplasm—that in contact with the cell wall and that in contact with vacuole—are called *cytoplasmic membranes.* Imbedded in the cytoplasm are numerous well differentiated bodies known as *plastids.* Plastids are usually centers of certain types of physiological activity. They are commonly classified on the basis of their colour into three groups : the *leucoplasts* which are colourless, the *chloroplasts* which contain the green chlorophyll pigments (also yellow pigments), and the *chromoplasts* which contain red or yellow pigments. *Chondriosomes* (mitachrondria), minute rod-like or granular bodies, are also found in the cytoplasm. The significance of these structures is not positively known, although a number of different roles have been ascribed to them.

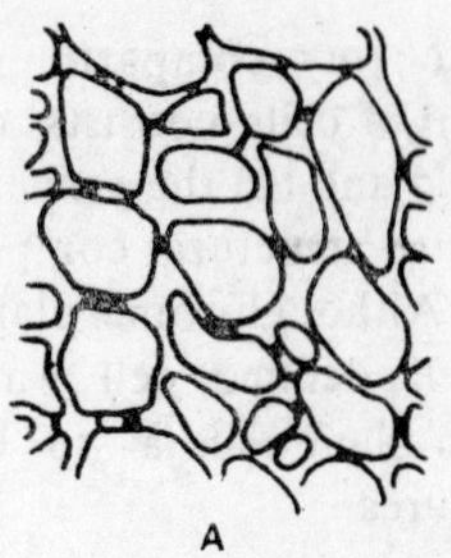

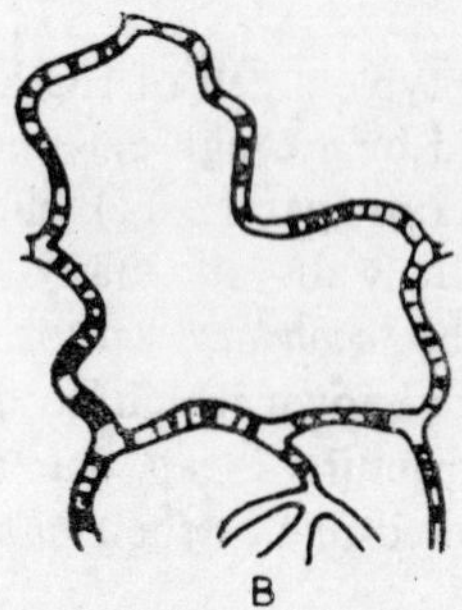

Fig. 1.1. Plasmodesms in cell walls of tobacco : (A) in sieve tubes and companion cells, (B) in epidermal cells of leaf. Redrawn from Livingston (1935).

Although the cell wall appears to imprison each protoplast and to effectively isolate it from the protoplasm of adjoining cells, actually there is probably a continuation of protoplasm from cell to cell. By certain techniques it can be demonstrated that minute pores extend from cell to cell through the cell walls. These pores often contain cytoplasmic strands which connect the cytoplasms of adjacent cells. These strands are termed *plasmodesms* (Fig. 1.1). Livingston (1935) has demonstrated the occurrence of plasmodesms in the walls of cells from a number of different tissues of the tobacco plant, and it is generally supposed that they are of widespread if not universal occurrence in the cell walls of higher green plants.

The bulk of the interior of most mature plant cells is occupied by a single large cavity, the *vacuole*, which is filled with *cell sap*. The cell sap is composed of water in which a great variety of substances are dissolved or colloidally dispersed.

There is no general agreement among cytologists regarding the exact classification of the parts of plant cells. The vacuole, for example, is frequently classified as a part of the protoplasm, because it first appears as minute droplets in the protoplasm of very young cells. Physiologically, however, the vacuole of the mature cell is as distinct an entity as the protoplasm or cell wall, and it is therefore considered as a separate part of the cell in this book. The following classification includes the principal parts of a mature plant cell. A few of these parts, as for example the various kinds of plastids, do not occur in every cell.

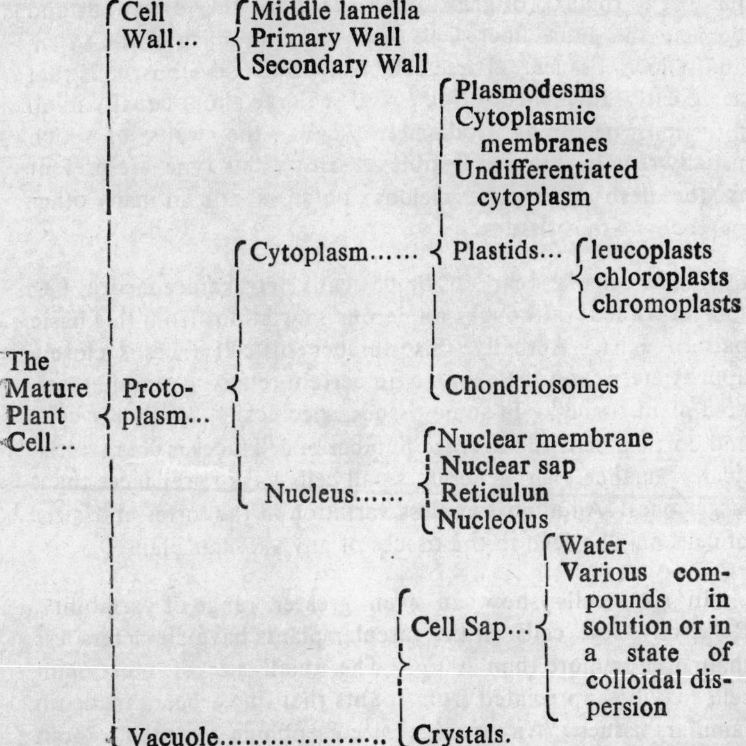

The Origin and Development of Cells. As soon as it became clear that all plant and animal tissues were composed of cells, the question of the origin of cells naturally arose. This proved to be a difficult problem for the pioneer investigators, and for many years there was much disagreement over the question. The now universally accepted principle that cells can arise only from pre-existing cells was first demonstrated beyond any reasonable doubt by Nageli just before the middle of the nineteenth century. In higher plants cell division occurs chiefly in certain restricted regions called *meristems.* The formation of new cells involves not only the division of pre-existing cells, but the subsequent enlargement and maturation of their cell progeny.

The Forms and Sizes of Cells. All newly formed cells do not differentiate morphologically in the same way. Some elongate parallel to the axis of growth more than in other directions and become the long fiber cells that make up much of the xylem and phloem tissues. These cells commonly develop walls that are greatly thickened. Some cells enlarge about equally in all directions becoming isodiametric cells, the walls of which usually remain thin and flexible. Cells of this type are present in the flesh of apples, melons, potatoes, and in many other parenchymatous tissues.

Plant cells appear to be basically tetrakaidecahedra, *i.e.*, fourteen-sided, although numerous deviations from this basic pattern exist. Actually this number of cell faces is closely approximated, on the average, in certain relatively undifferentiated plant tissues. In some tissues, specifically epidermis cells and cortical cells the average number of cell faces is characteristically smaller. In general, small cells have fewer faces than, larger ones. An almost endless variation in the forms and sizes of cells may be seen in the tissues of any vascular plant.

In size, cells show an even greater range of variability. Relatively few cells in the vascular plants have diameters less than 10 μ or more than 200 μ. The small size of most plant cells may be appreciated from counts that have been made on familiar tissues. An average sized apple leaf contains about

50,000,000 cells ; an object 1 square inch in area placed on the surface of many common leaves would cover about a million epidermal cells ; a section only 1 mm in thickness cut very near the tip of a root of the cowpea plant would contain about 128,000 cells. Some cells, however, are very much larger. Many of the cells making up the flesh of a watermelon are large enough to be seen without magnification and have a volume some 350,000 times that of the meristematic cells from which they developed. A cotton fiber is a single cell, and varieties of cotton are grown which produce fibers exceeding 4 cm. in length. The phloem fiber cells of *Boehmeria nivea* are known to reach a length of 55 cm. Such fiber cells have lengths that are many thousand times their diameters.

Origin and Development of the Cell Wall. In the vascular plants the division of a plant cell is preceded by the division of the nucleus. Nuclear division is a complicated process involving organization of the nuclear reticulum into unit chromosomes, splitting of those chromosomes, seperation of the split chromosomes, and finally reconstitution of a daughter nucleus from each set of daughter chromosomes. Just after the final stage ("telophase") of mitosis a membrane develops in the center of the cell and extends its margins until it makes contact on all sides with the existing cell wall, thus forming a septum which separates the protoplasts of the newly formed cells (Fig. 1.2). There is no evidence that this first membrane, the *middle lamella*, contains cellulose. It seems to be composed largely, if not entirely, of colloidal pectic compounds. As the cell begins to enlarge, a thin layer composed of cellulose and pectic compounds is deposited from the cytoplasm of each young cell on the two surfaces of the middle lamella. This new wall is joined firmly to the existing walls at the sides of the cell so that the protoplast of each daughter cell becomes encased within a continuous wall, the *primary wall*, of cellulose and pectic substances. As long as a cell wall retains a capacity for increase in area and for undergoing reversible changes in thickness it is classed as a primary cell wall. The walls of can bium, parenchyma, and collenchyma cells are all examples of primary cell walls.

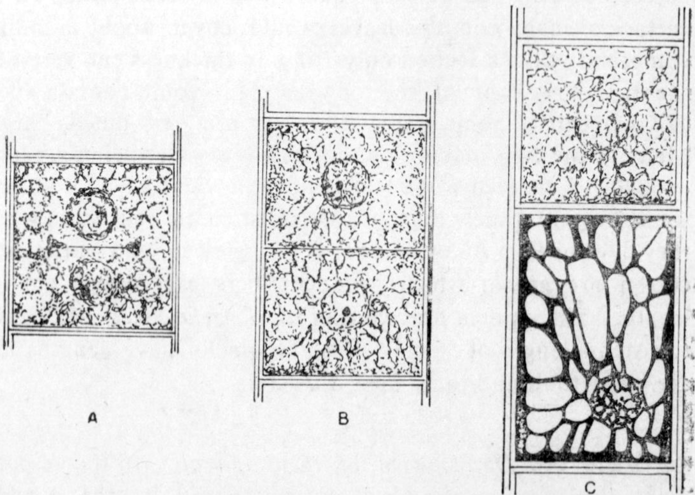

Fig. 1.2. Diagram illustrating formation of a plant cell wall. (A) Cell plate between two daughter nuclei, (B) cell plate (middle lamella) completely separating the daughter cells, (C) primary wall has been formed on each side of the middle lamella. One of the newly formed cells has undergone considerable enlargement.

After the enlargement of the cell ceases the deposition of cellulose on the inner surface of the primary wall may continue until the wall becomes conspicuously thickened. When this thickening is accompanied by a loss of the capacity for enlargement and a loss of elasticity, the added material is known as the *secondary wall.* In extreme cases secondary walls may increase in thickness until the wall occupies most of the interior of the cell. All cells have a middle lamella and a primary wall, but secondary walls are present only in certain types of cells. Phloem fibers, stone cells, tracheids and wood fibers are typical examples of cells with prominent secondary walls.

Increase in thickness of the wall usually takes place by the addition of definite layers of cellulose or other cell-wall constituents to the inner surface of the existing wall. In most walls these layers are too thin to be detected without swelling or otherwise treating the wall.

The Structure of the Cell Wall The structural framework of the plant cell wall is composed of cellulose and the basic unit of this framework is the cellulose molecule. Cellulose molecules are long chains formed by the condensation of many hundred β D-glucose molecules. Since the length of the chains is not constant, cellulose molecules are not units of definite molecular weight. Many of the cellulose molecules in the walls of living plant cells are long enough to contain at least 5000 glucose residues (a glucose residue is a molecule of β D-glucose minus a molecule of water), whereas others are much shorter.

The geometrical arrangement of the long cellulose molecules in the cell wall has been revealed by studies with the X-ray and polarized light. In certain regions the chain molecules are so closely parallel that they have acquired many of the properties of crystals. In other regions the chain molecules are less perfectly oriented giving rise to what is known as amorphous cellulose. The same cellulose molecule may be a part of a highly crystalline area, pass on into an amorphous region, and then continue on to enter another crystalline zone (Fig. 1.3). The transition between the crystalline regions

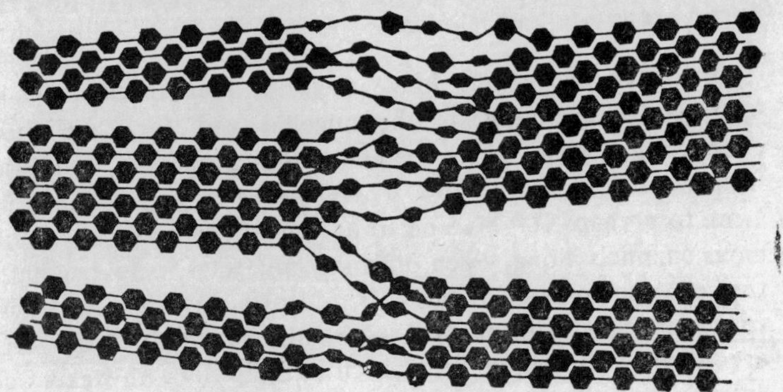

Fig. 1.3. Diagram of a small portion of a cellulose micelle, showing crystalline and amorphous cellulose.

(known as micelles or crystallites) and the amorphous areas is not abrupt so the wall is a fabric of very long chain molecules of cellulose which exhibit different degrees of parallelism in different parts of the wall.

Measurement of the X-ray diagrams of some of the natural plant fibers shows that the crystalline regions (micelles) have a rather definite thickness. A single micelle may be considered as rhombus about 5-6 mμ in thickness and at least 60 mμ in length. In terms of cellulose molecules this means that each crystalline micelle is a bundle of some 100-170 parallel molecules and that those portions of the molecule which lie within the micelle consist of at least 120 glucose residues. It is probable that in many cell walls long chain molecules formed by the condensation of other sugars are associated with the cellulose molecules.

The proportions of crystalline and amorphous cellulose in plant cell walls vary within wide limits. In the thickened walls of some plant fibers as much as 90 per cent of the cellulose is crystalline while in the enlarging walls of the young cells the crystalline cellulose makes up only a small proportion of that present. In some young cell walls crystalline cellulose appears to be entirely absent. Since the distinction between the crystalline and amorphous cellulose lies chiefly in the degree of orientation of the molecules it follows that stretching or drying a cell wall will increase, often very greatly, the proportion of crystalline cellulose.

Evidence of various kinds suggests that spaces of submicroscopic size separate the micelles and form an interconnecting system which extends throughout the cellulose' framework (Fig. 1.3 ; 1.4B). These spaces range in width from 1 mμ to perhaps 100 mμ and thus constitute relatively enormous channels along which diffusing molecules may penetrate the cellulose portions of the wall. In primary cell walls these spaces are filled with pectic compounds ; in woody tissues they are filled with lignin, and in cutinized walls with waxes and cutin. In walls that are almost pure cellulose, such as the secondary wall of cotton fibers, the spaces are filled with water.

The smallest *visible* units of cellulose walls are delicate thread-like strands of fibrils. These fibrils are aggregates of a great many molecules and micelles of cellulose both of which are oriented with their long axes parallel to the long axis of the fibril. In primary walls the fibrils form a fine anastomosing network (Fig. 1.4A), the meshes of which are filled with colloidal pectic compounds. In secondary walls the fibrils are often grouped into coarser branching strands which wind around the cell in a steep spiral the angle of which may vary in different layers and even in different parts of the same layer.

The minute details of cell-wall structure have been difficult to fathom because of the limitations imposed by even the finest

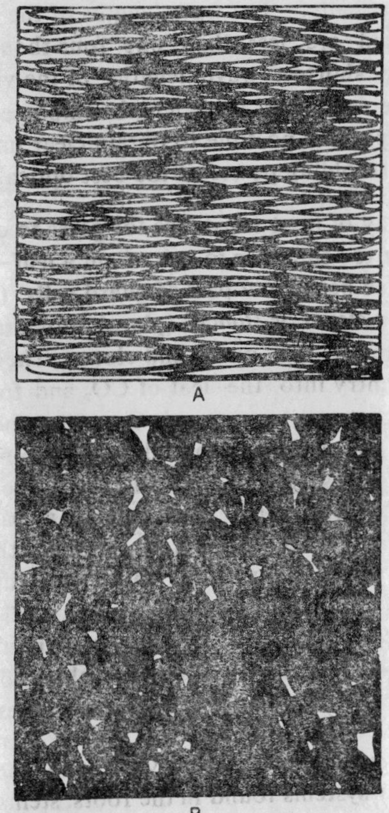

Fig. 1.4. **Diagrams illustrating submicroscopic structure of primary cell wall. (A) The meshwork of anastomosing cellulose fibrils as seen in surface view. (B) Cross section of a single fibril; showing *micelles* and intermicellar spaces.**

optical microscopes. The limits to the resolving power of an optical microscope are determined by the wave lengths of light used. The shorter the wave length, the greater the resolving power. It follows therefore that, if some method could be found to use wave lengths of radiation considerably shorter than those of the visible spectrum, it would be possible to see or photograph objects that were too small to be visible in the usual compound microscope. Such a method has been developed in the electron microscope. In this instrument beams of electrons are used instead of light, and the beams are focused magnetically. The wave lengths of the electron beams commonly used are less than one four-thousandth of the length of the shortest wave lengths of the visible spectrum. Because of the very short wave lengths employed in the electron microscope, it is possible to obtain clear images of objects far below the resolving power of the best optical microscopes.

The pathway of least resistance for gases to cross an epidermis—and thus to enter or to exit from a leaf—is through the adjustable space between a pair of guard cells. (Fig. 1.5). This pore is called a *stoma* or *stomate* (plural : stomata and stomates, respectively) and it can open and close. The stomatal pores allow for the entry into the leaf of CO_2 and for the release of the photosynthetically produced O_2. The inevitable loss of water vapour by transpiration also occurs mainly through the stomatal pores, as we will discuss. The stomata thus serve as a control, which helps strike a balance between freely admitting the CO_2 needed for photosynthesis and at the same time preventing excessive loss from the plant of the water vapour which evaporates from the walls of cells within the leaf and then diffuses out into the surrounding atmosphere. Air pollutants such as ozone (O_3) and sulphur dioxide (SO_2) also enter plants primarily through the open stomates.

Cells of vascular tissue. The xylem and the phloem make up the vascular systems found in the roots, stems, and leaves of plants (Fig. 1.5). In a tree trunk the phloem is one layer of the bark while the xylem constitutes almost all of the wood. Thus, the xylem can help support a tree or other plant. Conduction in the xylem often occurs only in the outermost annual

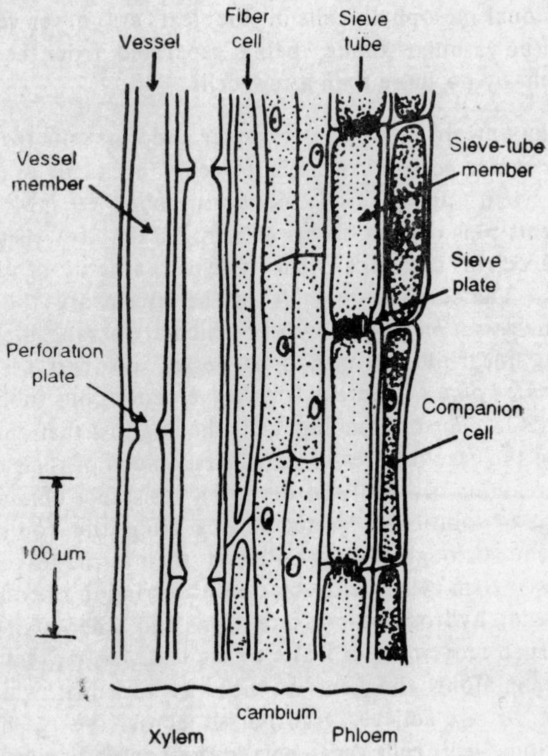

Fig. 1.5. Longitudinal section through part of a vascular bundle in a stem, showing elements of both the xylem and the phloem.

ring of a tree, while the nonfunctional phloem cells can be shed as pieces of the bark slough off. Having the phloem external to the xylem, as in a tree, is actually the general pattern for the stems of plants. As we follow the stem along a petiole and into a leaf, we observe that the xylem and phloem often form a vein, which sometimes conspicuously protrudes from the lower surface of the leaf. Reflecting the orientation originating in the stem or the trunk, the phloem is found beneath the xylem in the vascular tissue of a leaf, *i.e.*, the phloem is located on the side of the lower epidermis. The vascular system branches and rebranches as it crosses a dicotyledonous leaf, becoming smaller (in cross-section) at each step.

(Monocotyledons characteristically have parallel-veined leaves.) Individual mesophyll cells in the leaf are never very distant from the vascular tissue, being separated from the xylem or phloem by no more than a few cells.

The upward movement of water and nutrients from the soil to the upper portions of a plant occurs primarily in the xylem. The xylem sap generally contains about 10 mM inorganic nutrients plus organic forms of nitrogen that are metabolically produced in the root. The xylem is a tissue of various cell types. The conducting cells in the xylem are the *tracheids* and the *vessel members*, both of which are arranged end-to-end in long linear files. Contiguous vessel members are joined at *perforation plates*, thereby forming a continuous tube or vessel (Fig. 1.5). The conducting cells having lost their protoplasts and, in the case of vessel members, much of their end walls, the remaining cell walls border a low resistance channel for the passage of solutions. Water moves from the root, up along the stem, through the petiole and then to the leaves in these hollow xylem cells, with motion occurring in the direction of decreasing hydrostatic pressure. Ions and other solutes accompany such movement of water to the leaf. Some solutes leave the xylem along the stem on the way to a leaf, while others diffuse or are actively transported across the plasmalemmas of various leaf cells adjacent to the conducting cells of the xylem.

The movement of most *organic* compounds throughout the plant takes place in the other vascular system, the phloem. A portion of the photosynthetic products made in the mesophyll cells of the leaf diffuses or is actively transported across cellular membranes until it reaches the conducting cells of the leaf phloem. By means of the phloem, the photosynthetic products—which are then mainly in the form of sucrose—are distributed throughout the plant. The carbohydrates produced by photosynthesis and certain other substances generally move in the phloem toward regions of lower concentration, although diffusion is not the mechanism for the movement. The phloem is a tissue consisting of several types of cells. In contrast with the xylem, however, the conducting cells of the phloem contain cytoplasm (see Fig. 1.5). The conducting cells in the phloem are called *sieve cells and sieve-tube members*. They are joined end-to-end and thus form a transport system throughout the plant. Although these phloem cells often contain no nuclei at maturity, they remain metabolically active.

Root cells. Roots serve to anchor plants in the ground as well as to absorb water and nutrients from the soil and then to conduct these substances upward to the stem. In order to help understand uptake into the plant, we will examine the functional zones that occur along the length of a root.

At the extreme tip of a root is the root cap (Fig. 1.6*a*). It consists of relatively undifferentiated cells, frequently containing starch, which become scraped off as the root grows into new regions of the soil. Cell walls in the root cap often are mucilaginous in consistency, which might serve to reduce friction with soil particles. Behind the root cap is a meristematic region where the cells rapidly divide. Cells in this *apical meristem* tend to be isodiametric and have thin cell walls. Next

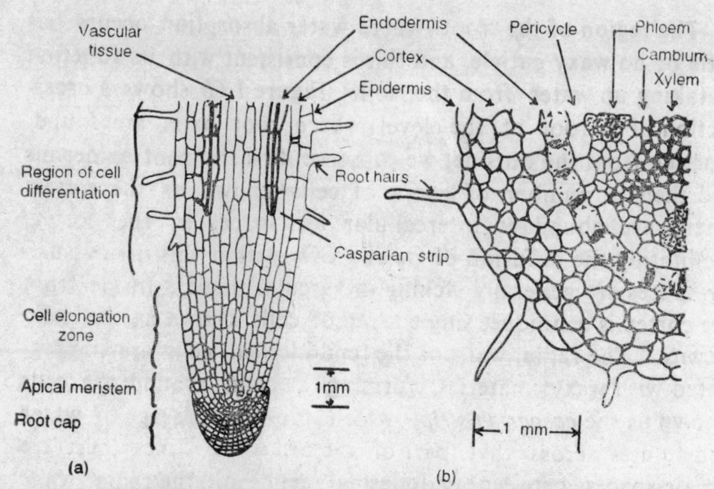

Fig. 1.6. Schematic diagrams of a root : (*a*) longitudinal section, indicating the zones which can occur near the root tip (*b*) cross-sectional view approximately 1 cm back from the tip, indicating the arrangement of the various cell types.

to this is a region of *cell elongation* in the direction of the root axis. Such elongation mechanically pushes the root tip through the soil, causing cells of the root cap to slough off by abrasion with soil particles. Sometimes the region of dividing cells is not spatially distinct from the elongation zone. Also, cell size and the extent of the zones vary with both plant species and physiological status.

The next region indicated in Figure 1.6*a* is that of *cell differentiation*, where the cells begin to assume more highly specialized functions. The cell walls become thicker and elongation is greatly diminished. The epidermal cells develop fine projections, radially outward from the root, called *root hairs*. These root hairs greatly increase the surface area across which water and nutrients can enter a plant. In fact, most of the water usually enters the root system in the region where the root hairs are located (generally a few millimeters or centimeters from the tip). As we follow the root upward, its surface becomes less permeable to water and its interior more involved with conducting water upward to the stem.

The region of the root where water absorption occurs has little or no waxy cuticle, and this is consistent with its function of taking up water from the soil. Figure 1.6*b* shows a cross-section of a root at the level where root hairs are found. Starting from the outside, we observe first the root epidermis and then a number of layers of cells known as the *cortex*. There are abundant intercellular air spaces in the cortex, facilitating the diffusion of O_2 and CO_2 within this tissue (such air spaces are generally lacking in vascular tissue). Inside from the cortex is a unique, single layer of cells known as the *endodermis*. The radial walls of the endodermal cells are impregnated with waxy material, forming a band around the cells known as the *casparian strip*, which prevents passage of water and solutes across that part of the cell wall. Since there are no air spaces between endodermal cells, and the radial walls are blocked by the waterproof casparian strip, water must pass through the lateral walls and enter the cytoplasm of endodermal cells in order to continue across the root. The endodermal cells represent the only instance in the entire pathway for water movement from the soil, through the plant, to the air, where it is necessary that the water enter a cell's cytoplasm. In the rest of the pathway water can move in cell walls or in the hollow lumens of xylem vessels.

Inside from the endodermis is a layer called the *pericycle*, which is typically one cell thick in angiosperms. The cells of the pericycle can divide and form a meristematic region, which in turn can lead to lateral or branch roots in the region just above the root hairs. Inside from the pericycle is the vascular tissue. The phloem generally occurs in two to eight or more strands located around the root axis (four strands, or a *tetrarch* arrangement, is illustrated in Fig 1.6*b*). The xylem often occurs along the root axis with sections radiating out between the phloem strands. Water does not therefore have to cross the phloem in order to each the root xylem. The tissue between the xylem and the phloem is the *cambium*, whose cells, upon division and differentiation, lead to xylem on the inside or phloem on the outside.

The endodermis surrounds the pericycle and vascular tissue for only a short length of the root. It does not begin until after the zone of cell elongation (Fig. 1.6*a*) and until a few phloem cells have differentiated. It breaks up just above the region of root hairs, where the root begins to widen appreciably as secondary xylem and secondary phloem are formed by the radial division of cambial layer cells. Thus, the main region for water entry into most roots is the zone of primary vascular tissue where there is an intact sleeve of endodermal cells. Toward the stem from this region the vascular tissue becomes larger in cross-section, the xylem is more devoted to conduction of water upward, and the phloem is needed to supply all the branch roots which have by then come together.

Our rather elementary discussion of leaves, vascular tissues, and roots leads to the following oversimplified but useful picture. The roots take up water from the soil along with nutrients required for growth. These are then conducted in the xylem to the leaves. Leaves of a photosynthesizing plant lose the water to the atmosphere along with a release of O_2

and an uptake of CO_2. Carbon from the latter ends up in photosynthetate translocated in the phloem back to the root. Thus the xylem and the phloem serve as the "plumbing" which connects the two types of plant organs that functionally interact with the environment. To understand the details of such physiological processes we must turn to fields like calculus, thermodynamics, and photochemistry for the analytical methods. Our next step is to bring the abstract ideas of these fields into the realm of cells and plants, which means we need to make calculations using appropriate assumptions and approximations.

Membrane permeability

With the knowledge of the general structure of membranes, let us turn to a quantitative analysis of the interactions between membranes and diffusing solutes. The rate-limiting step for movement of many molecules into and out of plant cells is diffusion through the plasmalemma. Because of the close packing and the interactions between the component molecules of a membrane, such diffusion is greatly restricted compared with the relative ease of movement in an aqueous phase like the cytoplasm, *i.e.*, the solute has great difficulty in threading its way between the molecules composing the membrane. The average diffusion coefficient of a small solute in a membrane is often about 10^6 times lower than in the adjacent aqueous solutions. Also, the membrane represents a different type of molecular environment than does an aqueous solution, so that the solubility of a species in the two phases must be taken into account in order to describe membrane permeability. As will be shown, Fick's first law can be modified to describe the diffusion of molecules across a membrane. Once past the barrier presented by the membrane, the molecules may be distributed throughout the cell relatively rapidly by diffusion, and over greater distances by cytoplasmic streaming.

Concentration difference across a membrane. The driving force for diffusion of uncharged molecules into or out of plant cells can be regarded as the negative of the concentration gradient of that species across the plasmalemma. Since the actual concentration gradient of some species j is not known in the plasmalemma (or in any other membrane for that matter), the driving force is generally approximated by the negative of the average gradient of that species across the membrane :

$$\frac{\partial c_j}{\partial x} \equiv -\frac{\Delta c_j}{\Delta x} = \frac{-c_j^i - c_j^o}{\Delta x} = \frac{c_j^o - c_j^i}{\Delta x} \qquad \ldots(1)$$

where c_j^o is the concentration of species j outside the cell, c_j^i is its concentration in the cytoplasm, and Δx is the thickness of the plasmalemma, which acts as the barrier restricting the penetration of the molecules into the cell (Fig. 1.7). The concentrations c_j^o and c_j^i appearing in Equation 1 can also represent values in the phases separated by any other membrane, such as the tonoplast. If c_j^i is less than c_j^o as we move in the positive x-direction (Fig. 1.7), $\partial c_j/\partial x$ is then

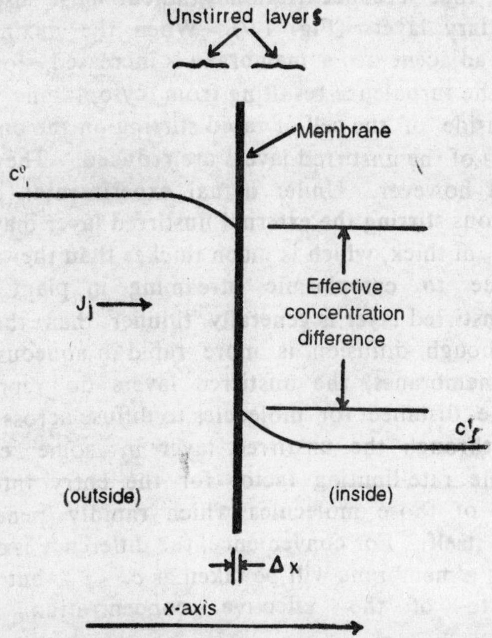

Fig. 1.7. The effect of unstirred layers on the concentration of species j near a membrane across which the solute is diffusing.

negative and so $-\partial c_j/\partial x$ is positive; the flux J_j would then occur in the positive x-direction ($J_j = -D_j\, \partial c_j/\partial x$). Also, if we designate a net flux into a cell as a positive flux—which is the commonly used convention—then the minus sign in $-\partial c_j/\partial x$ is incorporated into the concentration difference $(c_j^o - c_j^i)$ that we must use to describe the flux, viz., $J_j = D_j(c_j^o - c_j^i)/\Delta x$.

The magnitude of what may be called the "effective" concentration difference across a membrane is made somewhat uncertain by the existence of unstirred boundary layers (Fig. 1.7), as has been pointed out by J. Dainty. Boundary layers in which there is no turbulent mixing occur at the interface between any fluid (liquid or gas) and a solid, such as the unstirred air layers on either side of a leaf. A substance moves across these unstirred layers next to the membrane or other solid not by mechanical mixing but by diffusion, indicating that a concentration gradient must also exist in the boundary layers (Fig. 1.7). When the mixing in the solutions adjacent to a membrane is increased—for example, either by the turbulence resulting from cytoplasmic streaming on the inside of the cell or rapid stirring on the outside—the thicknesses of the unstirred layers are reduced. They are not eliminated however. Under actual experimental conditions with vigorous stirring the external unstirred layer may be from 10 to 100 μm thick, which is much thicker than the membrane itself. Due to cytoplasmic streaming in plant cells, the internal unstirred layer is generally thinner than the external one. Although diffusion is more rapid in aqueous solutions than in membranes, the unstirred layers do represent an appreciable distance for molecules to diffuse across. In fact, diffusion through the unstirred layer in some cases can become the rate-limiting factor for the entry into cells or organelles of those molecules which rapidly penetrate the membrane itself. For convenience, the difference is concentration across a membrane will be taken as $c_j^i - c_j^o$, but this is an overestimate of the effective concentration difference (Fig. 1.7).

The difference in concentration actually determining the diffusion of molecules across a membrane is the concentration just inside one side of the membrane minus that just within the other side. In Equation 1 the concentrations are those in the aqueous phases outside on either side of the membrane. Since membranes are quite different as solvents compared with aqueous solutions, the concentrations just inside the membrane can differ appreciably from those just outside in the aqueous solution. A correction factor must therefore be applied to Equation 1 to give the actual concentration gradient existing in the membrane. This factor is known as the *partition coefficient*, K_j. It is defined as the ratio of the concentration of a solute in the material of the membrane to that in equilibrium outside it in the aqueous phase (c_j^o or c_j^i). As a guide to the situation in the membrane, the partition coefficient is generally determined by measuring the ratio of the equilibrium concentration of some solute in a lipid phase—such as olive oil, which mimics the membrane lipids—to the concentration in an adjacent and immiscible aqueous phase (which mimics the solutions on either side of the membrane). This rather simple convention for obtaining partition coefficients is based on the high lipid content of membranes and the numerous experimental results which show that the relative ease of penetration of a membrane depends on the lipid solubility of the molecules Partition coefficients have a wide range of values, most being between 10^{-6} and 10. For example, an inorganic ion might have a K_j near 10^{-5} for membranes, while a nonpolar hydrophobic substance might have a value near 1. Usually K_j is assumed to be the same on the two sides of the membrane. These considerations lead us to write $K_j(c_j^o - c_j^i)$ for the concentration difference leading to diffusion across a membrane, where K_j is characteristic of a particular substance.

Permeability coefficient. We will next use Fick's first law ($J_j = -D_j \partial c_j / \partial x$) to obtain an expression that is convenient for describing the movement of a substance across a membrane. The negative concentration gradient will be replaced by $(c_j^o - c_j^i)/\Delta x$ (Eq. 1. Moreover, since D_j is the diffusion coefficient of species j within the substance of the membrane, we must use the actual concentration drop within the membrane, $K_j(c_j^o - c_j^i)$. These various considerations lead us to the following expression describing the diffusion of species j :

$$J_j = D_j \frac{K_j(c_j^o - c_j^i)}{\Delta x} = \frac{D_j K_j}{\Delta x}(c_j^o - c_j^i)$$

$$= P_j(c_j^o - c_j^i) \qquad \ldots(2)$$

where P_j is called the *permeability coefficient* of species *j*.

Since permeability coefficients have so many applications in physiology, we will define P_j explicitly :

$$P_j = \frac{D_j K_j}{\Delta x} \qquad \ldots(3)$$

which follows directly from Equation 2. The permeability coefficient conveniently replaces three quantities which can also be used to describe the diffusion of some solute across a membrane or other barrier. In this regard, the partition coefficients of membranes are generally determined by using a lipid phase like olive oil or ether, not the actual lipids occurring in membranes, and thus any such K_j will have some uncertainty. Also, measurements of Δx and D_j for membranes tend to be rather indirect. On the other hand, P_j is a single, readily measured quantity characterizing the diffusion of some solute across a membrane or other barrier. The units of P_j are length per time, *e.g.*, cm/sec.

Instead of calculating the flux of some species into a cell—the amount entering per unit area per unit time—we often focus our attention on the total amount of that species entering within a certain time interval. Let s_j be the amount of species *j* inside the cell, where a can be expressed in moles. If that substance is not involved in any other reactions, ds_j/dt represents the rate of entry of species *j* into the cell. The flux J_j is the rate of entry of the substance per unit area or $(1/A)(ds_j/dt)$, where A is the area of the cellular membrane across which the substance is diffusing. Using this expression for the flux, we can express Equation 2 as follows :

$$ds_j/dt = J_j A = P_j A(c^o_j - c_j^i) \qquad \ldots(4)$$

When the external concentration of species *j* (c_j^o) is greater than the internal one (c_j^i), species *j* will enter the cell and ds_j/dt will then be positive, as Equation 4 indicates. This equation is actually very useful for considering many experimental investigations of the diffusion of solutes into and out of plant cells.

Diffusion and cellular concentration. A question that often arises in cellular physiology is how the internal concentration of a penetrating solute changes with time. How long, for example, would it take the internal concentration of an initially absent species to build up to 70% of the external concentration ? In order to solve the general case, we must first express s_j in Equation 4 in a suitable fashion and then integrate the equation. It is useful to introduce the approximation—particularly appropriate to plant cells—that the cell volume does not change during the time interval of interest. In other words, because of the rigid cell wall, we will assume that the plant cell has a nearly constant volume (V) during the entry or exit of the solute being considered.

The average internal concentration of species j, c_j^i, is equal to the amount of that particular solute inside the cell (s_j) divided by the cellular volume (V), or s_j/V. Therefore, s_j can be conveniently replaced by Vc_j^i, and thus ds_j/dt in Equation 4 equals Vdc_j^i/dt when we can assume that the cell volume does not change with time. We are also assuming that species j is not produced or consumed by any reaction within the cell, so that the diffusion of species j into or out of the cell is the only process affecting c_j^i. For simplicity, we are treating the concentration of species j as if it were uniform within the cell. In a more complex case we might need a relation of the form of Equation 4 for each membrane-surrounded compartment. Moreover, we are presupposing either mechanical mixing or rapid diffusion inside the cell, so that the drop in concentration $(c_j^o - c_j^i)$ occurs essentially only across the membrane. Finally, let us assume that P_j is independent of concentration, a condition that is often satisfactorily met over a limited range of concentration. We can now proceed with the integration of Equation 4. Upon replacement of ds_j/dt by Vdc_j^i/dt, rearrangement in order to separate variables and the insertion of integral signs, Equation 4 becomes.

$$\int_{c_j^i(0)}^{c_j^i(t)} \frac{dc_j^i}{c_j^o - c_j^i} = \frac{P_j A}{V} \int_0^t dt \qquad \dots(5)$$

where $c_j^i(0)$ is the initial internal concentration of species j, *i.e.*, when $t=0$, and $c_j^i(t)$ is that at a later time t.

The volume outside certain unicellular algae and other membrane surrounded entities can be quite large compared with V. In such cases the essenal concentration (c_j^o) does not change appreciably. Such an approximation can also be appropriate for experiments of short duration or where special arrangements are made to maintain c_j^o at some fixed value. For those cases where c_j^o is constant, Equation 5 can be integrated to give the following expression :

$$\frac{P_j A t}{V} = \ln \frac{c_j^o - c_j^i(0)}{c_j^o - c_j^i(t)} \qquad \dots(6)$$

Starting with a known c_j^o outside a cell and determining the internal concentration both initially, *i.e.*, for $t=0$, and at some subsequent time t, we can calculate P_j from Equation 6 if A/V is known. Even when A/V is not known, the relative permeability coefficients for two substances can be determined. The above derivation can easily be extended to the case where c_j^o is zero. In that case, $P_j At/V$ equals in $[c_j^i(0)/c_j^i(t)]$, an expression that can be used to describe the diffusion of a photosynthetic product out of a chloroplast or some substance from a cell into a large external solution initially devoid of that species. Such studies can be greatly simplified by using radioactive tracers.

When diffusion occurs from all directions across a membrane surrounding a cell (or an organelle), the time to reach a given internal concentration can be directly proportional to the dimensions of the cell. For convenience of calculation in illustrating this point, let us consider a spherical cell of radius r. The volume V then equals $(4/3)\,\pi r^3$ and the surface area A is $4\pi r^2$. Hence, V/A is $r/3$. In the case of two spherical cells having the same permeability coefficient for some species, Equation 6 indicates that the time to reach a given concentration inside is *linearly* proportional to the radius, *i.e.*, t equals $(r/3P_j) \ln [c_j^o - c_j^i(0)]/[c_j^o - c_j^i(t)]$. For the same P_j, a given level inside a small cell is reached sooner than it would be reached inside a large cell. However, when diffusion across the membrane into the cell occurs from all directions, the ratio of the times to reach a given internal concentration for these two cells equals the ratio of the radii (assuming a uniform c_j^i). On the other hand, the time that it takes for a planar front to diffuse in one direction is proportional to the square of the distance traveled $(x_e^2 = 4D_j t)$.

A typical permeability coefficient for a small nonelectrolyte or uncharged molecule, *e.g.*, isopropanol, phenol, is 10^{-4} cm/sec for the plasmalemma while P_j for a small ion, *e.g.*, K^+, Na^+, might be about 10^{-7} cm/sec. The lower permeability co-efficients for charged particles are mainly because of the much lower partition coefficients of these solutes compared with nonelectrolytes which tend to be hydrophobic (recall that $P_j = D_j A/\Delta x$, ...). In other words, due to its charge, the electrolyte tends to be much less soluble in a membrane than is a neutral molecule, so that the effective concentration gradient driving the charged species across the membrane is generally smaller for given concentrations in the aqueous phases on either side of the barrier. By comparison, water can rapidly enter or leave cells, and often has a permeability coefficient of about 10^{-2} cm/sec. (Actually, it is hard to measure P_w because of the relative importance of the unstirred layers in this case.

Using Equation 6, we can calculate the time required for a substance initially absent from a cell $[c_j^i(0) = 0]$ to achieve an internal concentration equal to half the external concentration $[c_j^i(t) = \frac{1}{2}c_j^o]$. For spherical cells, V/A is $r/3$ as we mentioned above, and the conditions on the concentrations mean that $[c_j^o - c_j^i(0)]/[c_j^o - c_j^i(t)]$ equals $(c_j^o - 0)/(c_j^o - \frac{1}{2}c_j^o)$, which is 2. Substituting these relations into Equation 6 indicates that the time needed is $(r/3P_j) \ln 2$, where r is the radius of the spherical cell, P_j is the permeability coefficient of the solute being considered, and ln 2 equals 0.693. For a cell 50 μm in diameter—which is a reasonable dimension for spongy mesophyll cells in certain leaves (Fig. 1.5)—the time for an initially absent species to reach half of the external concentration is $\{(25 \times 10^{-4}\ \text{cm})/[(3)(10^{-4}\ \text{cm/sec})]\}$ (0.693), or 6 sec for a nonelectrolyte with a P_j of 10^{-4} cm/sec, and 1000 times longer (2 hr) for small electrolytes having a P_j near 10^{-7} cm/sec. Hence, the external membrane is an extremely good barrier for electrolytes, markedly hindering their entry into, or egress from the cell. On the other hand, small non-electrolytes can fairly readily diffuse in and out of plant cells, of course depending to a large extent on their relative lipid solubility.

Cell walls

The rigidity of the cell wall helps determine the size and the shape of the cell and ultimately that of the plant. This supportive role is performed in conjunction with the internal hydrostatic pressure, which causes a distension of the cell walls. The cell wall is also intrinsically involved in many aspects of the ion and water relations of a plant. Since it surrounds the plasmalemma of each cell, all fluxes of water and solutes into or out of a plant cell protoplast must cross the cell wall, usually by diffusion. Water which evaporates from a plant during transpiration comes directly from cell walls. The space surrounded by the cell wall in certain specialized cells can act as a channel through which solutions move. In the xylem, for example, the conducting cells have lost their protoplasts with the result that the pathway for the conduction of solutions in the xylem consists essentially of hollow pipes, or conduits, made exclusively of cell walls.

Cell walls vary from tenths to tens of microns in thickness, and generally are divided into three regions. The *primary cell wall* surrounds dividing meristematic cells as well as the elongating cells during the period of cell enlargement. The cell wall often becomes thickened by the elaboration of a *secondary cell wall* inside the primary one (Fig. 1.8), which makes the cell mechanically much less flexible. Hence, cells whose walls have under gone secondary unexning such as in xylem vessel members, are generally not capable of subsequent elongation. The cell wall also includes an amorphous region between contiguous cells called the *middle lamella*, the extent of which is not universally agreed upon. Although cellulose also occurs there, the middle lamella may be composed mainly of the calcium salts of pectic acids, the presence of which causes adjacent cells to adhere together.

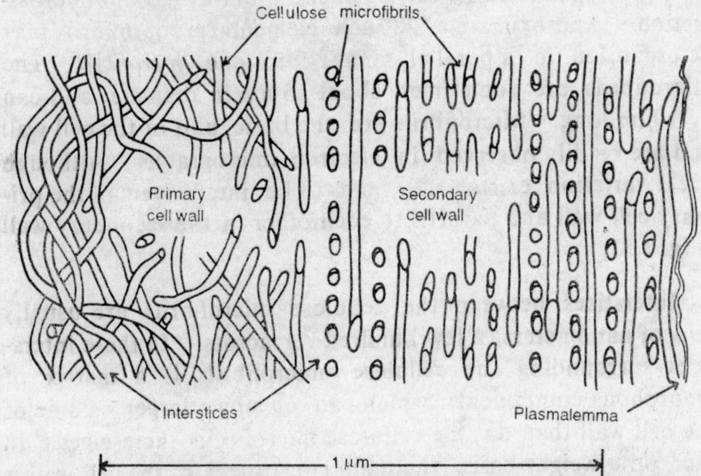

Fig. 1.8. Hypothetical thin section through a cell wall, indicating the cellulose microfibrils in the primary and secondary cell walls. The interstices are filled with noncellulosic material including appreciable water, the solvent for solute diffusion across the cell wall.

Chemistry and morphology. Cellulose is the most abundant organic component in the entire plant and animal kingdoms. It is the characteristic substance of the plant cell wall, constituting from 25 to 50% of the cell wall organic material. Cellulose is a linear (unbranched) polysaccharide consisting of 1.4-linked β-D-glucopyranose units :

The polymer is about 0.8 nm (8 Å) in its maximum width, 0.33 nm² (33 A°) in cross-sectional area, and often contains

8,000 to 12,000 glucose residues. Since the glucose residues have their rings in the same plane, the cellulose polymer is ribbon-like. These polymers are organized into various entities in the cell wall. Approximately 100 parallel chains of cellulose often form a structural unit known as an *elementary microfibril.* Elementary microfibrils can be about 4 nm by 8.5 nm in cross-section. Approximately 20 such elementary microfibrils may be organized in a parallel array, into one *microfibril.* The microfibrils are sometimes about 25 nm in diameter and can be 10μm long. Microfibrils are the basic unit of the cell wall and are readily observed in electron micrographs. Although great variation exists, they tend to be interwoven in the primary cell wall and parallel to each other in the secondary wall (Fig. 1.8).

Interstices between the cellulose microfibrils are usually tens of nanometers (a few hundred Å) across. In these interstices surrounding the cellulose microfibrils is a matrix of amorphous components which can occupy a larger volume of the cell wall than do the cellulose microfibrils themselves. In fact, on a weight basis, the main constituent of the cell wall is actually water.

The cell wall matrix contains lignins, noncellulosic polysaccharides such as pectin, a small amount of protein, bound and free water, appreciable calcium, other cations, and sometimes silicates. Lignins are complex polymers based on phenylpropanoid subunits (a 6-carbon ring, to which is attached a 3-carbon chain as in phenylalanine, a lignin precursor) plus certain other residues. Lignins constitute the second most abundant class of organic molecules in living organisms and are about half as prevalent as cellulose. They are quite resistant to enzymatic degradation and consequently are important in peat and coal formation. Moreover, since the plant itself cannot break down the lignin polymers, cells are unable to expand after extensive signification of their cell walls, as occurs in the secondary walls of xylem vessel members. Pectin consists primarily of 1,4-linked α-D-galacturonic acid residues, the carboxyl groups of which are normally dissociated and have a negative charge (galacturonic acid differs from galactose simply by having $-COOH$ instead of $-CH_2OH$ in the 6-carbon position). The negative charge of the dissociated carboxyl groups leads to the tremendous cation-binding capacity of cell walls. In particular, much of the divalent cation calcium (Ca^{++}) is bound, which may help link the various polymers together. Polymers based on 1,4-linked β-D-xylopyranose units (xylans), as well as on many other residues, can also be extracted from cell walls. They are loosely referred to as hemicelluloses. *e g.*, xylans, mannans, galactans, and glucans. In general, hemicelluloses tend to have a low molecular weight (in the ten thousands) compared with pectin (about five to ten times larger) or cellulose (up to 1000 times larger). The presence of the negatively charged pectins hinders the entry of anions into plant cells.

Diffusion across cell walls. How would we calculate the case with which molecules might diffuse across a cell wall ? A good place to start is with Equation 3 which indicates that the permeability coefficient of species j equals $D_jK_j/\Delta x$. We must first consider what we mean by D_j and K_j in a cell wall. The diffusion of solutes across a cell wall occurs mainly in the water located in the numerous interstices, which are often about 10 nm in diameter. Thus, the movement from the external solution up to the plasmalemma can be in aqueous channels through the cell wall. By the definition of a partition coefficient (solubility in barrier/solubility in water), we recognize that K_j would be unity for the water-filled interstices of the cell wall, it would be very low for the solid phases (cellulose, lignin, and the other polymers present in the wall), and the partition coefficient would have some intermediate value for the cell wall as a whole. Since diffusion takes place primarily in the cell wall water, we will let K_j be unity. If we use this convenient definition for K_j in cell walls, we must be careful how we define D_j. In particular, diffusion coefficients will have to be expressed on the basis of the whole cell wall, and not on the much smaller area presented by the interstices. The value of D_j in the water of the interstices could be very similar to that in a free solution, while its value averaged over the whole cell wall would be considerably less. Such effective diffusion coefficients of many small solutes are anywhere from ten to one hundred times less when averaged over the cell wall than are the D_js of the same species in an extended aqueous solution. Thus, the diffusion coefficients we will use here can be as high as about 10^{-6} cm^2/sec for small solutes diffusing across cell walls.

Next, let us calculate the permeability coefficient for a solute which has a diffusion coefficient of 10^{-6} cm^2/sec for that cell wall. We will assume a representative value of 2 μm for the cell wall thickness. Using Equation 3 ($P_j = D_j K_j / \Delta x$), we can thus estimate that P_j for the cell wall might be $[(10^{-6}\ cm^2/sec)(1)]/(2\times10^{-4}\ cm)$, or 5×10^{-3} cm/sec. Most of the permeability coefficients for small solutes crossing the plasmalemma range from 10^{-8} to 10^{-4} cm/sec. Hence, a cell wall generally has a higher permeability coefficient than does a membrane, which means that the wall is usually more permeable for small solutes than is the plasmalemma. For comparison, let us consider a permeability coefficient that might be appropriate for an unstirred liquid layer adjacent to a cell wall or membrane. Specifically, D_j for a small solute may be 10^{-5} cm^2/sec in water, K_j is unity in the aqueous solution and, let us assume, Δx is 30 μm for the unstirred layer. In such a case, P_j would be $(10^{-5})(1)/(30\times10^{-4})$ or 3×10^{-3} cm/sec, which is somewhat less than the above value for a cell wall.

Molecules have much greater difficulty diffusing a given distance across the plasmalemma, compared with the same distance in the cell wall or in the adjacent unstirred layer. For the above numerical values, $D_j K_j$ is 10^{-5} cm^2/sec in the aqueous solution, 10^{-6} cm^2/sec in the cell wall, while for a plasmalemma 10 nm (100 Å) thick, $D_j K_j$ is only 10^{-14} to 10^{-10} cm^2/sec. Membranes do indeed provide very effective barriers for the diffusion of solutes.

Secondary cell walls are often interrupted by localized pits. In such regions the primary cell walls also have local depressions where they are markedly decreased in thickness. A pit in the wall of a given cell usually occurs opposite a complementary pit in an adjacent cell, the cytoplasm of two adjacent cells being brought into close proximity at such a pit pair. The local absence of extensive cell wall substance facilitates the diffusion of molecules from one cell into the other. An easier way for molecules to move between plant cells is by means of the *plasmodesmata* (singular ; *plasmodesma*). These are fine, membrane-flanked, cytoplasmic threads that pass from a protoplast, through a pit, directly into the protoplast of a second cell. A tobacco leaf cell can have 10 to 30 plasmodesmata per 100 μm^2 of cell surface, each plasmodesma having a diameter near 0.2 μm. Plasmodesmata can be as small as 20 nm in diameter and their frequency in the hundreds per 100 μm^2. Usually they occupy about 0.1 to 0.5% of the surface area of the cell. The continuum of communicating cytoplasm created by such intercellular connections is commonly referred to as the *symplasm.* The presence of the plasmodesmata provides a particularly effective pathway for solute movement between adjacent cells since no permeability barrier in the form of either a cell wall or a membrane needs to be surmounted for diffusion of molecules from cell to cell in the symplasm.

Since movement from cell to cell through the symplasm is physiologically so important, let us next compare it with the concomitant diffusion across the plasmalemma. We will use Fick's first law as presented in Equation 2, *viz.*, $J_j = D_j K_j \Delta c_j / \Delta x = P_j \Delta c_j$. Let us consider two adjacent cells in which the glucose concentration differs by 1 mM. Glucose is rather insoluble in membrane lipids and would have a permeability coefficient of about 10^{-7} cm/sec for a typical plasmalemma. For movement from cell to cell across the two plasmalemmas in series, the effective P_j would be 0.5×10^{-7} cm/sec (inclusion of the cell wall as another series barrier for diffusion would only slightly reduce the effective P_j in this case. Thus, the flux of glucose across the plasmalemmas toward the cell with the lower concentration would be $(0.5 \times 10^{-7}$ cm/sec$)(1 \times 10^{-3}$ mole/1000 cm^3), or 5×10^{-14} mole/cm^2-sec. Let us next consider the flux for the symplasm. In the pores of the plasmodesmata, $D_{glucose}$ should be similar to its value in water, 0.7×10^{-5} cm^2/sec, while $K_{glucose}$ would be 1.0. We will let the plasmodesmata be 2 μm long. The flux in the pores would thus be $(0.7 \times 10^{-5}$ cm^2/sec$)(1.0)(1 \times 10^{-3}$ mole/1000 cm^3)/$(2 \times 10^{-4}$

cm), or 4×10^{-8} mole/cm^2-sec. If the pores occupy 0.002 of the surface area of the cells, then the rate of glucose diffusion through the plasmodesmata per unit area of the *cells* would be (0.002) (4×10^{-8}), or 8×10^{-11} mole/cm^2-sec. This is more than a thousandfold greater than the simultaneously occurring diffusion across the plasmalemmas. The permeability coefficients of the plasmalemmas for phosphorylated sugars like ribose-5-phosphate or ribulose-1,5-diphosphate would be less than for glucose, while the diffusion coefficients in the plasmodesmata would be about the same as for glucose. Thus, the discrepancy between fluxes of charged species through the plasmodesmata and those across the plasmalemmas would be even greater than the thousandfold difference calculated here for glucose. The joining of cytoplasms into a symplasm indeed facilitates the diffusion of solutes from one cell to another.

Stress-strain relations of cell wall. Cell walls of mature plant cells are generally quite resistant to mechanical stretching, especially when there is appreciable thickening of the secondary cell walls. Nevertheless, such structures will stretch when a stress is applied. Reversible elastic properties are generally described in terms of a measure of elasticity known as *Young's modulus*, which is the ratio of applied stress (force per unit area) to the resulting strain (fractional change in length) :

$$\text{Young's modulus} = \frac{\text{stress}}{\text{strain}} = \frac{\text{force/area}}{\Delta L/L} \quad \ldots(7)$$

Since $\Delta L/L$ is dimensionless, Young's modulus has the dimensions of force per unit area or pressure. A high value of this modulus of elasticity means that a large stress must be applied in order to produce a noticeable effect.

Young's modulus for dry cotton fibers (which represent nearly pure cellulose) is quite large, about 10^5 kilograms/cm^2, or 5% of that for steel. Because of both the complicated three-dimensional array of microfibrils in the cell wall and the presence of many other components, Young's modulus for a cell wall is considerably less than for pure cellulose. The modulus of elasticity for the cell wall of *Nitella*, for example, has been experimentally found to be about 7000 kg/cm^2.

The hydrostatic (turgor) pressure, P, acts uniformly in all directions in a cell. This internal pressure tends to cause the cell to expand and also leads to tensions (stresses) in the cell wall. The magnitude of these stresses varies with the physiological state of the plant as well as with direction in the cell wall.

A cylinder, closely approximating the large internodal cells of *Nitella* or *Chara*, is a useful geometrical model for evaluating the various cell wall stresses. Let the radius of the cylinder be r (as in Fig. 1.9). The force on an end wall is the pressure (P) times the area of the end wall, πr^2. The force is

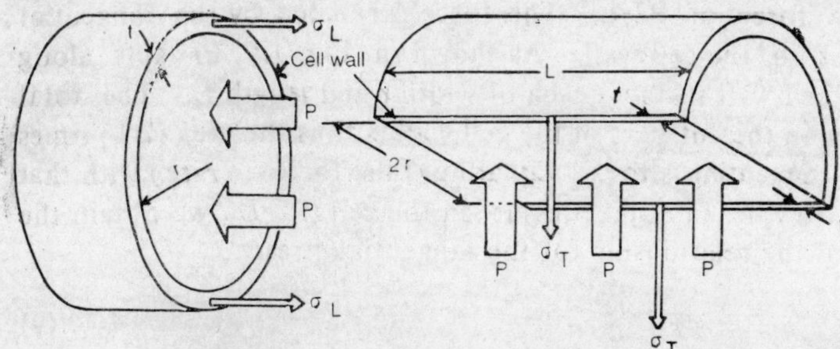

Fig. 1.9. Schematic sections of a hypothetical cylindrical cell resembling the internodal cell of *Nitella* or *Chara*, illustrating various dimensions and the stresses (σ_L and σ_T) existing in the cell wall. One way to calculate the stresses is to imagine that the cellular contents are removed, leaving only the cell wall, which has a uniform hydrostatic pressure P acting perpendicular to its inside surface. The projection of this P over the appropriate area gives the force acting in a certain direction while the reaction to this force is an equal force in the opposite direction in the cell wall. By dividing the force in the cell wall by the area over which it occurs, we can determine the cell wall stress.

then $P\pi r^2$. This force is balanced by a tension known as the *longitudinal stress*, σ_L, which occurs in the cell wall. The area over which σ_L acts is shown by the cut portion of the cell wall in the left part of Fig. 1.9. The longitudinal stress occurs in an annulus of circumference $2\pi r$ (approximately) and a width equal to the thickness of the cell wall, t. The force in the cell wall is thus $(\sigma_L)(2\pi r)(t)$. This force is an equal and opposite reaction to $P\pi r^2$, and so $\sigma_L 2\pi rt$ equals $P\pi r^2$ or

$$\sigma_L = \frac{rP}{2t} \qquad \text{...(8)}$$

The longitudinal stress acts parallel to the major axis of the cylinder and resists the lengthwise deformation of the cell.

A *tangential stress*, σ_T, also exists in the cell wall in response to the internal pressure and limits the radial expansion of the cell. In order to determine the magnitude of this stress, we will consider a cell split into half along its axis (right side of Fig. 1.9). The split part of the cell has an area of $2rL$ in the plane of the cut, which is acted on by the pressure P, leading to a force of $P2rL$. This force is resisted by the tangential stress in the cell wall. As shown in Fig. 1.9, σ_T acts along two cell wall surfaces each of width t and length L. The total force in the cut part of the cell wall is thus the area $(2tL)$ times the tangential stress. Equating this force $(\sigma_T 2tL)$ with that due to the internal pressure in the cell $(P2rL)$, we obtain the following relationship for the tangential stress :

$$\sigma_T = \frac{rP}{t} \qquad \text{...(9)}$$

The tangential stress in the cell wall given by Equation 9 is twice as large as the longitudinal stress (Eq. 8). Thus, this simple cylindrical model illustrates the important point that the stresses in a cell wall can vary with direction.

To estimate the magnitudes of the stresses and the resulting strains in the cell wall, let us consider a *Nitella* or *Chara* cell that is 1 mm in diameter with a cell wall 5 μm thick. In this case, r/t is (500 μm)/(5 μm), or 100. A reasonable estimate of P is 6 bars. (A bar is a convenient unit for expressing pressures occurring in plants and equals 10^6 dynes/cm^2 or 0.987 atmosphere. Using Equation 8, we can calculate that the longitudinal stress in the cell wall of such a cell is [(100)(6 bars)]/(2), or 300 bars. By Equation 9, the tangential stress would be twice as great, or 600 bars. One bar corresponds to 1.020 kg/cm^2 at sea level. The longitudinal stress in the cell wall is thus (300 bars)(1.020 kg/cm^2-bar), or 306 kg/cm^2, an appreciable tension. The strain produced by this stress can be calculated using the definition of Young's modulus (Eq. 7) and its particular value for the cell wall of *Nitella*, 7000 kg/cm^2. From Equation 7, the fractional change in length is then (306 kg/cm^2)/(7000 kg/cm^2), or 0.044, *i.e.*, only about 4% in this case. Hence, even with an internal pressure of 6 bars, which leads to a longitudinal stress in the cell wall of 300 bars, the cell wall (and consequently the whole cell) is not extended very much. The cell wall is indeed rigid and therefore well-suited for limiting the size of individual cells as well as for providing the structural support of a plant.

Before concluding this discussion of cell walls, it should be noted that the case of elasticity or reversible deformability is only one extreme of stress-strain behaviour. At the opposite extreme is plastic (irreversible) extension. If the amount of strain is directly proportional to the time that a certain stress is applied, and if it persists when the stress is removed, we have an example of viscous flow. The cell wall actually exhibits intermediate properties and is said to be *viscoelastic.* When a given stress is applied to a viscoelastic material, it turns out that the resulting strain is approximately proportional to the logarithm of time. Moreover, such extension is partly elastic (reversible) and partly plastic (irreversible). Underlying the viscoelastic behaviour of the cell wall are the crosslinks between the various polymers. For example, if a bond from one cellulose polymer to another is broken while the cell wall is under tension, the opportunity exists for a new bond to form in a less strained configuration, leading to an irreversible or plastic extension of the cell wall. The quantity responsible for the tension in the cell wall—which in turn leads to such visco-elastic extension—is the hydrostatic pressure within the cell.

VASCULAR TISSUE

The Procambium. In the differentiation of axis and leaf from promeristematic tissue, the formation of a uniseriate external layer is usually the first evidence of the complex structure seen in the mature organ. The first cells to mature, however, in either, leaf or axis, belong to the vascular tissue. In the promeristem, where all cells are essentially isodiametric and alike, continuing longitudinal divisions set apart in some areas strands of elongate, slender cells with dense cytoplasm. This meristematic tissue forms the primary phloem and xylem and is known as the *procambium.* (The term "procambium" is used here to indicate the meristematic tissue that gives rise to the morphological vascular units, not as applied to any elongate meristematic cells that are cambium-like in form). The form and arrangement of the procambium strands foreshadow the structure of the primary vascular skeleton. The first procambium appears as isolated strands very close to the apex in stem and root, commonly at a distance of only a few micra. It is continuous backward in the older tissues with older promeristem strands and mature vascular tissue, that is, development is acropetal.

The slender procambium strands increase in diameter by longitudinal cell division within themselves and by the addition of new cells upon their borders by transformation of adjacent promeristem cells. Other strands may appear a little later in other positions in the promeristem. Increase in size of the strands may be so great that some or all of the strands fuse,

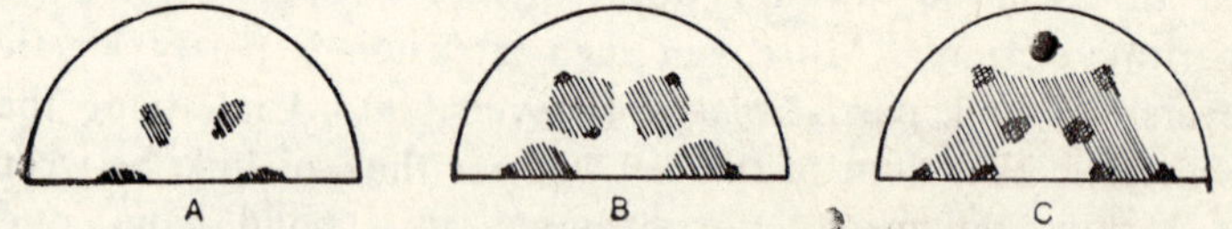

Fig. 1.10. Diagrams showing stages in the development of a procambium cylinder from promeristem. A, the first procambium has appeared in the promeristem in the form of strands ; B, the strands of procambium have increased in diameter, approaching one another, and the oldest portions (on the outside and the inside) have become phloem and xylem ; C, the strands of procambium have united laterally, and more procambium has matured as phloem and xylem. In a later stage all procambium will have become vascular tissue, forming a primary vascular cylinder.

and a hollow cylinder or a solid central core may be formed. Ultimately the procambium skeleton so built up has the form of the vascular skeleton of the region. But at no one time is the complete procambium mass present because the oldest procambium cells quickly mature into phloem and xylem cells, often long before the last of the procambium cells appear.

As new cells are formed in a strand, cell length increases greatly, and the ends become more tapering. Increase in length is related to the rate and extent of increase in length of the region in which the cells lie. In this way the tissue keeps pace with the increase in the surrounding tissues with few or no transverse divisions. In later maturing procambium cells, the transverse diameter of the cells may also increase progressively. Both these changes are reflected in the size and form of the vascular cells formed from them : the first sieve tubes, tracheids, and vessels are short and exceedingly slender ; the later matured cells increasingly longer and of somewhat greater transverse diameter. The difference in length is partly obscured by the stretching of first-formed cells.

As procambium develops, the diameter of the organ is increasing, and the promeristem cells are multiplying and enlarging. Spatial changes thus necessarily take place in the position of the strands in relation to each other and to the center of the organ. Further, within each strand, new-cell formation brings about spatial changes of the cells on the borders. In this way, as growth continues, the first phloem and the first xylem cells, which usually mature on the inner and outer margins of slender strands, are separated more and more widely as the strand becomes larger (Fig. 1.11).

The first-maturing cells in a young strand are phloem cells ; the first xylem cells soon follow. The first mature phloem cells are close to the apex in both stem and root, at a distance of a few micra, frequently within one millimeter.

The order of maturing, longitudinally, of the phloem cells in a strand is acropetal, like that of the procambium ; that of the xylem is both acropetal and basipetal, from one or more points of beginning. Definiteness and uniformity of development transversely in a procambium strand are important characteristics of primary vascular tissue. The points of beginning and the order of cell maturation are constant for certain organs and for the major plant groups. The first xylem and phloem cells mature usually on the inner and outer margins of the strands and are separated radially by procambium in stems and leaves (Fig. 1.10) and tangentially by promeristem in roots. Order of development from the point of beginning is radial (as to the axis) and lateral in stems and leaves, radial only or chiefly in roots. After the first vascular cells mature, divisions in the central promeristem of the bundle tend to be in such planes that the later matured cells stand more or less in radial rows. This is especially true of herbs, vines, and woody plants of the more advanced families.

Centripetal and Centrifugal Growth. The development radially in a procambial strand of mature vascular cells progressively from the point of beginning toward the center of the axis is known as *centripetal growth* (Fig. 1.11B), and that away from the center of the axis, *centrifugal growth* (Fig. 1.11A). The development of phloem is probably always centripetal, whereas that of xylem is sometimes centripetal and sometimes centrifugal. Less commonly, the first-formed cells are situated, not at the edge of the procambium, but nearer the center, and development is then both toward and away from the center of the axis (Fig. 1.11C). In xylem when the development is toward the center of the axis, *centripetal xylem* is formed, and the xylem group or strand is said to be *exarch* (Fig. 1.11B); when it is *away from* the center of the axis, *centrifugal xylem* is formed and the xylem unit is said to be *endarch* (Fig. 1.11A).

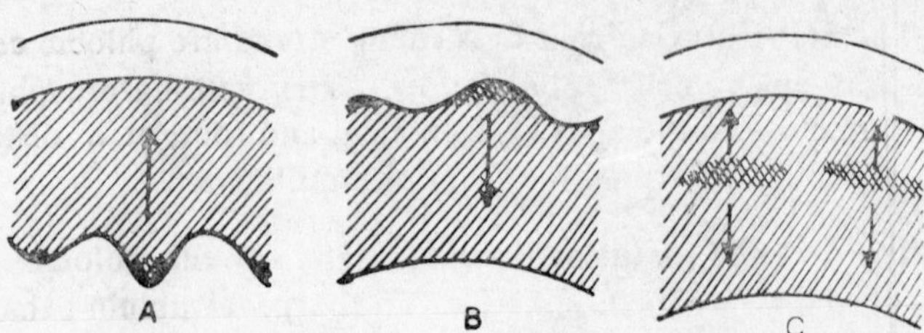

Fig. 1.11. Diagrams showing order of development of primary vascular tissue transversely in an organ. A, centrifugal ; B, centripetal ; C, both centrifugal and centripetal. These diagrams may also illustrate types of xylem from the standpoint of order of development: A, endarch ; B, exarch ; C, mesarch.

When development is such that both centripetal and centrifugal xylem are formed—even though the amount of one type be very small—the xylem is *mesarch* (Fig. 1.11C). (Development must always be considered as *relative to the center of the axis*, not to the center of the vascular bundle in question.) The terms "centripetal growth" and "centrifugal growth" refer only to sequence of appearance and of maturation of procambium cells in definite directions and do not imply growth in the sense of successive new-cell formation. The distinction of xylem as exarch, endarch, and mesarch is of morphological and phylogenetic importance because each type is characteristic of definite organs or parts of organs and is restricted more or less in its distribution in the larger plant groups. For example, the root is always exarch ; the stem of seed plants is endarch ; the axis of the club mosses is exarch : mesarch xylem is common in the ferns and infrequent elsewhere.

The First Vascular Elements : *Protophloem and Protoxylem.* The first cells of the phloem to mature are known as the *protophloem* ; those of the xylem, the *protoxylem.* These cells differ markedly from later formed cells of the same tissues in cell type ; in size and shape, being very slender and long because stretched ; and, in xylem cells, in the structure of the wall. Since they are formed very early in the ontogeny of the tissue in which they lie, they are subject to stresses due to increase in diameter and particularly to increase in length of the organ. To these stresses the surrounding meristematic cells accommodate themselves by new-cell formation and by increase in size, but the cells of the protophloem and protoxylem are mature and no longer subject to growth changes. In the region in which protophloem and protoxylem lie, the greatest change occurring is increase in length. The stresses brought about by this elongation tend to stretch the already matured cells. To such longitudinal stretching protoxylem and protophloem cells are, to a limited extent, adaptable. The cells are long and slender with walls which are thin and of cellulose, reinforced in protoxylem by bands of lignified secondary wall (Fig. 1.12). These bands, in the form of rings and spirals, prevent to some extent the collapse of the thin, plastic walls and the consequent closure of the lumen after stretching.

To those phloem and xylem cells which are thus capable of stretching, the terms "protophloem" and "protoxylem" are restricted. The parts of phloem and xylem which are not

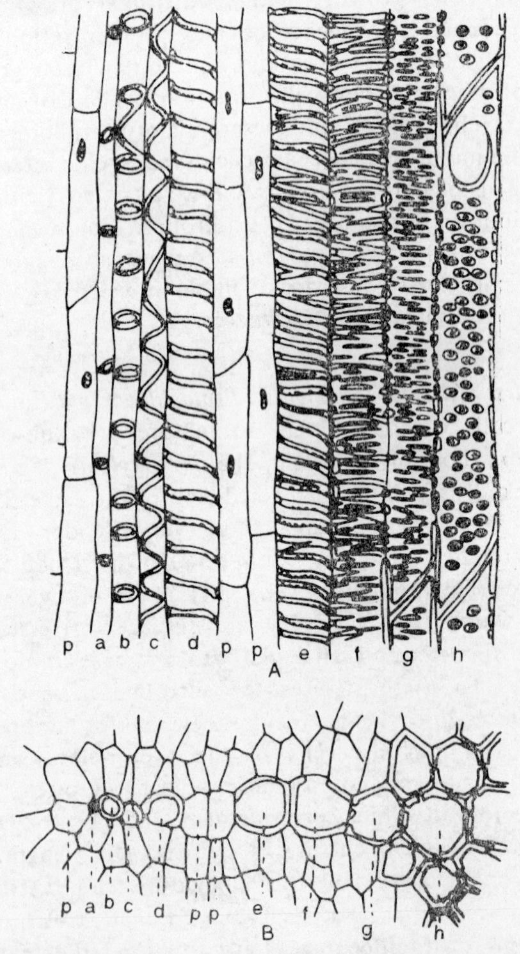

Fig. 1.12 Protoxylem and metaxylem in transverse, B. and longitudinal, A, section in *Lobelia*, *a*, *b*, annular elements ; *c*, *d*, *e*, spiral elements ; *f*, scalariform element ; *g*, scalariform-reticulate element ; *h*, pitted vessel ; *p*, parenchyma cell.

stretched by elongation of the region in which they lie are known as *metaphloem* and *metaxylem* respectively. Thus primary phloem consists of protophloem and metaphloem; primary xylem, of protoxylem and metaxylem. In secondary phloem and secondary xylem no protophloem and protoxylem exist, for secondary tissues develop only after elongation ceases. A strand of xylem may consist wholly of protoxylem—the condition in most very small vascular bundles—or wholly of metaxylem, as in some slow-growing roots and rhizomes, but commonly both types of xylem are present, the proportions varying with the plant, the organ, and the rapidity of growth of the region in question. The phloem varies similarly.

It is obvious that in exarch xylem the metaxylem lies internal to the protoxylem (Fig. 1.11B, the protoxylem heavily shaded); in endarch xylem, external to the protoxylem (Fig. 1.11A); in mesarch xylem, both toward the inside and toward the outside of the protoxylem (Fig. 1.11C).

Primary Phloem. The first vascular cells to mature in a given region are sieve elements of the phloem—sieve cells in gymnosperms and lower groups, sieve tubes in angiosperms. These cells are of the protophloem type. Their structure is difficult to determine because they closely resemble the procambium cells from which they are derived. Their diameter is very small; they live only a few days; they are soon greatly stretched, crushed between adjacent cells, and absorbed.

In phloem, as also in xylem, the first-formed cells are longest, because they develop while the region in which they lie is elongating most rapidly and they are therefore most stretched. Later formed cells are progressively shorter until metaphloem is formed. Similarly, the earliest of these cells have in general the smallest transverse diameter, and this diameter increases progressively in later formed cells. The change in diameter of these cells is related to the similar change in the procambium cells from which they mature. The first-formed phloem and xylem cells mature while all procambium cells are very slender.

The very first mature phloem cells of a region can be found only when the procambium is still in early stages. Later all evidence of their earlier presence has disappeared. Parenchyma cells and fibers mature later from the surrounding procambium cells and obliterate the spaces they occupied. The first protophloem probably consists only of sieve cells or sieve tubes distributed among procambium cells. Companion cells are rare or absent. Later, the procambium cells about these first-maturing cells differentiate as parenchyma cells and fibers. The fibers may be aggregated in clusters or masses which soon appear homogeneous because the early formed sieve tubes have disappeared, or they may be mingled with phloem parenchyma. These phloem fibers constitute what have been in many plants called pericyclic fibers. They, with the phloem parenchyma about them, make up a part, perhaps sometimes all, of the tissues commonly distinguished as pericycle. Mature conducting cells of protophloem are present in very young meristematic tissue—in stem and leaf tips are close to the apex as 0.3 mm ; in root apices, about 1 mm.

The metaphloem is a complex tissue with well-developed cells of all types—sieve cells or sieve tubes, companion cells, parenchyma cells, and sclerenchyma in the form of fibers or sometimes sclereids. Parenchyma cells are added to the sieve cells or sieve-tube elements as soon as metaphloem begins to be formed. The various cell types are essentially the same as those of the secondary phloem of the same plant. The metaphloem, like the protophloem, may be early crushed. In woody plants and in herbs with well-developed secondary vascular tissues, the soft cells of the primary phloem are often completely destroyed within a short time after maturing by the development of secondary tissues beneath them. The remnants of the flattened cells may exist for some time or be quickly absorbed like those of the protophloem.

Primary phloem is of only temporary functional importance where secondary phloem is developed in considerable amount. Where little or no secondary phloem is formed, the primary phloem persists throughout the life of the organ and may be of the greatest importance physiologically. Structurally, such phloem is highly specialized, consisting usually only of sieve tubes of the highest type and their accompanying companion cells. Such is the condition in the phloem of most monocotyledons, and of such dicotyledons as *Cucurbita*, *Ranunculus*, and *Podophyllum*. In these genera the phloem of the bundles is in part secondary, but the primary and secondary phloem differ little histologically except that the protophloem may lack companion cells. In other herbaceous plants, such as *Solanum*, *Aster*, and *Lobelia*, where definite cylinders of secondary vascular tissue occur, but where the amount of secondary phloem is small, the primary phloem persists as functioning tissue through the life of the stem. In such forms the sieve tubes and companion cells often occur in small groups and are remarkable for their small size. An entire group may be of no greater diameter than one of the adjacent parenchyma cells.

Primary Xylem. Protophloem probably always consists only of one kind of cell, but protoxylem is a complex tissue made up of tracheids, vessels, and parenchyma cells. Fibers are probably always absent. Parenchyma cells surround the earliest formed tracheal cells and appear to mature with or soon after the conducting cells. (The term "protoxylem" is commonly applied only to the more or less isolated tracheids and vessels but is better used for the continuous tissue, including the parenchyma cells).

The water-conducting cells of the protoxylem are the characteristic cells of this tissue because of the peculiar adaptation of their walls to the stretching which they normally undergo. The thin, plastic, primary walls of these empty cells are strengthened by the addition of a lignified secondary wall in the form of narrow rings and spiral bands. These bands apparently help to keep the conducting channels open during the elongation of the cells. The first-formed protoxylem cells have small amounts of secondary wall in the form of rings spaced at intervals along the cell (Fig. 1.12*a*,*b*). Such cells are called *annular cells*, or *annular elements*, or, more specifically, *annular tracheids* or *annular vessels*. (The term "element" is often used in speaking of cells of the protoxylem when the cells in question may be either tracheids or vessels.) Cells formed a little later than annular cells possess secondary walls in the form of spiral bands ; such cells are called *spiral cells*, or *spiral elements* (Fig. 1.12*c,d,e*). The flatness of the coils and, to some extent, the number of the bands increase in the cells formed successively later and later. The proportionate amount of secondary wall thus increases in the successively formed cells. The first-formed cells are subjected to an excessive amount of elongation, those formed later and later to less and less. Structurally, the cells formed at any given period are adapted to the amounts of elongation occurring during their formation and functioning period. Annular cells can be stretched to a greater extent than spiral cells ; cells with steep spiral coils more readily than those with flat coils ; those with one spiral more freely than those with several. Where the structure of the secondary wall suggests a spiral lying in flat coils, the coils fused at the corners of the cell, or where a more definitely ladder-like appearance is produced by transverse bar-like thickenings running from corner to corner, the cell is known as *scalariform* (Fig. 1.12). (The scalariform cell of protoxylem is best called a scalariform protoxylem cell, since the term "scalariform cell" is a loose descriptive term and may refer to a scalariform-pitted tracheid or to a scalariform-perforated vessel. A scalariform protoxylem cell may, of course, be also a scalariform vessel.)

The scalariform protoxylem cell is, obviously, capable of little if any stretching. Where the bands of secondary wall are more extensively and less uniformly tied together, forming a network of secondary thickening, a *reticulate cell* is formed (Fig. 1.12*g*). Where the secondary wall is still more extensive and the thin spots in the wall are definite and uniform in size and shape, the cell is *pitted* (Fig. 1.12*h*). Reticulate and pitted cells cannot be stretched. Primary xylem consisting of these cells is, therefore, *metaxylem*. Scalariform cells fall on the line between protoxylem and metaxylem. No sharp line separates any one type of cell from the next type, or protoxylem from metaxylem.

The water-conducting cells of protoxylem may be either tracheids or vessels. They are tracheids in pteridophytes and gymnosperms and in some of the less highly specialized angiosperms. In other flowering plants both kinds of cells may occur together, and in herbaceous forms and some woody forms the water-conducting cells of the protoxylem are largely vessels. Because protoxylem cells are made abnormally long by stretching, and the end walls are not readily seen, these cells, vessel-like in appearance, have been called "protoxylem vessels" regardless of their tracheidal or tracheal nature. This incorrect usage probably grew out of the idea of the seventeenth and eighteenth centuries that the animal-trachea-like appearance of spiral cells indicates a conducting tube of indefinite length. (The use of the term "vessel," or "trachea," in this sense is responsible for the opinion formerly held that "vessels occur in the gymnosperms only in the protoxylem.") It often cannot be readily determined whether a given cell of the protoxylem is a vessel or a tracheid ; such a cell may, then be called, for example, "an annular cell," or "an annular element" ; it should not be called a vessel unless there is proof of the perforation of its end walls.

Proportion and Arrangement of types of Protoxylem Elements. In the protoxylem of a given vascular bundle, annular, spiral, and scalariform cells may all exist, and each in any proportion ; or one or two of the types may be lacking. Spiral cells, especially those with closely coiled bands, constitute a large proportion of most protoxylem strands. Where growth of the axis is rapid, a high percentage of annular elements is found ; where slow, very few or no such cells occur. As a strand of protoxylem forms, not only is the wall of the successively formed cells of different character, but the cells themselves each become somewhat larger than the preceding ones. The order of development may thus usually be determined, when several or all cells are mature, on the basis of size alone. Often there are variations and exceptions to the sequence in size, but the structure of the wall is constant in its relation to order of development.

Elongation of Protophloem and Protoxylem. Though protophloem and protoxylem are structurally adapted to stretching, and the earliest formed cells especially so, the elongation to which they are subjected is often so great that their plastic capacity is surpassed and they are ruptured. The protophloem cells are rapidly absorbed after destruction, but the torn remains of the protoxylem tracheids and vessels persist permanently. Annular cells, being the first cells formed, are most frequently destroyed. Probably nearly all annular cells that are not destroyed are so distorted soon after their formation as to become nonfunctional. The stretching of these cells at first separates the supporting rings, and the thin, distended wall between them soon sags or collapses. The further pulling of the walls tilts the rings and turns them up on edge (Fig. 1.13B). Still further elongation fragments the cells, and isolated rings of secondary wall, often with pieces of the primary wall attached, are to be found where the cells have been destroyed (Fig. 1.13B). In spiral cells the spiral is pulled out so that it

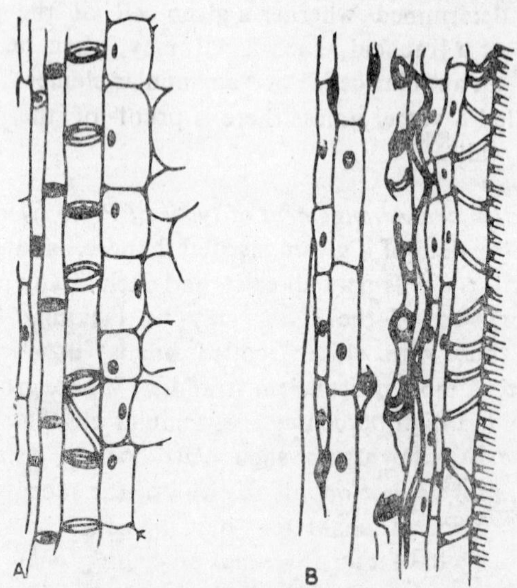

Fig. 1.13 Protoxylem in longitudinal section. A, from fruit of *Arisaema*, the elements are all annular and there has been little stretching and no rupture; B, from stem of *Equisetum*, there has been extensive stretching of annular and spiral cells, with rupture of the former and the formation of a protoxylem lacuna. The space occupied by the first-formed cell has been occluded except where the rings lie.

becomes steeply wound or is even straightened out (Fig. 1.13B). In such cells also the primary wall collapses.

Protoxylem Lacunae. Where a number of protoxylem cells lying together have been thus destroyed, a canal-like cavity, known as a *protoxylem lacuna,* may be formed (Fig. 1.13B). In some instances such an intercellular space is much enlarged by the pulling away of the surrounding parenchyma cells. Exaggerated protoxylem lacunae formed in this way are found in many herbaceous plants, and especially in the monocotyledons (Fig. 1.14C, D) and the horsetails (*Equisetum*). On the other hand, the growth of tissues surrounding the developing bundle at the time of the rupture of the protoxylem cells may be such that no lacuna is formed, but the surrounding parenchyma cells press in upon the flattening xylem cells and fill up the space occupied by them (Figs. 1.13B, left annular element ; 1.14A, B). In annular or loosely spiral cells, parenchyma cells thus may be found pushing into the hollows formed by the sagging wall, or tylosis-like, into the lumen when the wall is ruptured. Rings from annular cells sometimes become pressed, upright, between parenchyma cells (Fig. 1.13B). Under such conditions the position of the first-formed protoxylem cells may be difficult to ascertain in transverse sections.

Ontogeny of Protoxylem Elements. The stretching and tearing of the walls of the tracheary elements of protoxylem takes place in the thin, delicate primary wall. It has been claimed that the rings, spirals, and bars of the secondary wall are formed by the tearing apart of a previously formed, continuous secondary wall. Early studies demonstrated that in these tracheary elements of protoxylem the protoplast forms secondary wall only in those areas where rings, spirals, etc., are later present. Recent work has confirmed the earlier interpretation. The presence of spiral bands stretched until straight is further evidence that stretching does not break up spirals and form rings.

Arrangement of Cells in Primary Vascular Tissues. The cells of primary tissues of all kinds and in any organ may be arranged uniformly according to some pattern or may be without regularity of arrangement. The cells of secondary tissues,

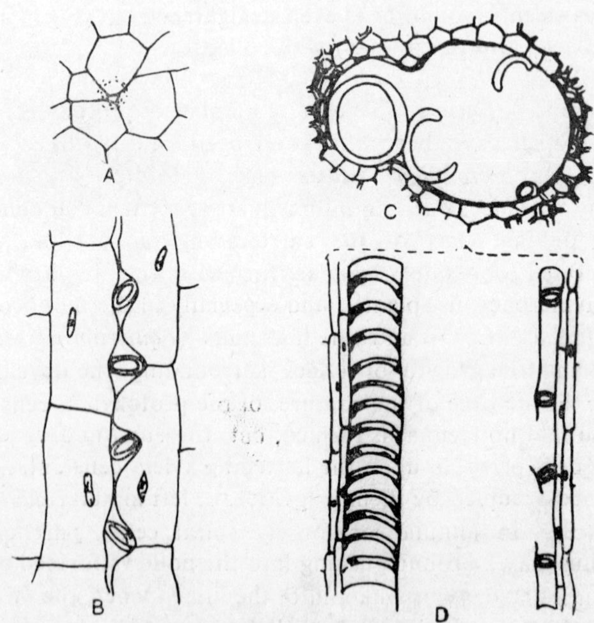

Fig. 1.14. Protoxylem in transverse and longitudinal section, showing behaviour of surrounding tissue. A, B, from *Lobelia*, showing a stretched annular element, the primary wall collapsed between the rings, the adjacent parenchyma cells pressing into the spaces, in cross section the space originally occupied by the cell nearly filled in ; C, D, from *Zea*, the surrounding parenchyma pulled away from the protoxylem elements, leaving a large lacuna.

because of the method of their origin from meristems of the cambium type, tend to be arranged in rows or sheets. It is frequently stated that primary tissues are characterized by irregular, and secondary by regular arrangement, and this distinction is applied to vascular tissues especially, but the orderly arrangement of secondary tissues may be broken up by unequal development of the constituent cells. Also in the primary vascular tissues of many herbaceous and some woody angiosperms, the primary xylem and phloem, especially the protoxylem and metaphloem, stand in radial rows. This arrangement further suggests secondary origin because cells of the secondary tissues stand in rows continuous with them.

Distinction of primary and secondary tissues can be made—except for protoxylem and part of the metaxylem—only on method of origin, by procambium or by cambium. Since the cambium itself is developed from the procambium as the formation of primary tissues ceases, only an approximate limit can be set between primary and secondary phloem and between primary and secondary xylem.

DIFFUSION

The chemical elements which constitute the bulk of the body of any plant are among the commonest ones on the surface of the earth. They occur in the environment of plants as relatively simple inorganic compounds and enter plants in the form of such compounds. From the relatively small number of compounds which enter it from its environment the green plant fabricates the numerous complex organic compounds which are essential to its continued existence as a living system.

The movement of substances into a plant from its surroundings is accomplished largely by the process known as *diffusion*. Substances enter a plant partly through its aerial organs and partly by way of the root system. From the atmosphere, carbon dioxide and oxygen gases diffuse into plants, principally through the stomates. From the soil, water and the cations and anions of inorganic salts pass into the plant at least partially by diffusion, although other more complex mechanisms are also involved in the entrance of substances into roots.

Similarly the loss of substances from a plant into its environment is accomplished principally by diffusion. Large quantities of water vapour pass out of leaves and other aerial organs of plants into the atmosphere by this process. At times oxygen gas and at times carbon dioxide gas diffuse from plants into the atmosphere. Certain volatile compounds also escape from the aerial organs of many plants by diffusion. Similarly the roots lose carbon dioxide and other compounds into the soil by diffusion.

Likewise some of the movement of substances from one part of a plant to another is accomplished by diffusion. This is true both of the gases which move through the intercellular spaces and the water and solutes which move within the cells. However, much of the translocation of materials from one plant organ to another is accomplished by more complicated mechanisms than diffusion. Within any living plant cell diffusion of substances from one part of a cell to another is also continually in progress.

Diffusion is the mechanism for most, if not all steps by which CO_2 from the air reaches the sites of photosynthesis in chloroplasts. For instance, CO_2 diffuses from the atmosphere up to the leaf surface and then diffuses through the stomatal pores. After entering the leaf, diffusion of CO_2 occurs within intercellular air spaces. Next, CO_2 diffuses across the wet cell wall, crosses the plasmalemma of a leaf mesophyll cell, and then diffuses through the cytoplasm to reach the chloroplasts. Finally, CO_2 enters a chloroplast and diffuses up to the enzymes that are involved in carbohydrate formation. If the enzymes fixed all the CO_2 in their vicinity, and no other CO_2 were to *diffuse* in from the atmosphere surrounding the plant, photosynthetic processes would stop. Diffusion is likewise involved in many other aspects of plant physiology—here we will pay particular attention to the diffusion of substances across membranes. The mathematical formulation necessary for understanding both diffusion across a membrane.

Fick's first law. Fick in 1855 was one of the first to examine diffusion on a quantitative basis. For such an analysis we need to consider the concentration (c_j) of some species j in a solution. (The subscript j simply calls attention to the fact that we are considering only one species out of the many that could be present). We will assume that the concentration of species j in some region is less than in a neighboring one. A net migration of molecules then occurs by diffusion from the concentrated to the dilute region Such a molecular flow down a concentration gradient is analogous to the flow of heat down a temperature gradient from a warmer region to a cooler one.

The analogy is actually rather good (especially for gases), since both processes depend on the random thermal motion of the molecules.

To express diffusion quantitatively, we will consider a diffusive flux or flow of species j, J_j. We will restrict our attention to diffusion involving planar fronts of uniform concentration. (Fortunately, as we shall see, this has widespread application to situations of interest in biology). We will let J_j be the amount of some species j crossing a certain area per unit time, *e.g.*, moles of particles per cm^2 in a second. From reasoning based on an analogy with heat flow, Fick deduced that the "force," or causative agent, leading to the net molecular movement depends on the concentration gradient. A gradient is simply a measure of how a certain parameter changes with distance, *e.g.*, the gradient in concentration of species j in the x-direction is $\partial_j/\partial x$. In general, *the flux of some substance is proportional to the appropriate force*, a relation which we will use repeatedly in this text. In the present instance the driving force is the negative of the concentration gradient of species j, which we will represent by $-\partial c_j/\partial x$ for diffusion in one dimension. (To appreciate why a minus sign occurs, we should recall that the direction of net diffusion is toward regions of lower concentration). We can now write down the following relation showing the dependence of the flux on the driving force :

$$J_j = -D_j \frac{\partial c_j}{\partial x} \qquad \ldots(10)$$

Equation 10 is commonly known as Fick's first law of diffusion, where D_j is the *diffusion coefficient* of species j. The partial derivative is used in Equation 10 to indicate the change in concentration in the x-direction of Cartesian coordinates at some moment in time (constant t) and for specified values of y and z. (For the cases that we will consider, the flux in the x-direction actually has the same magnitude at any value of y and z).

The diffusion coefficient is often expressed in cm^2/sec. For instance, for c_j in moles/cm^3 (hence, $\partial c_j/\partial x$ in moles/cm^4) and J_j in moles/cm^2-sec, D_j is in cm^2/sec. Since D_j varies with concentration and temperature, it is properly called a *coefficient* in the general case.

We will now consider the magnitude of D_j for solutes in water. Some diffusion coefficients determined for 10 mM solutions at 25° are as follows : sucrose, 0.52×10^{-5} cm^2/sec ; glucose, 0.67×10^{-5} cm^2/sec ; glycine, 1.1×10^{-5} cm^2/sec ; calcium (as the chloride), 1.2×10^{-5} cm^2/sec ; and potassium (as the chloride), 1.9×10^{-5} cm^2/sec. Hence, the diffusion coefficients of small ions and other molecules in aqueous solutions are approximately 10^{-5} cm^2/sec. A globular protein with a molecular weight of 15,000 might have a diffusion coefficient near 10^{-6} cm^2/sec at 20°, while D_j for a 10^6 molecular weight globular protein is usually close to 2×10^{-7} cm^2/sec. Fibrous proteins often have approximately twofold lower diffusion coefficients than do globular proteins of the same molecular weight. The lower value of D_j for fibrous proteins is due to the greater frictional interaction between water molecules and fibrous proteins compared with the more compact globular ones.

To illustrate the time-distance consequences of Equation, we will quantitatively consider the diffusion of small molecules in an aqueous solution. How long, on the average, would it take a small solute with a D_j of 10^{-5} cm^2/sec to diffuse 50μm, the distance across a typical leaf cell ? From Equation the time required for the population of molecules to shift this distance is $(50 \times 10^{-4}\ cm)^2/[(4)\ (10^{-5}\ cm^2/sec)]$ or 0.6 sec. Thus, diffusion is a fairly rapid process over subcellular distances. (In the cytoplasm of living cells there is a mechanical mixing due to *cytoplasmic streaming*, which can lead to even more rapid movement than by diffusion). Next, let us consider the diffusion of the same substances over a distance of 1 meter. The time needed is $(100)^2/[(4)(10^{-5})]$ or 2.5×10^8 sec, which is eight years ! Diffusion is indeed not a rapid process over long distances. Thus, inorganic nutrients in xylary sap would not ascend a tree by diffusion at a rate necessary to sustain life. On the other hand, diffusion is often sufficient for the movement of solutes within leaf cells and especially inside organelles such as chloroplasts and mitochondria. In summary, diffusion in a solution is fairly fast over short distances (less than about 100 μm) but extremely slow for very long distances.

Diffusion of gases in the air surrounding and within leaves is necessary for both photosynthesis and transpiration. For instance, water vapour evaporating from the cell walls of leaf cells diffuses across the intercellular air spaces to reach the stomata and from theré diffuses across the unstirred air layer into the atmosphere CO_2 diffuses from the atmosphere through the open stomata to the surfaces of leaf cells while the photosynthetically evolved O_2 traverses the same pathway in the reverse direction, also primarily by diffusion. The experimentally determined diffusion coefficients of these three gases in air at 20° are approximately as follows : for water vapour, 0.25 cm^2/sec ; for carbon dioxide, 0.16 cm^2/sec ; and for oxygen, 0.20 cm^2/sec. The diffusion coefficients in air are approximately 10^4 times greater than the D_j describing the diffusion of a solute in a liquid, indicating that diffusion coefficients depend markedly on the medium. In particular, many more intermolecular collisions occur per unit time in a liquid phase than in a less dense gas phase. Thus, a molecule can diffuse further in air than in an aqueous solution before being influenced by other molecules. Most cells in animals are bathed by fluids, so that both a hemoglobin-type molecule and a circulatory system are necessary in order to transport O_2 to their cells and CO_2 away. Plants, on the other hand, often have conspicuous intercellular air spaces by which they can take advantage of the large values of D_{O_2} and D_{CO_2} in a gas phase.

As indicated by Graham's law, the diffusion coefficients of gases in air are generally inversely proportional to the square root of the molecular weights. This proportionality also often holds for chemically related small substances in solution. (For large substances, the proportionality is often with the cube root of molecular weight. The molecular weight of H_2O is 18, and for CO_2 it is 44. Water vapour therefore should have a diffusion coefficient $\sqrt{(44)/(18)}$, or 1.56 times higher than that of CO_2, as the above values for D_j in air indicate. In addition to this dependence of D_j on molecular weight and, by implication, on molecular size, the diffusion coefficient depends inversely on the viscosity of the medium.

Because of the complicated anatomy of a leaf, the actual pathway for the movement of gas molecules in the intercellular air spaces can be rather tortuous. Nevertheless, calculations using Equation can give useful estimates of the diffusion times for the many processes involving gaseous movements in a leaf. An upper limit for the diffusion pathway in such intercellular air spaces of a leaf might be 1000 μm. Using Equation and the diffusion coefficients given in the previous paragraph, we can calculate that the times needed to diffuse 1000 μm for water vapour, carbon dioxide, and oxygen in air are from 10 to 16 milliseconds. This illustrates that diffusion of molecules in a gas is a fairly rapid process. However, when illumination and temperature are not limiting, the rate of photosynthesis in plants is generally limited by the amount of CO_2 diffusing up to the chloroplasts, while the rate of transpiration is controlled by diffusion of water vapour from the cell walls within the leaf to the outside air. In the latter case, the diffusion limitation helps prevent excessive water loss from the plant and therefore is quite useful, physiologically speaking.

Diffusion of Gases. If a small vial of bromine gas is broken under a bell jar which has previously been evacuated of air, the entire jar quickly becomes filled with the brownish vapour of bromine. The distribution of the bromine gas throughout the bell jar has been accomplished by the kinetic activity of the bromine molecules and is a simple example of the process o diffusion. If the vial of bromine be broken under a bell jar which has not been evacuated of air, the time required for the bromine gas to occupy the bell jar completely will be longer than when diffusion of the gas occurs into a vacuum. Under such conditions the freedom of movement of the bromine molecules is impeded by the presence of molecules of the gases of the air, and the diffusion process is retarded. If the pressure of the air within the bell jar be increased to two atmospheres (which is equivalent to doubling the concentration of all the gases in the jar), the rate of diffusion of the bromine gas through the air would be still less than when the jar was occupied by air at atmospheric pressure.

Many other simple examples of the diffusion of gases might be cited. If a bottle of ammonia, ether, peppermint oil, or of any other readily volatile substance with a characteristic odor be opened indoors, within a very short time the distinctive odor of that substance can be detected in all parts of the room. Such a dispersal of gas molecules is accomplished at least in part by diffusion, although air currents often assist in speeding up such a distribution of molecules. Except in the rare case of diffusion into a vacuum, diffusing molecules move between the molecules of other substances.

The following somewhat fanciful analogy may aid in visualization of the kinetics of the diffusion process in gases. Suppose two adjoining rooms to be connected by a closed double door. Imagine also that one of these rooms contains a large number of tennis balls traveling in various directions along straight pathways at different rates of speed. The average distances between the tennis balls are supposed to be relatively great in proportion to their diameters. Each tennis ball represents a molecule. The individual balls will be constantly bumping into each other and into the walls of the room. Because of the large number of balls present innumerable collisions will occur every second. Each time a ball strikes a wall of the room it will bounce off along a different linear pathway. Similarly, whenever two balls collide, each will be deflected out of its course along a different route, to which it will hold undeviatingly until it is again deflected from its path by another collision. The course of each ball will thus be a zigzag progression through space, each short segment of its path being terminated by a collision which changes its direction. All of the balls do not move at the same speed at any given time, but the speed of each fluctuates from moment to moment as a result of the numerous collisions in which it participates. The average speed of the entire group will remain constant, however, as long as the temperature remains unchanged. As a result of this haphazard mutual buffeting, the tennis balls will remain equally distributed throughout the room.

Suppose now that the double doors connecting the two rooms are thrown open. As a result of their haphazard movement some of the balls close to the door will pass into the empty room. The first ones to do this will travel without interruption until they bump into one of the walls, at their rate of progression will not be impeded by collisions with other molecules. When they hit against one of the walls of the room they will bounce back into its interior along a new pathway. As more and more of the balls pass into the originally empty room their rate of progress becomes slower since the greater their concentration in the room, the greater the number of collisions per unit of time. As soon as any appreciable number of tennis balls has invaded the empty room, as a result of their random movements, some will pass back through the doorway into the room which originally contained all of them. As long, however, as the concentration (number per unit volume) of the tennis balls is greater in one room than in the other their random movements will result in more passing through the door into the room in which their lesser concentration prevails, than in the opposite direction.

In a relatively short time the concentration of tennis balls will have become equal in both rooms. In other words, they have "diffused" from one room into the other. After equality of concentration has been established the number of balls passing through the door in one direction in any interval of time will be exactly equal to the number moving in the opposite direction. When this condition of dynamic equilibrium is attained, diffusion, in the sense the word is used in this discussion, is no longer occurring.

If, in the hypothetical illustration just described, one room were filled with white tennis balls and the other with red tennis balls, diffusion would occur simultaneously in both directions. The red balls would "diffuse" toward the room in which their initial concentration was zero while the white balls would "diffuse" in the opposite direction. At equilibrium the concentration of the white balls would be equal throughout the two rooms, and this would likewise be true of the red balls.

Even after a dynamic equilibrium has been attained in any system, haphazard kinetic activity of the molecules continues. This is sometimes referred to as diffusion. The term diffusion will be used only to characterize situations in which there is gain in the number of molecules of a certain kind in one part of a system at the expense of other parts. According to this concept diffusion can occur only when the concentrations of the diffusing substance is not uniform throughout the system, and the process can continue only as long as differences of concentration are maintained.

The phenomenon of diffusion is exhibited by the molecules of liquids, solutes, and even solids, as well as by those of gases. Diffusion of ions also occurs, and even colloidal particles diffuse, but only at very slow rates.

Diffusion phenomena should be clearly distinguished from *mass movements*, in which the moving units are not single molecules, but more or less extensive assemblages of molecules. Winds and air currents generally are examples of mass movements of gas molecules. The draft of warm air ascending a chimpsy is another. All of the phenomena just listed, and many others, result primarily from differences in the density of the gases in various parts of a system. Heavy gases are more strongly attracted by gravitational forces than light gases. Hence, as in a chimney, cold (relatively heavy) air, displaces hot (relatively light) air, foreing the latter to rise. Similar phenomena on a grand scale are the principal cause of winds and air currents. They are representative of the physical process called *convection*. Such phenomena also occur in liquids. Mass movements of gases and liquids can also be caused in many other ways.

Diffusion Pressure. That diffusing gases sometimes result in the development of measurable pressure can be readily shown by certain simple experiments. In Fig. 1.15 is depicted an apparatus which can be used to illustrate a number of aspects of the phenomenon of diffusion of gases. This apparatus consists essentially of a vertically arranged glass tube, to the upper end of which is attached, by means of rubber stopper, a cylindrically shaped porous clay cup. This cylinder is hollow and its thick walls are pierced by numerous minute capillaries. These pores are fine enough to prevent mass movement of gases at any appreciable rate, but the porous clay is not a membrane in the usual sense of the word. The lower end of the vertical glass tubes dips in some coloured water. If such a porous clay cylinder, containing air at atmospheric pressure, be enclosed within a bottle containing hydrogen gas, also at atmospheric pressure, a rapid bubbling of gas will occur from the lower end of the tube through the dye solution into which it dips. When the bottle is removed a rapid and sudden rise of liquid up the glass tube will ensure. This is followed by a slow subsidence in the level of the liquid in the glass tube, until it falls to that of the water in the beaker.

The explanation of this sequence of events is as follows. For reasons which are discussed later, the *rate* of diffusion of hydrogen is greater than that of nitrogen or oxygen under

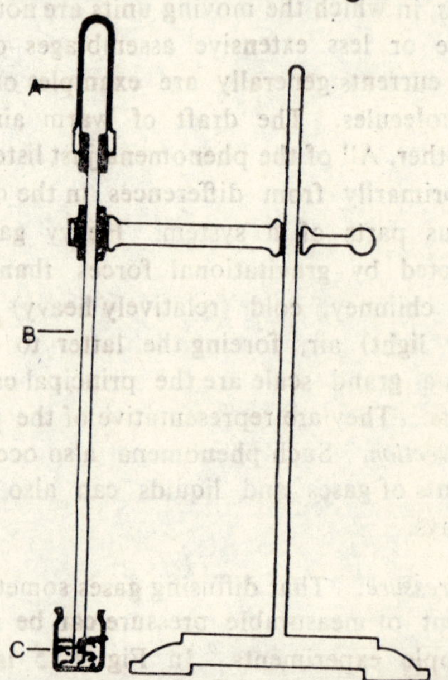

Fig. 1.15. Apparatus for demonstrating the development of pressure during diffusion of gases : (A) porous clay cup, (B) glass tube, (C) vessel of coloured water.

comparable conditions. When the porous clay cup is first enclosed within the bottle of hydrogen, rapid diffusion of hydrogen gas occurs through the pores of the cup. This raises the total gas pressure within the cup, since outward diffusion of oxygen and nitrogen occurs at a much slower rate. Hydrogen gas diffuses into the cup in spite of the fact that this results in a greater total pressure inside the cup than in the surrounding atmosphere. The *direction* of the diffusion of the hydrogen gas is controlled entirely by its own differences in concentration and is unaffected by the presence of other gases. The greater gas pressure inside the cup results in the outward bubbling of a mixture of all three gases at the lower end of the vertical glass tube. When the bottle is removed from around the porous clay cup, the direction of the diffusion of hydrogen is reversed, since the concentration of hydrogen gas is now greater inside of the cup. Since, while enclosed in the bottle of hydrogen, some of the nitrogen and oxygen diffused out of the cup and some was lost by the bubbling of gases from the lower end of the glass tube, this rapid outward diffusion of hydrogen results temporarily in a lower total gas pressure inside the cup than that originally present. Hence the solution rises rapidly in the glass tube. Finally, nitrogen and oxygen diffuse slowly inward, because of the reduced concentration of these gases inside of the cup, and the liquid slowly falls in the glass tube to its original level.

That the process of diffusion may result in the development of pressure also can be shown very strikingly by means of another simple experiment. If a pure rubber balloon containing only a little air be suspended in a closed bottle of carbon dioxide gas, it will gradually become distended. Rubber membranes are quite readily permeable to the molecules of carbon dioxide but virtually impermeable to those of oxygen, nitrogen, or other gases of the atmosphere. Since, temperature remaining constant, the pressure exerted by any gas is directly proportional to its concentration (number of molecules per unit volume), it follows that the diffusion of the molecules of gases may be interpreted in terms of the differences in the *partial pressure* exerted by that gas in different parts of a system. Because of the greater partial pressure of carbon dioxide in the surrounding atmosphere, diffusion of this gas through the walls of the balloon continues until the partial pressure of the carbon dioxide is the same on the two sides of the rubber membrane. Since the initial partial pressure of carbon dioxide inside the balloon was virtually zero, and that in the bottle equivalent to 1 atm., the inward diffusion of carbon dioxide gas results in a considerable distension of the balloon.

In the analysis of diffusion phenomena it is often clarifying to speak of the partial pressure of a gas as its *diffusion pressure*. Solutes and liquids may also be considered to possess a diffusion pressure, although the existence of such a physical quantity is less evident in such systems than in gases. Hence the concept that diffusion is the movement of the molecules of a substance from a region of its greater to a region of its lower diffusion pressure is usually more satisfactory than the interpretation of diffusion phenomena in terms of concentration differences.

Principle of Independent Diffusion. In the first experiment described in the preceding section it was shown that while one gas (hydrogen) was diffusing in an inward direction through the pores of a clay cylinder, other gases (oxygen and nitrogen) were simultaneously diffusing in an outward direction through the same pores. This exemplifies one of the most important principles governing diffusion phenomena—namely, that the direction in which any substance will diffuse is controlled entirely by its own differences in diffusion pressure, and is not influenced by either the direction or rate of diffusion of other substances in the same system. Hence, in any given system, as, for example, two adjacent plant cells, a number of substances may be diffusing in one direction across the intervening membranes, while simultaneously other compounds may be diffusing in the opposite direction across the same membranes. Each one of these individual substances will diffuse in the direction determined by its own differences in diffusion pressure, and at a speed which is determined by the factors that are influencing the diffusion of that particular substance.

Factors Influencing the Rate of Diffusion of Gases. 1. *Density of the Gas.* Different gases diffuse at different rates even when influenced by the same set of environmental factors. Hydrogen, for example, diffuses more rapidly than any other gas. The same, Thomas Graham who conducted some of the earliest studies on the properties of colloidal systems also was one of the first investigators to study quantitatively the phenomenon of gaseous diffusion. He discovered the principle, often called "Graham's Law of Diffusion," that the relative speeds of diffusion of different gases are inversely proportional to the square roots of their relative densities. By relative density is meant the weight of a given volume of gas as compared with the weight of the same volume of hydrogen. The relative density of oxygen is 16. Hence the rate of diffusion of hydrogen is proportional to $1/\sqrt{1}$ while that of oxygen is proportional to $1/\sqrt{16}$. Hydrogen gas will therefore diffuse four times as rapidly as oxygen gas under the same conditions of temperature and pressure. The relative density of carbon dioxide is 22 ; hence by similar reasoning it is apparent that hydrogen gas will diffuse nearly five times as rapidly as carbon dioxide gas.

2. *Temperature*. Increase in temperature increases the speed of diffusion. This is due, at least in part, to the correlated increase in the kinetic activity of the molecules of the diffusing substance. Actual measurements of the Q_{10} of diffusion generally yield values between 1.2 and 1.3. Such values are characteristic of any purely physical process such as diffusion.

3. *Diffusion-Pressure Gradient*. The fact that the *direction* of the diffusion of gases is from a region of their greater diffusion pressure to a region of their lesser diffusion pressure has already been emphasized. The *speed* of diffusion is also influenced by differences in diffusion pressure. In general, the greater the difference in diffusion pressures between the two regions, the more rapidly diffusion will occur. The rate of diffusion is influenced, however, not only by the difference in diffusion pressures, but also by the distance through which the diffusing molecules must travel. These two factors are components of what may be called the *diffusion-pressure gradient* or *concentration gradient*. This concept may be clarified by a specific

illustration, such as that depicted in Fig. 1.16. In both parts, A and B, of this biagram, oxygen is represented as diffusing

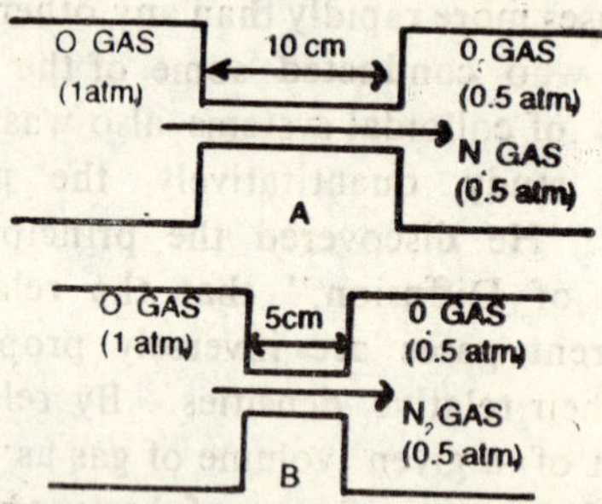

Fig. 1.16. Diagram illustrating that the length of a diffusion gradient is a factor influencing its steepness.

from a region in which its diffusion pressure is maintained at 760 mm. Hg (atmospheric pressure) into a region in which its diffusion pressure it just half as great. The length of the connecting tube is, however, just twice as great in A as in B. Under the conditions postulated the rate of diffusion from the region of greater diffusion pressure to the other will be just twice as rapid in B as in A. The diffusion-pressure gradient is equivalent to the difference in the diffusion pressures between the delivering and receiving ends of the diffusion system divided by the length of the distance between. The greater or "steeper" this gradient the more rapidly diffusion will occur. The steeper the gradient, the more rapid the change in diffusion pressure per unit of length along the axis of the diffusion gradient. The steepness of a diffusion-pressure gradient may be changed by varying either factor, the difference in diffusion pressures, or the length of the gradient.

4. *Concentration of the Medium through Which Diffusion Occurs*. In general the more concentrated the medium, *i.e.*, the more molecules per unit volume in the medium through which the diffusing molecules must pass, the slower the rate of diffusion. Bromine gas, shown earlier in this chapter, diffuses more rapidly through a vacuum than through air.

Diffusion of Solutes. The molecules or ions of a solute possess sufficient kinetic energy to move from place to place within the limits of a solution. The simplest method of demonstrating the diffusion of a solute is to introduce a crystal of copper sulfate, or some other compound which is coloured when make in solution, into the bottom of a tall glass cylinder filled with water. The cylinder should then be placed in an environment of equable temperature where it will be free from disturbance. The diffusion of the molecules or the ions which pass into solution in the water can be followed by the slow change in colour of the water. One of the most striking facts illustrated in such experiments is the extremely slow rate of diffusion of solutes through water. This results partly from the fact that in such an experiment the steepness of the diffusion gradient decreases with time, but principally from the fact that the densely packed molecules of the liquid enormously impede the diffusion of the dissolved molecules or ions.

The true rate of diffusion of solutes is probably even less than the rates indicated in such experiments, because in all such setups convection currents may develop in the water and aid in the distribution of the solute throughout the body of the liquid. The diffusion of solutes is often demonstrated by employing a gel rather than a liquid as the medium into which diffusion occurs. Such a technique avoids errors introduced by the development of convection currents.

The *direction* of the diffusion of any solute occurs in accordance with its own differences in diffusion pressure, regardless of the rate or direction of diffusion of other solutes in the same system. The *rate* of diffusion of solute particles is governed by principles essentially similar to those which control the rate of diffusion of gases and is controlled by the following factors :

1. *Size and Mass of the Diffusing Particles.* Small molecules or ions diffuse more rapidly than large ones. A hydrogen ion, for example, diffuses many times more rapidly than a glucose molecule. Similarly, highly hydrated ions diffuse more slowly than those which have fewer water molecules bound to them, since the association of water of hydration with a molecule or ion in effect increases its size. The mass of the particle will

also be a factor influencing the speed of its diffusion. As between two particles of the same size, but different masses, the heavier particle will diffuse more slowly.

2. *Temperature*. The kinetic activity, and hence the rate of diffusion of solute molecules, increases with increase in temperature.

3. *Diffusion Gradient*. The steeper the diffusion gradient, the more rapidly solute particles diffuse.

4. *Solubility*. In general, the more soluble a substance is in a liquid, the more rapidly it will diffuse through that liquid. This influence of solubility upon diffusion rates can be interpreted principally in terms of its effect upon the diffusion gradient, since obviously steeper gradients can be built up if the solute is very soluble in the liquid than if it is only slightly soluble.

2

Soil Water

Water plays several different roles in the ecological relations of plants and animals. As a material entering the organism water is important as a necessary and abundant constituent of protoplasm, and the plant or animal body as a whole generally continus a large percentage of water—sometimes 90 per cent or more. Water is essential also as one material taking part in the photosynthetic reaction, through which energy becomes available either directly or indirectly to all living beings. Water is necessary as a solvent for food and as an agent for the chemical transformation of the materials within the boby. Plants obtain mineral nutrients from the soil after they are in aqueous solution. Although most animals take in their food in solid form, this material must be dissolved before it can be absorbed by the blood and tissues. The intestines could be stuffed full of solid food and yet the animal would starve if no water were available for its digestion. Furthermore, water serves as a vehicle for transport or circulation within the bodies of organisms. Minerals are carried up the stem of the plant by the transpiration stream. Water is the principal constituent of the circulatory, excretory, and reproductive fluids of animals, and it is necessary as a transfer agent at respiratory and olfactory surfaces. Water also acts as a regulator of temperature for plants and animals.

For these essential purposes the proper concentration fo water must be maintained inside the organism and, at the same time, the transfer of water must be suitably regulated. If the organism could live hermetically sealed in a capsule, the retention of the necessary amount of water would be easy, but exchange of this material between the organism and the outside world is necessary. The crucial need for a proper water balance is demonstrated by the consequences of even small water losses in some instances. During starvation, for example man may lose as much as 40 per cent of his body weight including half of the proteins and nearly all the glycogen and fat without serious danger, but if 10 per cent of the water content of the human body is lost, serious disorders result. If as much as 20 per cent of the water is lost, death follows.

The concentration of water divides the environment into aquatic and terrestrial habitats. At first sight, one might say that more than enough water exists in the aquatic environment and that water is a problem only on land. Closer scrutiny shows, however, that water is tending to enter or to leave the organism too fast in almost every situation. In certain terrestrial habitats the water supply may be excessive for many organisms, as in some tropical rain forests where the air is often 100 per cent saturated and the ground is completely permeated. A visitor to such a habitat sees moisture condensing on every surface and hears the steady dripping of water from the vegetation.

In the ocean in spite of the thousands of cubic miles of sea water a scarcity of water would nevertheless exist for many plants and animals ! For these organisms the concentration of water is too low relative to the abundance of salt for proper osmotic equilibrium. Fresh-water environments, viewed similarly in relation to osmotic balance, tend to contain too much water. This contrast between marine and inland aquatic situations is, it a sense, analogous to the difference between dry and humid climates on land. The maintenance of the proper water balance is thus a problem that must be considered in all habitats for all types of organisms.

WATER PROBLEM IN THE AQUATIC ENVIRONMENT

Since the water in natural environments contains a varying amount of dissolved materials and usually a different amount from the fluids of the organism, the resulting differences in osmotic pressure raise a problem in regard to water exchange. In order to determine the tendency of water to enter or to leave the organism we must know something of the composition of the surrounding medium.

Composition of Natural Waters

Representative values for the amounts of the more numerous ions found in sea water and in the water of ponds and lakes are shown in Table 2.1. A larger amount of each of the common ions occurs in sea water than in typical fresh water. Hard fresh water contains more dissolved salts than soft fresh water, particularly with respect to the calcium and the carbonate ions. A comparison of the *relative* amounts of each type of ion in each kind of water may be made by recalculations in which sodium is given a value of 100 in each case (Table 2.2). In sea water chloride and sodium

Table 2.1. Composition of some typical natural waters.

	Na	*K*	*Ca*	*Mg*	*Cl*	SO_4	CO_3	*Total (g/litre)*
Soft fresh water	0.016	—	0.010	—	0.019	0.007	0.012	0.065
Hard fresh water	0.021	0.016	0.065	0.014	0.041	0.025	0.119	0.30
Sea water	10.7	0.39	0.42	0.91	19.3	2.69	0.073	34.9

Table 2.2. Relative Abundance of Ions (by Weight) in Natural Waters

	Na	*K*	*Ca*	*Mg*	*Cl*	SO_4	CO_3
Soft fresh water	100	—	62.5	3.3	119.0	43.8	75.0
Hard fresh water	100	76.0	310.0	66.8	195.0	119.0	567.0
Sea water	100	3.6	3.9	12.1	181.0	20.9	0.7

are the first and second most abundant ions, respectively, whereas in typical hard fresh water carbonate is the most abundant with calcium second. In typical soft fresh water the calcium and carbonate ions are relatively less concentrated than sodium and chloride.

Samples of sea water from different localities are surprisingly constant in respect to the relative abundance of the major constituents. So much is this the case that for many purposes it is not necessary to measure the concentration of more than one ion. The standard procedure for determining the salinity of sea water is to find the amount of chloride ion present by titration and then to calculate the total salinity and the concentration of the other ions from the known ratios. The individual salts occurring in fresh water by contrast, vary widely—often with far-reaching consequences that will be considered later.

At the moment our concern is with the maintenance of the proper osmotic balance between the inside and the outside of the organism. Since osmotic pressure is determined by the total concentration of molecules and ions in solution, values for the aquatic medium will vary according to the total salt content of the water body (Table 2.3). Dissolved materials in soft fresh water vary in amount from practically zero to about

Table 2.3. Total salt content of some typical natural waters.

Soft fresh water	>65.0 ppm	0.065%	0.0065%
Hard fresh water	300.0	0.300	0.03
Devils Lake, N.D.	11,000.0	11.0	1.1
Great Salt Lake	170,000.0	170.0	17.0
Baltic Sea		8.0	0.8
Ocean water		35.0	3.5
Red Sea		40.0	4.0

65 parts per million. Lakes and ponds with hard water exhibit a variety of higher values for salt content, but the value given in the table, 300 ppm, is representative. Salt lakes may contain exceptionally large amounts of dissolved substances.

In contrast to the extreme variability in the salt content of inland waters, the open ocean tends to remain highly uniform in salinity over huge areas (Fig. 2.1). A fish could swim up the middle of the Atlantic Ocean, for example, from the Cape

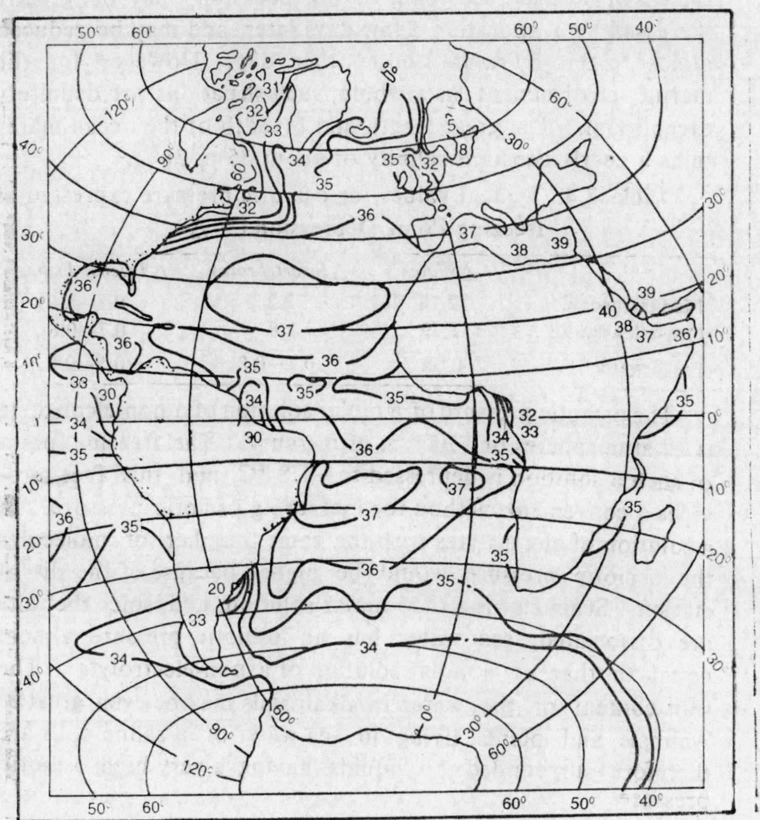

Fig. 2.1. The surface salinity of the Atlantic Ocean.

of Good Hope almost to Iceland without encountering a salinity, lower than 35% or higher than 37% (parts per thousand). Ocean areas with unusually high evaporation, like the Mediterranean Sea and the Red Sea, exhibit salinities that run up to 40%. On the other hand, the salinity of the Baltic Sea is reduced to 8% or less due to the large inflow of fresh water.

In river estuaries and in coastal areas receiving large amounts of run-off, salinity may vary over the whole range from nearly fresh water to full sea water. In smaller habitats, such as rock pools filled by ocean spray, the salinity may be typical of ordinary sea water on one occasion, may be greatly increased by evaporation a few days later, and may be reduced almost to zero by rainfall on another day. However, for the marine environment as a whole, such situations are definitely exceptional. The great length and breadth of the ocean maintains a nearly constant salinity of about 35 to 36%.

Table 2.4. Typical values for osmotic pressure expressed as freezing Point Depression (= △).

	△*Water*	△*Invertebrates*	△*Teleost Fishes*
Mediterranean	2.14	2.2	0.8—1.0
Coastal Atlantic	1.79	1.8	0.7—0.8
Fresh water	0.03	0.1–0.8	0.5—0.6

The osmotic pressure of a molar solution of a non-electrolyte is 22 atmospheres or 1700 cm of mercury. The freezing point of such a solution is depressed to—1.84°C, and this fact provides a convenient method for measuring osmotic pressure. In a solution of electrolytes with the same number of molecules the osmotic pressure would be higher because of the dissociation. Sea water is a 0.55 molar solution, and, since the salts are dissociated, sea water has an osmotic pressure almost equal to that of a molar solution of a non-electrolyte. The salt content of the water in alkali soils may be even greater. Animals and plants living in sea water or in saline soils are therefore surrounded by liquids having a very high osmotic pressure.

WATER PROBLEM IN THE TERRESTRIAL ENVIRONMENT

When we turn to dry land as an environment, the water problem becomes extremely acute. Everyone is familiar with the quick death which awaits aquatic organisms brought out on land. Many terrestrial forms can endure the land environment only if they remain in damp places. This is true for a great many species of plants and also for animals without special protection. A salamander that has escaped from its terrarium and run across the dry floor will soon be dead if it is not rescued and put back in a humid atmosphere. Earthworms frequently crawl out on a concrete sidewalk during a spring rain. If the sun comes out and dries the sidewalk, many of the animals will be killed before they can find their way back to the moist ground.

The successful invasion of the dry land is dependent upon the possession of methods for securing and retaining sufficient water and at the same time allowing adequate exchange with the environment. Terrestrial plants and animals as well as aquatic forms must maintain a proper *water balance* between water income and loss. Great variation exists in the moisture content of land organisms and also in the amount of water intake and output that is necessary to provide for metabolic processes. Insects subsisting on dry food probably hold the record for the minimum water exchange. At the other end of the scale are many types of higher plants that are extravagant water users. Some species of plants absorb and transpire more than 2000 grams of water for every gram of dry matter produced by assimilation. An oak tree may transpire as much as 570 liters of water in a single day.

How many taxonomic groups have succeeded in colonizing dry land ? Actually, only a very few of the prominent kinds of plants and animals have accomplished this. In the plant kingdom it is chiefly the vascular forms that are able to live in really dry places, but some low-growing bryophytes, algae, and fungi—particularly the rock lichens—form noteworthy additions to the list. Among the vertebrates, only the reptiles, birds, and mammals can live freely exposed to dry air. Although several phyla of invertebrates are reported in the land fauna, only the insects, spiders, and snails occur in important numbers in dry habitats. By contrast, there are a great many plant and animal groups that are exclusively aquatic.

Occurrence of Water in Land Environment

The only ultimate source of water for the terrestrial environment is condensation—chiefly in the form of rain. If precipitation were evenly distributed, it would cover the earth to a uniform depth of 1 m each year. Quite the contrary is the case. A glance at Fig. 2.2 will reveal the irregularity in the geographical distribution of rainfall. In the desert areas the rainfall is less than 25 cm per year and is inadequate for most organisms. In the black areas of the chart an excessive rainfall of 200 cm or more is reported each year. The mean annual rainfall on the south side of the Himalayas is recorded as 1232 cm. At the other extreme is Iquique, Chile, with an average precipitation of 0.125 cm per year. Among the Olympic Mountains in the state of Washington the annual rainfall totals as much as 380 cm on the windward side of some ridges and as little as 25 cm on the leeward side.

Seasonal variations in the distribution of rain may be of even greater importance than the geographical aspect. In many regions the division of the year into a rainy season and a dry season is of more ecological significance than the change in temperature between summer and winter. In the climatographs shown in Fig. 2.3 two situations are contrasted. In the Chicago region representing a temperate climate, an average of 5 to 9 cm of rain falls in every month of the year, but the average monthly temperature varies from about −4°C in the winter to about 22°C in the summer. In the tropical climate of Barro Colorado Island the average monthly temperature changes by not more than 1 or 2°C throughout the year, whereas the precipitation drops to less than 3 cm per month in the dry season and rises to more than 40 cm per month in the wet season.

Wherever the amount of precipitation varies greatly during the course of the year, the time of occurrence of rain in relation to the temperature cycle has great effect on the vegetation. Only those species will thrive whose varying needs for water in the different life stages are satisfied by the seasonal distribution of available moisture. In many of the grass-covered or forested areas of the temperate zone the major portion of rain occurs in the summer, but in other regions, of which southern California is an example, most of the precipitation occurs in the winter. In the latter situation we find a special type of vegetation with broad evergreen leaves known as the sclerophyllous

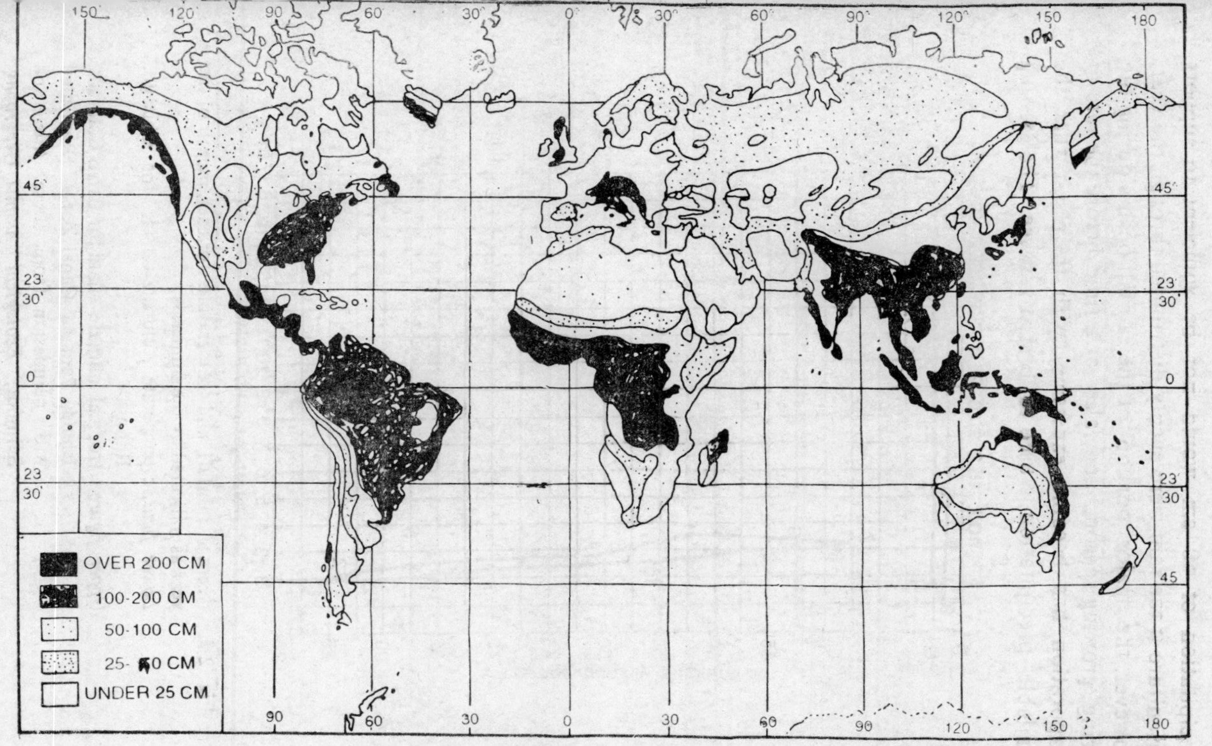

Fig. 2.2. Geographical distribution of mean annual rainfall.

forest. In the prairie provinces of Canada the average annual precipitation of 50 cm would not be sufficient to support agriculture if the rain were evenly distributed through the year. However, the major portion of the rainfall occurs during the spring growing season, and also at a time before the rate of evaporation has become excessive, with the result that the available precipitation is used to best advantage. The existing

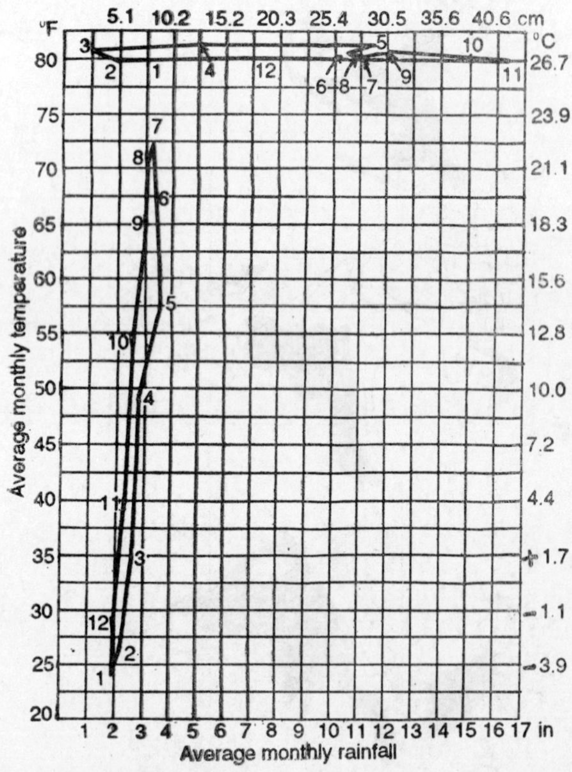

Fig. 2.3. Average rainfall and temperature for each month of the year as indicated by the numerals.

Lower figure : temperate climate—data for Chicago, Illinois.

Upper figure : tropical climate—data for Barro Colorado Island, Panama Canal Zone. (Reprinted with permission from Hesse, Allee, and Schmidt, *Ecological Animal Geography*, 1951, John Wiley and Sons, New York.)

combination of ecological conditions thus permits western Canada to be an important wheat-growing region.

The foregoing discussion brings into relief the fact that in considering moisture conditions on land the actual amount of the precipitation is only part of the story. The other part, and often the more important part, is the loss of water. Of the rain that falls upon the surface of the soil, part runs off immediately, part sinks in, and another portion is lost by evaporation. The amount of moisture at any one time and place depends upon the relative rates of the supply of water to the soil and evaporation from it—another example of an ecological factor whose value depends upon an equilibrium. In evaluating the water factor in the terrestrial environment both supply and loss processes must be taken into account.

Moisture in the soil. Since the terrestrial environment is so varied, it is not surprising that the water factor is very complex, and the moisture in the soil and in the air will first be considered separately. When rain water enters the soil, it fills the spaces between the particles. The volume that can be filled is known as the *pore space* and commonly varies between 60 per cent for heavy soils and 40 per cent for light soils. When the rain ceases, a certain portion of the water in the soil soon drains out, and this is known as the *gravitational water*. The portion of the soil water that is held by capillary forces around and between the particles is the *capillary water*. That part of the pore space which is not filled by gravitational or capillary water may be occupied by water vapour. Another from of soil water, termed the *hygroscopic water*, occurs as an extremely thin film on the soil grains but this cannot move as a liquid. A small portion of the soil water is chemically bound with soil materials and is known as *combined water* The total amount of capillary, hygroscopic, and combined water plus the water vapour constitutes the *field capacity* and is the maximum amount of water that the soil can hold after the gravitational water has drained away (Fig. 2.4).

The maximum potential supply of water available for the vegetation is represented by the full field capacity of the soil.

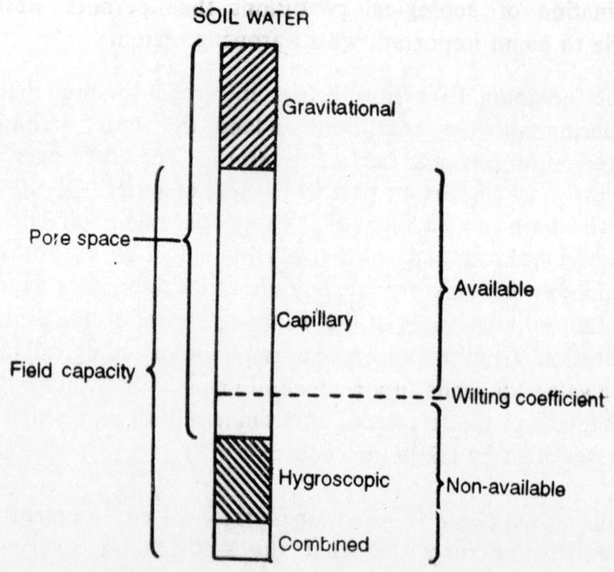

Fig. 2.4. Generalized diagram of the forms of soil water. The relative proportions and the value of the wilting coefficient differ according to the nature of the soil.

As plants draw on the soil water, they gradually reduce the amount held in capillary spaces. Below a certain moisture content plants tend to wilt in the hottest part of the day. When the capillary water has been still further depleted, a point is reached at which the plants will not recover from wilting until more water is added to the soil, regardless of other environmental conditions. At this point absorption of water by the plants has become too slow to replace the water lost by transpiration The moisture then remaining in the soil is designated as the *permanent wilting percentage*, or the *wilting coefficient*. Since little difference has been found in the abundance of soil water when wilting occurs for plants of various species growing in the same soil, the permanent wilting percentage is primarily a characteristic of the soil, and as such has great ecological significance. The amount of water between the full field capacity, as a maximum, and the wilting coefficient, as a minimum, represents the *available water*. As indicated in Fig. 2.4, a considerable amount of water may still remain in the soil after the wilting coefficient has been reached. A fraction of the capillary water, the hygroscopic water, and the combined water, as well as the water vapour, cannot be obtained by the plant, and these constitute the *non-available water*.

Soils vary considerably in the relative proportions of the different categories of soil water. In some instances the amount of non-available water may actually be larger than the available water. The fact that a portion of the water in the soil is unavailable to land organisms and that water in the sea is unavailable to aquatic organisms not adapted to its osmotic pressure contributes further to the general concept of the universality of the water problem. From the ecological point of view the actual amount of water present in any habitat is not as important as its availability.

Moisture in the Air. For many terrestrial organisms the moisture in the soil is chiefly important as constituting the principal source of water, whereas the moisture in the air is mainly significant as controlling the loss of water. For this reason the *absolute humidity*, or total amount of water in the air, is generally of far less ecological consequence, than the *relative humidity*. The ecologist therefore focuses his attention on the relative humidity or the amount of moisture in the air as a percentage of the amount which the air could hold at saturation at the existing temperature. Since the capacity of air for water vapour increases with temperature, the relative humidity of the atmosphere is reduced in any situation in which an increase of temperature occurs without an accompanying increase in the total moisture content of the air.

In the terrestrial environment the ecological effect of the water factor is consequently strongly influenced by the temperature factor. Of two regions having the same rainfall the warmer is the drier in the ecological sense (Fig. 2.5). The climate of a locality with a mean annual precipitation of 50 cm would be characterized as humid if the mean annual temperature were—7°C or less. On the other hand, another locality with the same rainfall would be regarded as having a

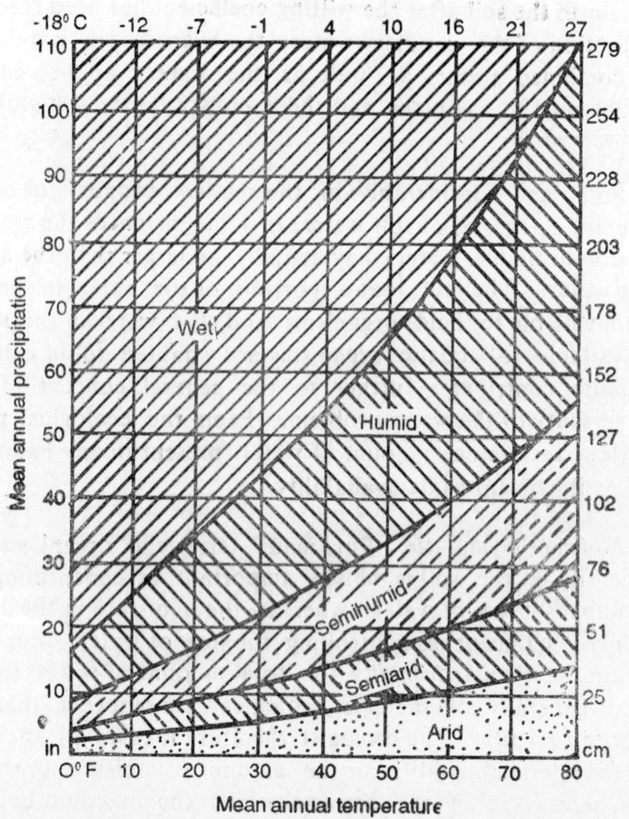

Modified from McDougall, 1925

Fig. 2.5. Schematic representation of the influence of rainfall and temperature on climate.

semiarid climate if the annual temperature were 21°C or more. The contrast in climate between the Canadian prairie and the Mexican desert, both with about 50 cm of rain but with very different temperature conditions is an excellent illustration of this principle.

The geographical variation in relative humidity is very great. Relative humidities of 80 to 100 per cent characterize the tropical rain forest. Regions reporting values of less than 50 per cent are regarded as having dry climates, and those with values of less than 20 per cent are extremely arid. It is of interest to note in passing that in cold winter weather the relative humidity inside our houses is rarely higher than 35 per cent.

At any one locality the relative humidity may remain relatively constant for long periods of time or may vary widely. On many oceanic islands the humidity is very nearly the same throughout the year. In other localities, characterized by wet and dry seasons, the humidity fluctuates widely from one part of the year to another. In certain situations as on the plains and in desert regions considerable changes in moisture content may occur during the course of each day. Records made in a short-grass prairie in the United States showed a variation from a relative humidity of less than 30 per cent in the early afternoon to more than 95 per cent in the middle of the night (Fig. 2.6). Obviously animals and plants living in such habitats must be equipped to withstand these rapid and extensive changes in the water factor, and their lives must be attuned to them. In deserts where daytime humidities are extremely low and evaporation is excessive, the greater relative humidity at night helps to relieve the critical moisture condition. Many desert animals take advantage of this situation by going abroad only during hours of darkness ; in some desert plants the stomata open only at night when transpiration loss is at a minimum.

The foregoing has been a brief oescription of the amount of moisture in the air in terms of its relative humidity. As already implied, the chief ecological significance of the relative humidity is its effect on the rate of water loss. Terrestrial animals and plants lose water directly to the air by evaporation and transpiration. The water supply in their substratum is also reduced by direct evaporation from the soil and indirectly by the transpiration of the vegetation. Evaporation takes place very rapidly from soil because of the great surface area presented by the fine particles and causes the drying of the upper layers. Water is lost from the deeper layers chiefly by absorption by roots and attendant transpiration.

The magnitude of water loss by transpiration is often not appreciated. Perhaps some idea can be obtained from the fact that a hectare (2½ acres) of mature oak trees will transpire as much as 25,000 liters of water per day. This volume would be equivalent to a layer of water 0.25 cm deep over the whole area. In other words, if transpiration continued at this rate for 10 days, a rainfall of at least 2.5 cm would be required to restore the water. One might think that the maximum rate of water loss would occur from saturated soil or from a pond. Actually the rate of water loss per unit area from the ground due to the transpiration of the vegetation may be nearly twice as great as that from a free water surface due to evaporation.

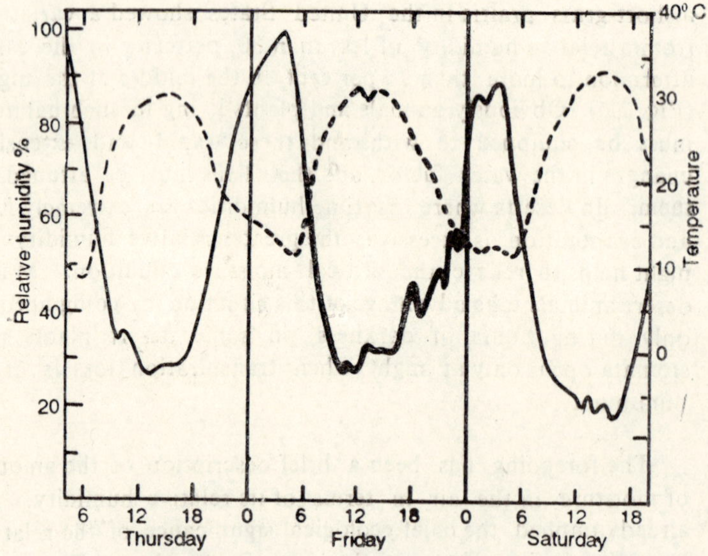

Fig. 2.6. Hygrothermograph record or temperature (*broken line*) and relative humidity (*solid line*) in the short-grass plains of central United States during days in early July.

Three atmospheric conditions that greatly modify the rates of evaporation and transpiration are the saturation deficit, the temperature, and the wind velocity. The *saturation deficit* is the difference between the actual vapour pressure and the maximum possible vapour pressure at the existing temperature.

The saturation deficit thus gives more information of ecological significance than the relative humidity alone since the saturation deficit provides a measure of the capacity of the air to take up additional moisture. An increase in the saturation deficit produces a rise in evaporation rate.

An increase in temperature similarly speeds up the evaporation process A good indication of the magnitude of this influence is obtained from the amounts by which reservoirs are lowered in regions of different temperatures. During one year the water level in a reservoir in Ontario dropped 38 cm, a reservoir in California lost 2.4 m, and one in Egypt was lowered 3 6 m through evaporation alone. The distribution of plants often reveals the influence of temperature on direct and indirect water loss. On mountain slopes vegetation zones are found at different altitudes according to the exposure to the heat of the sun. Moisture-demanding species are restricted to higher levels on the side of the mountain toward the equator than they are on the cooler opposite side. The variation in the amount of rain required for a good growth of short grass furnishes another illustration of the effect of temperature. In Montana a precipitation of about 35 cm is sufficient, but in northwestern Texas a rainfall of 53 cm is needed to produce the same amount of grass. The explanation of this fact is easily apparent when it is realized that evaporation in the 6 summer months in Montana amounts to 84 cm, whereas in Texas it averages 137 cm.

The wind velocity also exerts a major effect on the loss of water. With a gentle zephyr of only 8 km per hr the transpiration of plants is increased 20 per cent over that which they exhibited in still air. A wind velocity of 16 km per hr increases the transpiration by 35 per cent and one of 24 km per hr increases it by 50 per cent. The desiccating action of warm dry winds sometimes prevents the invasion of windward mountain slopes by vegetation that grows perfectly well on the leeward sides of the same slopes where water loss is much reduced. The harmful effect of excessive evaporation caused by the wind may also be seen in one's own garden. Here winter killing is occasionally due not to very low temperatures, but to a combination of wind and high temperatures ! If a warm, dry wind blows for too long a period when the ground is frozen, the plants may lose water faster than their roots can obtain it and they will succumb as a result.

Our discussion of moisture in the soil and in the air has indicated that living organisms—particularly the higher plants—may exert a profound reciprocal influence on the water conditions in the terrestrial environment. Water is a highly modifiable factor in land habitats, and living elements may act either to augment or to diminish the amount of moisture present. Vegetation tends to catch the rain, so that, if the precipitation comes in short showers, the rain may never reach the soil. The plant cover also tends to reduce the water content of the deeper layers of the soil as a consequence of its transpiration, but at the same time it adds moisture to the air. On the other hand, the presence of forest vegetation favours the reduction of evaporation at levels near the ground because it lowers the temperature, shows down the wind, and sometimes causes increased condensation.

The amount of water present at any point in a land habitat and the rates of gain and loss are thus seen to be the result of the equilibrium of processes both climatic and biological. The interrelations of the water factor and the vegetation are frequently very complicated and are the subject of elaborate studies beyond the scope of this book. In addition to presenting the general picture, the present discussion reveals the interdependency of vegetation and moisture as another outstanding example of the organism and the environment acting as a reciprocating system.

MICROCLIMATES

In the foregoing discussion of the variation in water loss under various conditions of humidity, temperature, and wind the values given were often those characteristic of extensive areas. Very frequently, however, the values of these factors in the immediate surroundings of the organism differ markedly from the regional values. In other words, the immediate climate of the organism is often sharply different from the average climate of the region as reported by standard meteorological records. The realization of this fact has led to the development by ecologists of the concept of the *microclimate*. In considering the effect of climatic influences on plants and animals it is absolutely essential to measure the environmental factors as they actually reach the organism. The effective climate, upon which the life of the organism depends, is its microclimate. The characteristics of the microclimate may, or may not, be similar to the regional climate.

Local variations in climatic factors of concern to many organisms may involve distances of several meters in the surroundings, but the microclimates of forms living in the soil, in rock crevices, under the bark of trees, or in similar confined habitats often have dimensions measured in fractions of a centimeter. The concept of microclimates is particularly important in relation to moisture because relative humidity varies widely within short distances in any irregular habitat. In swampy woodland, for example, measurements have shown that a high rate of evaporation characteristically exists near the tops of the trees. At half this height the evaporation rate was reduced to 30 per cent in one instance. Lower down, near the damp soil, the animals and plants were subject to an evaporation rate of only 7 per cent of that near the treetops. We are consequently not surprised to find that this vertical change in moisture, as well as vertical gradients in light and in other factors, contributes toward the establishment of a characteristic vertical layering by species of the vegetation and sometimes also of the arboreal animal life in forests. Among animals the difference in evaporation rate with distance above the ground is of particular importance to soft-bodied insects. Many of these are known to move higher in the vegetation at night, but during the day they are forced by excessive evaporation to retreat to lower levels.

Contrasting values for the rate of water loss in two microclimates were reported in an interesting experiment performed in the Middle West. Two sets of green ash seedlings were planted in pots ; one set was placed in the midst of a sumac thicket, and the other set was put out on the open prairie only a short distance away. At the end of 8 days the first group of seedlings was found to have lost an average of 0.38 grams of water per sq cm of leaf surface. The seedlings set out in the adjoining prairie had lost 0.88 grams of water per sq cm, or more than twice as much during the same period.

A traditional method of finding the approximate direction of north in the woods is by observing on which side of tree trunks the growth of green "moss" is thickest. In the North Temperature Zone in habitats in which the atmospheric moisture is near the critical point, algae and other non-vascular plants are characteristically more abundant on the north surfaces of tree trunks, because the higher evaporation due to insolution effectively retards growth on the south side. However, in many situations differences in exposure to wind may outweigh the effect of the sun in causing excessive water loss. Thus, although the "north side" rule does not always hold, direct measurements show that the growth of epiphytic moes is a very sensitive indicator of microclimate, and the striking differences in the growth of algae on opposite sides of tree trunks can be seen by anyone who walks through the woods. An average value for the moisture condition of the forest would give no inkling of these small-scale ecological differences of crucial importance to the species concerned.

The burrows of desert animals furnish another example of a microclimate in which the physical conditions are radically different from those of the region as a whole. When the kangaroo rat retreats within its burrow, it breathes air two to five tinnes more humid than that outside. Measurements indicate that, if this animal remained out of its burrow continuously, the rate of evaporation from its lungs would exceed the rate at which it could obtaid water, and the resulting water loss would eventually cause death. Temperatures in desert burrows have been found to be as low as 28°C at the same time that values for the surface of the soil surpassed 71°C. Such critical differences in temperature and in other factors in microclimates will be considered further in subsequent chapters.

MEETING WATER PROBLEM ON LAND

Land plants can live only where sufficient water reaches them either through the air or through the soil. Some species can get along with relatively small amounts of water, and these plants, inhabiting deserts and other arid regions, are termed *xerophytes.* Plants eke out an existence on a reduced water supply in a variety of ways. Certain annual plants are adapted to maintain themselves in desert areas by remaining in the seed stage during the dry season and then passing rapidly through their life cycles when a rainy period occurs. Other desert plants, like the cacti, store water in succulent organs in sufficient quantities to tide them over long intervals of drought. In addition, various species of non-succulent perennials, such as the creosote bush, sage brush, and many grasses are included among the xerophytes because they possess unusual ability to endure long periods of permanent wilting—in some instances running into years. Lichens and some mosses similarly can survive a surprising degree of drying out. Plants with water relations intermediate between those of the xerophytes and the hydrophytes are classified as *mesophytes.*

Since land animals can move about, they can actively seek pools of water from which to drink or can obtain water in the form of snow or dew. Many are able to endure long periods between visits to the water supply. Some terrestrial animals, such as frogs and toads, have the ability to absorb water through their skins from damp surroundings, whereas others find sufficient liquid water in the tissue fluids of their food. Another method of solving the problem of securing water in dry land habitats, and one employed widely by the insects, is the use of *metabolic water*, that is, the water resulting from the chemical breakdown of food materials. The clothes moth and the meal worm have the ability to obtain metabolic water in sufficient quantity to enable them to live their whole lives without any access to free water in the environment. Experiments indicate that the for the survival of a plant in relation to its water need may occur, during adult life. Plants differ greatly in the amounts of water that they must absorb from the soil and transpire. Those with a high water requirement, or *transpiration ratio*, are limited to habitats where the supply of moisture is adequate.

Differential ability to withstand drought will also play its role in controlling geographical distribution. If the soil moisture drops below the permanent wilting coefficient, most annuals and other tender plants will die within a matter of hours or days unless the soil water is renewed. On the other hand, succulents and plants that can aestivate are able to withstand varying periods of drought extending into years with some species. Thus the duration of drought periods is of crucial importance in determining whether a region may be inhabited by different plant types.

Local differences in the moisture factor were clearly shown to influence the species composition of the forest vegetation in a study made in Indiana. A series of observations in a line running from north to south over a ridge revealed the fact that soil moisture was 20 per cent higher on the north side of the ridge than on the south side. Moreover, the rate of evaporation was more than twice as great at stations on the south slope than at those on the north slope. Quadrat counts of the number of trees on the two sides of the ridge gave the tabulated results. Although other ecological factors undoubtedly played some part, moisture conditions were chiefly responsible for the very pronounced differences in the occurrence of the various species on the two sides of the ridge.

Species	*North Slope*	*South Slope*
Maple	202	31
Beech	112	0
White oak	0	80
Black oak	0	70
Hickory	4	44

Another example of the control of distribution by the moisture factor, and one on a much larger scale, is found in the regional occurrence of the natural vegetation in the United States. A glance at Fig. 2.7 will reveal the fact that in the central part of the country, leaving out of account the irregularities due to the mountains, the zones of principal vegetation types tend to run north and south. In the eastern part of the continent the typical natural vegetation is forest (in areas where man has not interfered). West of the forest zone is a belt of tall-grass prairie, and beyond that, running from Montana and North Dakota down to Texas is a zone of short-grass rangeland.

Still farther westward lies the desert belt. These fundamental zones are indicated schematically in Figure 2.8.

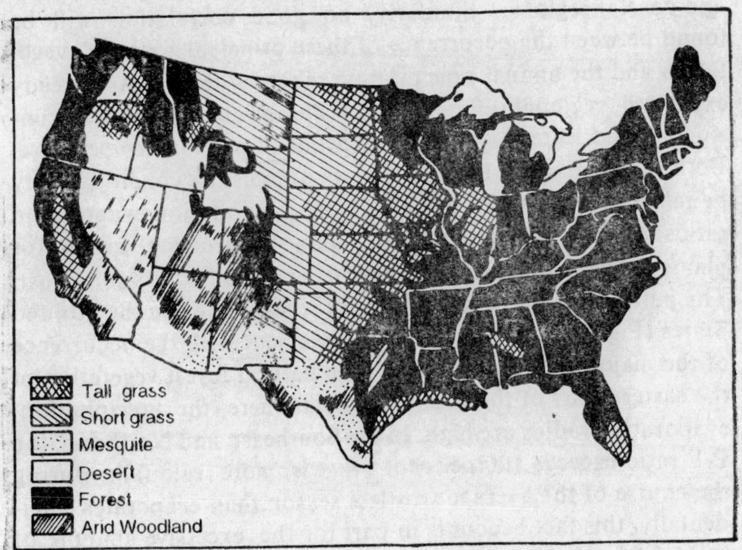

Modified from the Graphic Summary of American Agriculture, U.S. Department of Agriculture.

Fig. 2.7. The distribution of the natural vegetation in the United States.

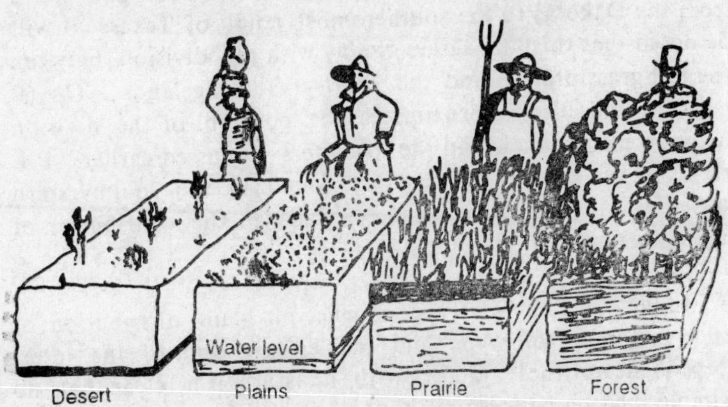

Fig. 2.8. Schematic representation of the four fundamental vegetational zones in central United States (from *This is Our World* by Sears, 1937, Univ. of Oklahoma Press).

This zonation is obviously not controlled primarily by temperature since the temperature belts are seen to run generally east and west, if we again leave out of account the mountain and coastal regions. Similarly, no good correlation will be found between the occurrence of these principal zones of vegetation and the annual precipitation taken by itself. As already explained, regional differences in evaporation must also be considered, and water loss in turn depends in part on temperature. Both the supply and the loss aspects of moisture conditions may be taken into account by calculating-precipitation evaporation ratios for the "frostless season," *i.e.*, the growing season for plants between the last spring frost and the first autumn frost. The pattern of the distribution of these ratios in the United States (Fig. 2.9) shows a general agreement with the occurrence of the major vegetation zones. The natural forest vegetation of the eastern part of the country occurs where the precipitation-evaporation radios are high. In the Southeast and Northeast, the P-E ratio exceeds 100 per cent ; that is, more rain falls during the course of the average frostless season than evaporates. Incidentally, this fact accounts in part for the excessive amount of soil leaching in these regions. In the Middle West precipitation-evaporation ratios between 80 and 100 per cent exist. Values below 100 per cent indicate that the potential rate of water loss is greater than the rate of water supply. The line representing a P-E ratio of 60 per cent runs almost due south from the Dakotas to the southernmost point of Texas. It will be noted that this line agrees closely with the division between the tall-grass prairies and the short-grass rangelands. The 60 per cent line also approximates the position of the division between the pedocals and the pedalfers discussed earlier. P-E ratios of less than 20 per cent are found in the southwestern states, and their occurrence corresponds to the distribution of the desert vegetation.

In view of the modifying influence on plant distribution of such factors as soil type and temperature and of the interdependence of these factors with moisture it is clear that no simple relation between the type of vegetation and any single physical influence can be expected. However, the broad corres-

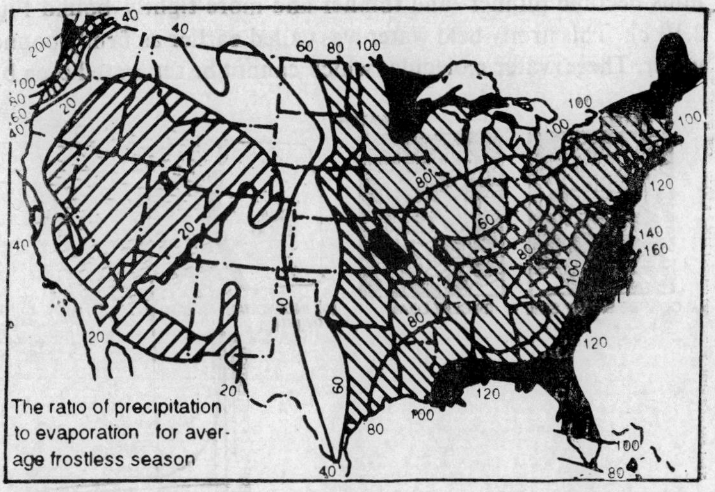

Modified from Livingston and Shieve, 1921

Fig. 2.9. The precipitation-evaporation ratio indicated as a percentage for the average frostless season in the United States.

pondence between the zones of vegetation and the availability of moisture will suffice to emphasize the major role of the water factor in the region under discussion in controlling both the soil and the plant life that develops with it. The application of the P-E ratios to these distributional problems has again brought into relief the fact that the balance between supply and loss is the most crucial aspect of the moisture factor for the vegetation.

SOIL WATER POTENTIAL

Water in soil may exist in three states *viz.*, solid (ice), liquid (soil solution) and gas (vapour). The water that enters into the soil either by rain or irrigation may pass through the surface soil layers to the water table and is known as 'ground water'. The water which is not drained deep into the soil profile is either retained in the soil pores and channels or on the surface of the soil particles. As can be seen from Fig. 2.10 *a* and *b*, soil particles are surrounded by the thin water films. These films of water are bound to soil particles by strong molecular forces of adhesion and cohesion. As soil dries out, these water films become thinner and thinner and more tightly bound Fig. 2.10 *c*). This firmly held water was called earlier as firmly bound water. These water molecules which cannot be separated even by

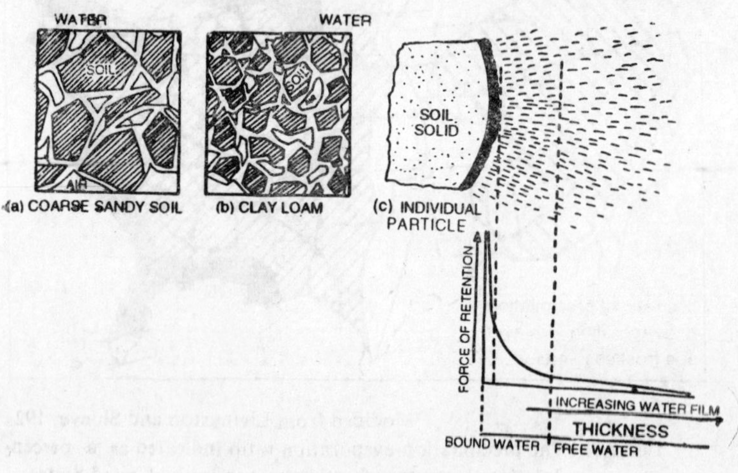

Fig. 2 10. Water retention by soil.

centrifugal action develop forces several tens of thousand times that of gravity. This layer of firmly bound water has a thickness of the order of few tens of molecular coats and in its turn, links and orients water molecules by which it is surrounded. The latter form of water was earlier known as loosely bound water. It is, however, not possible to draw a boundary line between firmly bound and loosely held water.

The distribution of water in the soil mass depends on the nature of the solid phase, *i.e.* the composition, the size distribution and the arrangement of soil particles. The hydrophilic nature of soil particles also determines the thickness of the water film and the force of retention on the solid surface. The large specific surface, both internal as well as external, of the colloidal soil particles is available for water adsorption. In view of the complex nature of soil particles, their surfaces and pore-size distribution, several other types of water have been recognized in soil, *viz.* vapour, hygroscopic, pellicular, capillary, gravitational, water in solid state, quasicrystalline and chemically bound water.

The water vapour fills the pore spaces which are not filled with liquid water. The hygroscopic water is the water that condenses on the soil particles. This water is adsorbed when dry soil particles come in contact with humid air. The pellicular water forms on the particles under the influence of one molecular forces of adhesion. Capillary water is the water held by the forces arising due to presence of thin bore capillary size pores and the air-water interfaces. Gravitational water is the free water that is easily drained by the force of gravity. The chemically combined water is in the crystal structure of clay minerals which cannot be removed even by drying the soil at 110°C. The amount of water present in a soil at a given time can be expressed quantitatively. The water content of a sample of soil is defined as the amount of water lost when dried to constant weight at 105°C. The water lost is expressed either as mass of water per unit mass of soil or volume of water per unit volume of soil. Recent studies have established some functional relationships between the soil water content and the different forces with which water is held by the soil or the energy status of water in the soil.

Because different terms used earlier in classifying soil water are too arbitrary to describe the state of soil water quantitatively, the energy concept of soil water is described using two fundamental approaches, one from the mechanical point of view based on the potential theory of classical physics where the concept of soil water potential will be discussed in detail and the second from the thermodynamical point of view where the chemical potential of soil water will be discussed. In the first approach the forces are considered with respect to mechanical equilibrium assuming that the soil-water system is under isothermal condition and there is no significant movement of ion species. The thermodynamic approach, however, accounts for these effects.

POTENTIAL CONCEPT

The most important concept in nature is the concept of energy because it is a fundamental entity common to all forms of matter in our physical world. Water retention and movement in soil is a consequence of energy effects. Commonly speaking, the capacity of doing work is called 'energy'. Thus, closely associated with energy is the concept of work. In a physical sense, the work is done only if a force applied on an object causes its displacement, *i.e.* the work is done when an object is moved by a force which has a component acting in the direction of motion. Quantitatively, in its simple mechanical form, work is defined as the force times the distance through which the force acts, *i.e.*

$$W = F \times h$$

(Work)=(Force) × (Distance)

Consider an example of the work done in lifting the mass, *m*, of water to a height, *h*, from *b* to *a* (Fig. 2.11) above the water level (reference level). It will be necessary at this point to define the force. According to Newton's second law of motion, when a body is acted upon by a constant force, its resultant acceleration is in the direction of the applied force and is given by

$$a = \frac{F}{M}$$

or

$$F = m \times a$$

(Force)=(Mass) × (Acceleration)

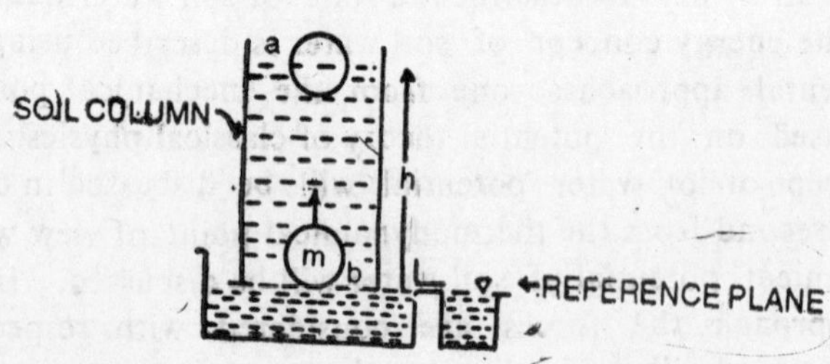

Fig. 2.11.

Therefore, by Newton's second law of motion, the force required to lift the mass, *m*, is equal to its own weight

$$F = m.g$$

where *g* is the acceleration of gravity. Substituting the weight *mg* for F in Eq. (1)

$$W = mgh$$

If a 5 *g* mass of water is raised vertically to a distance of 1 *m*, the work done will be

$$W \times 5 \times 980 \times 100 = 4,90,000 \text{ dynes-cm or ergs.}$$

We may state here that any type of potential should have similar definitions or expressions as written above. We may show this fact by defining the more familiar term the 'electrical potential' that is related with the transfer of charge. In physics the electrical potential, V_e of a charge *e* placed in the field is defined as the amount of work done or potential energy stored, per unit positive charge in carrying any charge *e* from the reference point to the point in question, *i.e.*,

$$V_e = \frac{E_p}{e}$$

where V_e = electrical potential, volt ; and *e*=charge moved into the field, coulomb (1 coulomb = 3×10^9 electrostatic units of charge).

In MKS system the unit of electrical potential is volt. If two points differ in potential by 1 volt, 1 joule of work is required to move 1 coulomb of charge from one point to another,

$$1 \text{ Volt} = \frac{\text{Joule}}{\text{Coulomb}}$$

Soil Water Potential

Because the velocity of flow of water in soil is often very low its E_k is negligible and its potential energy, E_p is primarily responsible for determining the state and movement of water in soil. The potential energy of soil water varies considerably from point to point. This difference of potential energy between two points gives rise to the tendency for water to flow. The water flows from the point where the potential energy is higher to where it is lower in the direction of greatest decrease in potential energy. The rate of decrease of E_p with distance is the moving force causing flow. The potential energy is acquired by the water whether the work is done against gravitational, elastic, osmotic, electric or magnetic forces.

The term soil water potential denotes the specific potential energy of soil water to that of water under standard reference state. Before we discuss the potential energy of water, we may first examine the major forces that act on water in the soil. They are :

1. Gravity, inertial or body forces.
2. Hydrostatic or forces associated with pressure gradients.
3. Osmotic forces related with solutes.
4. Adsorptive forces-long range electrical forces and the short range London van der Waal's forces.

If temperature of the system is not constant, an additional force field arising due to temperature will also be acting. For simplicity, we assume that the soil system in question is under isothermal conditions.

For any force acting on soil water, a corresponding potential component can be specified. For gravity there will be a gravitational potential, ϕ. Similarly, other potentials that are associated with above mentioned forces will be pressure potential, ψ, osmotic potential, ω, adsorptive potential, α, etc. Since potential is a scalar quantity, it can be expressed as algebraic sum. The total soil water potential, Φ, can thus be expressed as summation of different potential components :

$$\Phi = \sum_k \Phi_k$$

or

$$\Phi = \phi + \psi + \alpha + \omega + \ldots,$$

where Φ=total soil water potential, erg g^{-1} ; ϕ=gravitational potential, erg g^{-1} ; ψ=pressure potential, erg g^{-1} ; α=adsorptive potential, erg g^{-1} ; and ω=osmotic potential, erg g^{-1}.

The total soil water potential can thus be defined as[1] "the amount of work that must be done per unit quantity of pure water in order to transport reversibly and isothermally an infinitesimal quantity of water from a pool of pure water, at a specified elevation and at atmospheric pressure, to the soil water at the point under consideration."

Reference State

The selection of reference level in all the above examples is purely arbitrary. In any soil water system the zero energy level is known as reference state. The conventional characteristic of reference state of water are the following :

1. *Pure water* : The water at reference state has no dissolved gases or solids and free of any other impurities.

2. *Atmospheric pressure* : Generally, one can use barometric pressure but it continues to change depending upon weather conditions. Therefore, standard pressure of standard atmosphere is used.

1 atmosphere $= 1.013 \times 10^6$ dynes cm^{-2} = Pressure exerted by a column of water of height $h = 1035$ cm (with $\rho_w = 1\ g\ cm^{-3}$ and $g = 980$ cm sec^{-2}).

In metric system, the unit of pressure in atmosphere is replaced by bar ($= 10^6$ dynes cm^{-2}).

3. *Zero hydrostatic pressure* : The reference level is considered to be free flat surface with zero hydrostatic pressure.

4. *Zero elevation* : All points which are measured in the soil water system, should be from this reference level.

WATER MOVEMENT IN SOIL

When water, whose source is either rainfall, or irrigation, s applied to soil, it enters the soil pores replacing the air contained in them. If sufficient quantity of water is available, the entire pore space may be filled with water and the excess water would move downward by a physical process known as saturated flow. The flow of water in the upper saturated soil layers could be compared to the flow of water through the pipes. Such a type of flow results primarily from gravitational and pressure potential gradients and is comparatively rapid. If, however, the application of water is limited—as happens when rain is scanty, or irrigation water is in short supply—the rate of such movement is drastically curtailed. If evaporation is taking place at the soil surface and the plants are absorbing water from wetted soil zones, the soil water from the profile is rapidly depleted. Under such conditions, the soil pores will be only partially filled with water and the soil water potential decreases further. The flow of water under these conditions will occur in response to the differences in soil water potentials in the different regions of the soil profile. Water moves from a region of high potential to depleted soil water zones (low potential). Such a type of water movement is termed as unsaturated flow. Although movement of water in soil is mainly in liquid phase, its movement in vapour phase also occurs due to vapour pressure gradients caused by temperature and salt concentration gradients. The movement of water in the vapour state may take place within or over the soil surface. From the soil surface, the water is lost to the atmosphere by diffusion and convection. This phenomenon is known as evaporation. Due to differences in the vapour pressure, the water vapour in the soil diffuses from one region to another. Thus, water movement in soil may be of three types : (1) saturated flow ; (2) unsaturated flow ; and (3) vapour flow.

Under field conditions, the movement of water is complex. During irrigation and subsequent redistribution of water in the soil profile and its loss through evaporation and deep drainage, one may encounter all the kinds of water movement phenomena described above. The nature of the soil profile, its texture, structure and pore size distribution determine to a large extent how water penetrates and is retained in the soil. In a uniformly moist homogeneous soil, water moves at nearly the same rate in all the directions. Water introduced at the centre of a vertical soil column is observed to move almost as far upwards as downwards and laterally. In this case, the gravitational potential gradient is negligibly small in comparison with the soil suction gradient. This circular pattern of water is also observed in furrow irrigation.

When the water moving downward through a well-aggregated soil encounters a clay layer this layer is wetted readily because of the lower suction in the wetting front in the coarse soil. However, the downward movement is appreciably slowed down due to greater resistance to water in passing through fine pores in the clay layer. Water tables generally build up over the layer of clay. On the other hand, when the water moving downward through the fine textured soil encounters a layer of course sand the downward movement is temporarily stopped until the soil suction in the wetting front is sufficiently reduced. This is due to the presence of large macropores in the course sand which are too large to take in water from the micropores of the soil. However, when the soil above the course sand is almost saturated, the water moves into and through such sand layers rapidly. Though the soil above the course sand gets very wet, water tables are not built up because the underlying coarse material will conduct away all excess water rapidly, acting as a drain for the overlying soil. Similar is the case with respect to the soil overlying gravels. Therefore, soil overlying sand and gravel absorb 2·3 times as much water as the same soil material if it did not overlie sand or gravel. In the presence of sand lens, greater water movement is observed where the sand lens is in direct contact with free water supplied from the top.

The aggregates also behave in the same way as the sand layer. However, the amount of water moving through the aggregates is limited by a relatively small number of contacts between the aggregates. It can be seen that the presence of coarse soil aggregates considerably retards the downward movement of water in soil. Similarly, coarse soil, which infiltrates water more rapidly than fine soil, transmits water more slowly than fine soil during the later stages of drainage due to the higher conductivity of fine soil relative to the coarse soil under unsaturated conditions.

Incorporation of crop residue creates favourable porosity conditions for better water penetration. In vertical mulching if the mulched channel is open to the surface for free entry of water, such a channel would help in allowing the water to enter deep into the soil, thereby reducing the surface runoff. If the channel is not open to the surface such a phenomenon is not observed. This also shows that favourable surface conditions help water entry while under unfavourable surface conditions, even moderate rainfall does not infiltrate and causes runoff and soil erosion.

If the flow of water in a saturated soil can be compared with the flow of water through tubes or pipes. Poiseuille's equation can be used to quantify it. The soil pores, however, do not resemble uniform smooth pipes or tubes. In fact, soil pores are highly irregular and tortuous. Therefore, Poiseuille's law may be of little help in describing the nature of water flow in such a complex porous medium. The numerous constrictions or necks in the pore space and dead-ends restrict the flow of water. As a result, the flow velocity varies considerably from point to point. However, if the soil pore space can be imagined as a bundle of uniform capillaries of radius *r* and with their surface area in the direction of the potential gradient, we may still apply Poiseuille's equation and can expect better approximation of the discharge through soil. To visualize this, consider a cylindrical soil mass, of macroscopic radius *a* containing *n* number of pores of radius *r* (Fig. 2.12). The overall flow rate will be equal to the sum of the separate flow rates through the individual tubes.

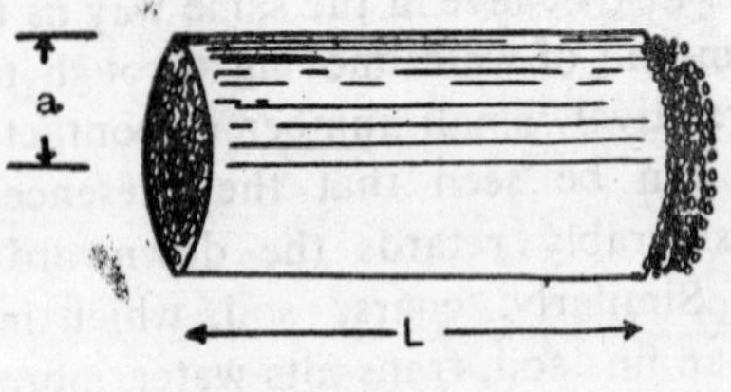

Fig. 2.12.

DARCY'S LAW

The flow of water through porous materials such as sand was experimentally studied, more than a century ago (in 1856), by a French engineer Henry Darcy. He investigated the behaviour of vertical sand filter beds for the water supply to the town of Dijon in France. He measured the quantity of water filtering through the sand bed (Fig. 2.13) and showed that the volume of water passing in unit time depended upon several factors, namely, the cross-sectional area of the bed, the thickness of the bed, the height of the water on the filter bed and the proportionality constant, K. The value of K as observed by him depended upon the nature of the bed. The famous experimental Darcy formula is

$$Q = K\frac{A(h+L)t}{L}$$

where Q = volume of water passing through sand filter, cm^3 ; L = length of the sand filter bed, cm ; A = cross-sectional area of the sand filter bed, cm^2 ; h=height of water on the filter bed, cm ; t=time, sec ; and K=proportionality constant cm sec^{-1}.

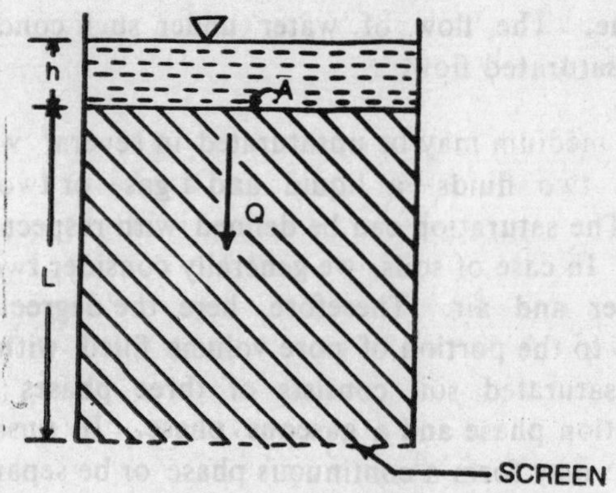

Fig. 2.13. The Darcy sand filter.

Substituting q=Q/At, Eqn. can be rewritten as

$$q = K\frac{(h+L)}{L}$$

where q=volume of water passing through unit cross-sectional area of material in unit time and is termed as flux, cm^3 cm^{-2} sec^{-1} or cm sec^{-1} ; h+L=hydraulic head difference between the inlet and outlet ends of the coloumn, cm ; and $(h+L)/L$=hydraulic gradient, dimensionless.

From the foregoing discussion, it is clear that the discharge rate (the volume of water flowing through the soil column in unit time t) is directly proportional to the cross-sectional area, A, and to the hydraulic head difference and inversely proportional to the length of the soil column, L. The proportionality constant, K, is known as hydraulic conductivity. Hydraulic conductivity is thus the constant of proportionality in Darcy's law as applied to the viscous flow of water in soil. It is the flux of water per nuit gradient of hydraulic potential.

FLOW OF WATER IN UNSATURATED SOIL

We have so far considerered the flow of water in a soil where sufficient water was available to fill the pores and the movement of water was in accordance with Darcy's law. We may now apply Darcy's law to a soil where the pore space is not completely filled with water, that is, the effectively conducting pore space is not the total pore space but a partially saturated one. The flow of water under such conditions is known as unsaturated flow.

A porous medium may be unsaturated in several ways. It may contain two fluids—a liquid and a gas—or two liquids and a gas. The saturation can be defined with respect to one of the fluids. In case of soils, we generally consider two fluids, namely, water and air. Therefore, here, the degree of saturation refers to the portion of pore volume filled with water. Thus an unsaturated soil consists of three phases: a solid phase, a solution phase and a gaseous phase. In unsaturated soil the fluid may form a continuous phase or be separated by soil particles. We shall consider both air and water forming continuous phases. It is, however, possible for the air bubbles to be entrapped among the soil particles.

Let us consider a flow system shown in Fig. 2.14 similar to Darcy's filter bed. In the system described in this case (Fig. 2.14), the pores are only partially filled with water, that is, the soil column is unsaturated and water flows under the

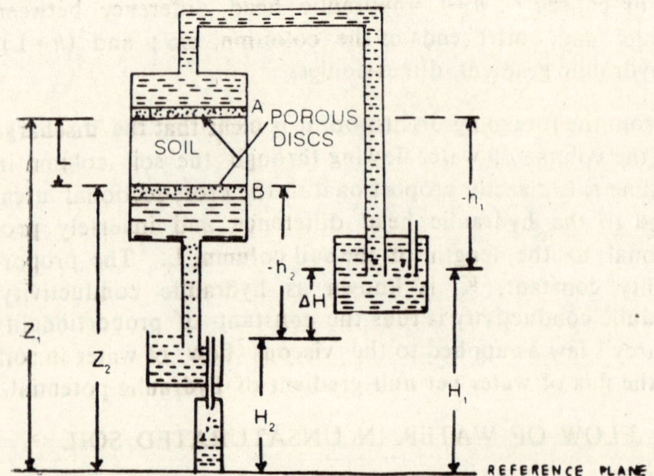

Fig. 2.14. Unsaturated flow system.

influence of negative hydrostatic pressure head or suction head difference. In such a situation, the hydraulic head at the inflow (A) and outflow (B) boundaries of the soil column will be as :

Hydraulic head, H		*Pressure head, h*	*Gravitational head, z*
H at A	=	$-h_1$	$+z_1$
H at B	=	$-h_2$	$+z_2$
ΔH	=	$-(h_1-h_2)$	$+(z_1-z_2)$

Also, $\Delta H = H_1 - H_2$.

Here, h_1 and h_2 are the pressure heads with a negative sign and z_1 and z_2 are the gravitational heads.

If h_1 and h_2 are maintained constant and the column length is small, the flow will soon attain a steady value. According to Darcy's law, the flux through the column will be given by

$$q = K\frac{\Delta H}{L}$$

Since the soil column is not uniformly saturated, the gradient along the column may not be constant. In a long soil column, the suction will vary along the axis of the column and so will the water content and the hydraulic conductivity. If the steady flow velocity is maintained everywhere in the column, the gradient of potential should vary along the axis of the sample ; being high where the conductivity is low and *vice versa*. With the increase in suction, the suction gradient will also increase and the conductivity will decrease. The distribution of water content, soil suction head and hydraulic conductivity in such a column is shown in Fig. 2.15. However, in the case of is stratified soil column, consisting of coarse sand overlying fine sand (see Fig. 2.16), with a steady downward flow, the distribution of soil suction with depth shows abrupt discontinuity at the junction between the two layers. The soil water suction in the fine sand which is in contact with the water table depends upon the amount of water that has entered it. Since the downward movement of water in fine sand will be appreciably reduced at the junction between the two due to the

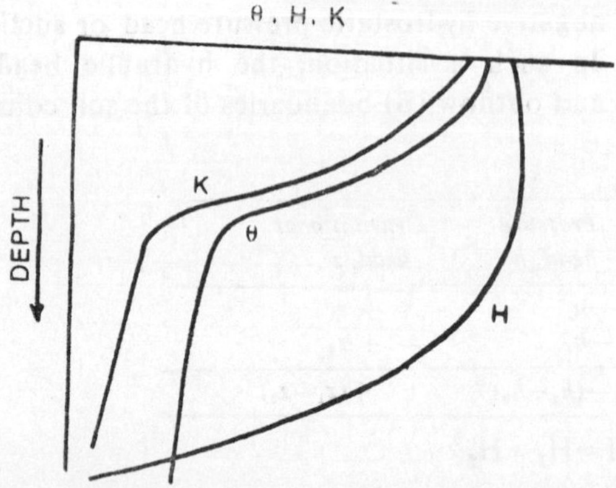

Fig. 2.15. Variation of hydraulic conductivity (K), suction head (H), and water content (θ) in an unsaturated soil column under steady flow.

presence of fine pores, which offer a high resistance to water passage, the soil suction in coarse sand remains low compared to fine sand showing a high gradient at the junction. When steady flow is approached a linear distribution of suction as shown in Fig. 2.16 is observed in both the layers and the flux is equal to the effective hydraulic conductivity of the column which is determined by the unsaturated transition zone. If, however, the fine sand overlies a coarse sand the picture is different, as shown in Fig. 2.17. In this case, suction increases at the junction and the zone above the junction has a higher suction in the fine sand than the limiting case. The water will

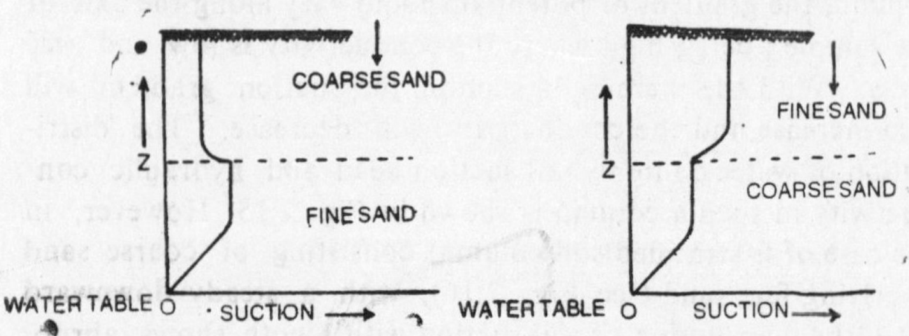

Fig. 2.16. Suction distribution in stratified column during downward steady flow above the water table.

Fig 2.17. Suction distribution in stratified column during downward steady flow above the water table.

enter the coarse soil only when the suction is reduced to the value determined by the pores in the coarse soil at the junction.

A cultivated soil consists of aggregates containing discrete groups of smaller pores within the aggregates (also called crumb pores or micropores) separated by a more continuous system of large pores between them, that is, the intercrumb pores or macropores. A well-aggregated field soil which has drained to attain a uniform moisture content after saturation can be compared to a bed of aggregates whose intercrumb pores have drained to a uniform water content, leaving the crumb pores saturated. Such a pore system with a bimodel pore size distribution is characterized by a double sigmoid water characteristic curve. The different stages in desaturation of aggregated soil are shown in Fig. 2.18.

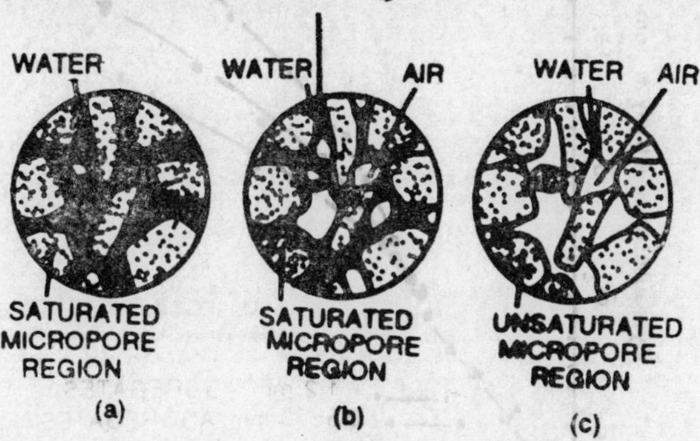

Fig. 2.18. Various stages in the desaturation of an aggregated soil.

In Fig. 2.18 all the pores—crumb pores and intercrumb pores—are saturated and water flow takes place in the larger macropore channels, although water can also move through the aggregates. It is the size and distribution of these macropore channels which determine the hydraulic conductivity of the soil. With a small suction, the macropores will drain leaving the micropore region still saturated (Fig. 2.18*b*). The liquid continuity in the macropores is broken and the water moves through the aggregates which form the continuous path for water movement by means of lenses of water near their points of contact. The hydraulic conductivity of the aggregates decreases rapidly as the area of connecting lenses becomes smaller with decrease in water content. With further increase in suction, the micropores themselves begin to get desaturated (Fig. 2.18*c*), causing further reduction in hydraulic conductivity. The experimental results shown in Fig. 2.19 agree with the above reasoning. It appears that a well-aggregated surface soil could be helpful in soil water conservation. The hydraulic conductivity of the bed of aggregates may be very low even if the crumb pores are nearly saturated. Therefore, during periods of evaporation from the soil surface a mulching effect is obtained. It should, however, be noted that a well-aggregated soil may not attain complete saturation in practice because of entrapment of air inside the aggregates.

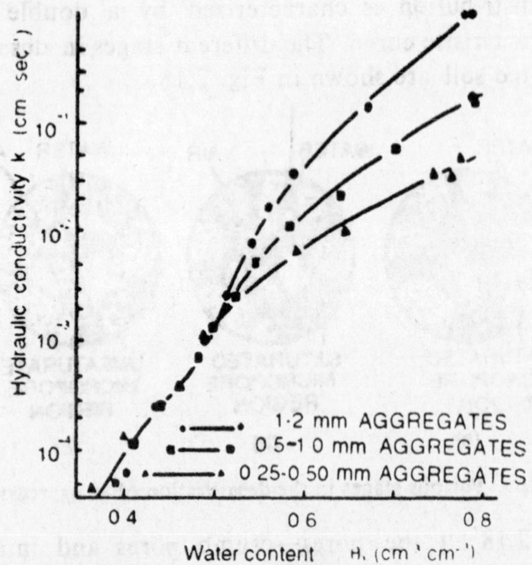

Fig. 2.19. Relationship between hydraulic conductivity and moisture content of various-sized aggregates.

During heavy rainfall or flood irrigation, water in the macropore channels may be saturated, leaving the centre of the micropore region filled with air. During the uptake of water by soil crumbs, the entrapped air may be compressed to such an extent that the uptake may practically stop. There is also the possibility of the explosion of aggregates and the breakdown of soil structure when the pressure inside the crumb exceeds the breaking point.

A number of experiments have been conducted to determine the variations of hydraulic conductivity with soil water content or suction. The generalized results (K versus τ) are presented in the Fig. 2.20. It is seen that the hydraulic conductivity is sharply reduced with an increase in soil suction.

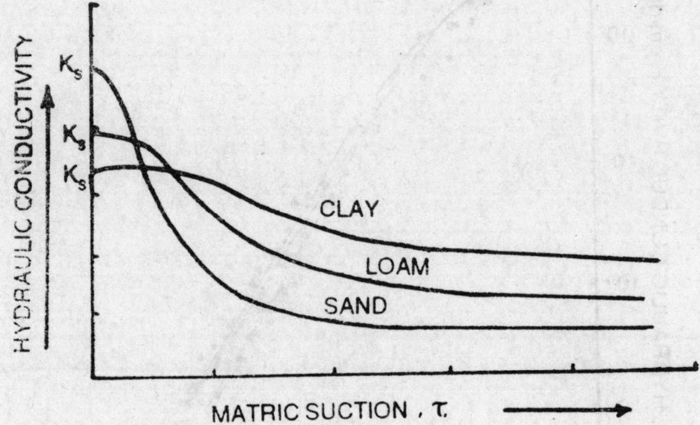

Fig. 2.20. Dependence of hydraulic conductivity on matric suction in soils of different textures.

The shape of the curve for the three soil textures is different. The saturated hydraulic conductivity of sand is the highest and that of clay is the lowest. However, with an increase in soil suction, the trends are reversed. The unsaturated hydraulic conductivity of sand decreases more rapidly with the increase in suction while that of clay decreases less sharply. At a higher suction value, the conductivity of sand is much lower than that of clay soil. In general, the hydraulic conductivity is very sensitive to water content or suction and ranges over a factor of 10^5 for the water contents measured in the field. Similar results are also shown in Fig. 2.21 for three fine textured soils.

The foregoing results indicate that under unsaturated conditions, the flux is not linearly related to the hydraulic gradient as in the case of saturated low. Darcy's equation is thus non-linear under unsaturated conditions, in the sense that the hydraulic conductivity depends on soil water content, θ, or soil suction, τ. Since K is independent of flux and/or suction

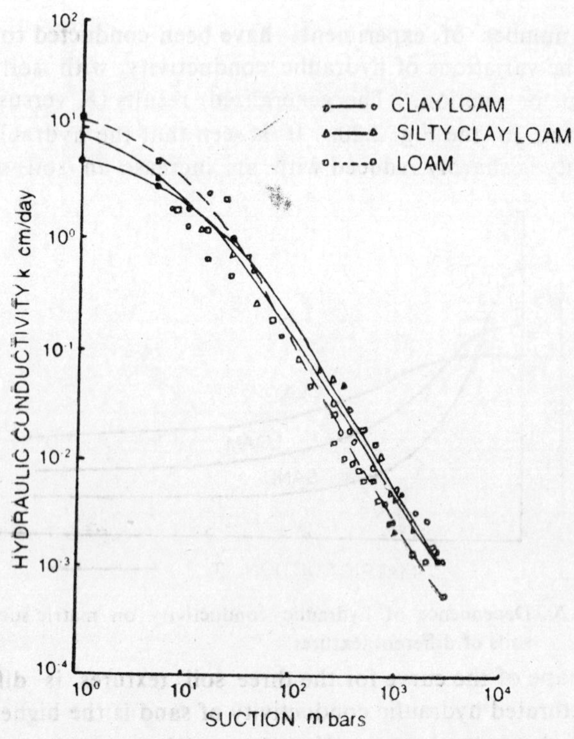

Fig. 2.21. Hydraulic conductivity as a function of suction.

gradient, a plot of flux versus hydraulic gradient at a fixed average water content θ will give a straight line passing through the origin. A family of such straight lines (Fig. 2.22) can be obtained from careful experimentation such as shown in Fig. 2.14. The slope of the line (Fig. 2.22) is the hydraulic conductivity at the indicated water content. Since the hydraulic conductivity is no more constant but highly dependent upon soil water content or suction, attempts have been made to modify Darcy's law for unsaturated soil by introducing functional relationships between hydraulic conductivity and water content or matric suction. For unsaturated soils Darcy's law has been written as :

$$q = K(\theta)i$$

where i = hydraulic gradient.

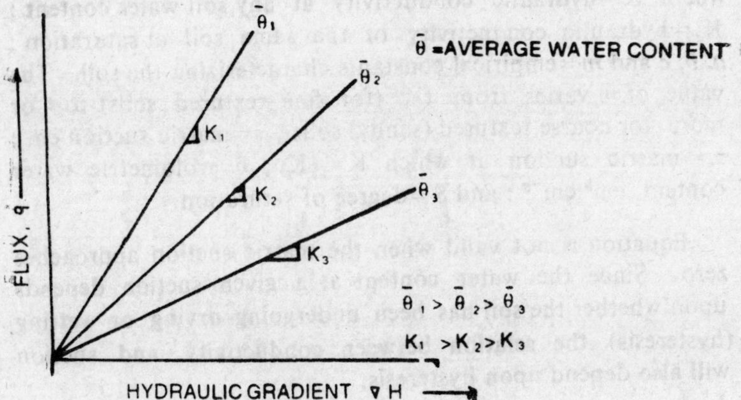

Fig. 2.22. Relationship between flux and hydraulic gradient with water content as parameter.

In the vector form, Equation can be rewritten as

$$q = -K(\theta)\nabla H$$

or

$$q = -K(\tau)\nabla H$$

These equations show that K is not constant but changes with water content or matric suction. If the soil water content is known at a particular point, the flux at that point will be [from Equation].

q	=	K	×	i
Flux at a point		Hydraulic conductivity at that point		Hydraulic gradient at that point

No universal relationships are available for hydraulic conductivity and soil water content or suction. Nevertheless, several empirical relationships have been proposed. The following equations have been found useful in practice as it is difficult to predict precisely the unsaturated hydraulic conductivity from fundamental soil properties :

$$K = a\tau^{-m}$$

$$K = a(b+\tau^m)^{-1}$$

$$K = K_s[1+(\tau/\tau_c)^m]^{-1}$$

$$K = K_s(b+\tau^m)^{-1}$$

$$K = K_s(1+c\tau^m)^{-1}$$

$$K = a\theta^m$$

$$K = K_s S^m$$

where K=hydraulic conductivity at any soil water content; K_s=hydraulic conductivity of the same soil at saturation; *a*, *b*, *c* and *m*=empirical constants characterizing the soil. The value of *m* varies from 1-2 (for fine textured soils) to 4 or more for coarse textured (sandy) soils; τ=matric suction cm; τ_c=matric suction at which $K=\frac{1}{2}K_s$; θ=volumetric water content, $cm^3\ cm^{-3}$: and S=degree of saturation.

Equation is not valid when the matric suction approaches zero. Since the water content at a given suction depends upon whether the soil has been undergoing drying or wetting (hysteresis) the relation between conductivity and suction will also depend upon hysteresis.

INFILTRATION

In water management and conservation studies, accurate information on the rate at which different soils will take in water under different field conditions is required. The rate of water entry into soil varies widely between different soil types and also within a single soil type, depending upon soil water content and management practices.

When water is applied to the soil surface through irrigation or rain, it subsequently enters the soil. If the supply rate is more than the rate of entry into the soil, the excess will accumulate over the surface and may flow as runoff. The time rate at which water will percolate into the soil through its soil-atmosphere interface is known as infiltration. Quantitatively, infiltration rate is the volume of water entering into the soil per unit area in unit time, when the soil is subjected to a shallow depth of ponding at the surface.

In hydrology, the term infiltration capacity is frequently used and is defined as the maximum rate at which rain can be absorbed by a soil in a given condition, *i.e.*, it signifies soil infiltrability. It is this characteristic that determines how much of the falling rain will flow directly over the ground surface into streams and rivers and how much will enter the soil and will be retained there for some time before being either passed downward as percolation or returned to the atmosphere by the process of evapotranspiration. The terms infiltration capacity of the soil and infiltration rate are the same.

Generally, under continued ponded conditions, the infiltration has a high rate in the beginning, decreasing rapidly and then more slowly until it approaches a constant rate (Fig. 2.23). As water moves deeper into the soil, filling the pores, the hydraulic gradient goes on decreasing and, hence the infiltration rate. Dispersion of aggregates, blocking of the cracks and channels, swelling of colloids also reduce the infiltration rate. Soil pores are also blocked and the infiltration rate is sharply reduced when water is turbid.

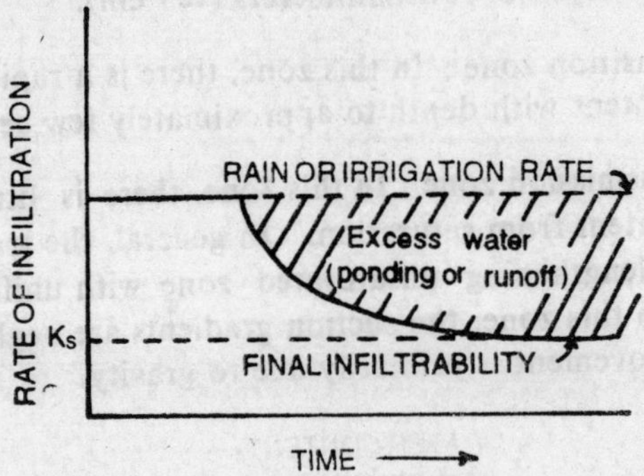

Fig. 2.23. Rate of infiltration as a function of time, under an irrigation of constant intensity lower than the initial value, but higher than the final value of soil infiltrability.

The infiltration process is of great practical importance and it affects many aspects of hydrology and agriculture. It influences runoff, and determines the water content of the soil. It is related to evapotranspiration. It is possible to modify infiltration through suitable soil and crop management practices.

PROFILE MOISTURE DISTRIBUTION DURING INFILTRATION

If we study the moisture distribution during infiltration in a uniform profile, it will be seen that the surface of the soil is saturated to a depth of a few centimeters (Fig. 2.24). Below this saturated zone is a transmission zone through which water is being transmitted from the saturated zone to the wetting zone at the bottom. The transmission zone is a lengthening zone with uniform water content near saturation. In the wetting zone, the moisture content decreases with depth till it merges into the wetting front which forms a sharp boundary between the wet and dry soil. The wetting zone and wetting front move downward continuously during the process of infiltration. Thus, during infiltration, the wetted portion of the homogeneous soil profile comprises of five zones.

1. Saturated zone : In this zone, the soil is saturated and extends to a depth of few millimeters ($\simeq$1 cm).

2. Transition zone : In this zone, there is a rapid decrease of water content with depth to approximately few centimeters.

3. Transmission zone : In this zone, there is little change in water content from saturation. In general, the transmission zone is a lengthening unsaturated zone with uniform water content. In this zone, the suction gradients are negligible and the water movement is primarily due to gravity.

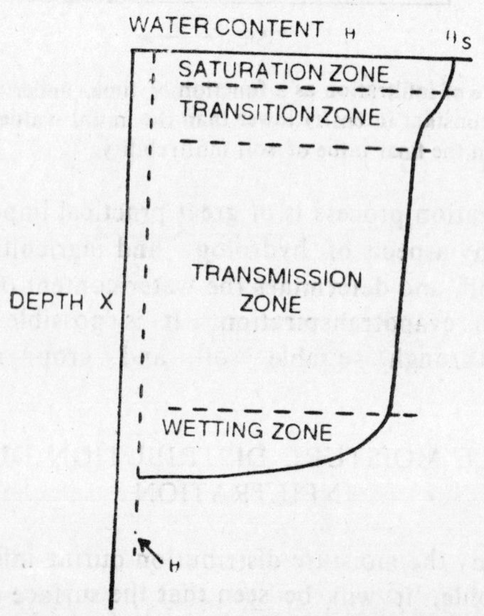

Fig. 2.24. The four zones of the infiltration profile.

4. Wetting zone : In this zone, the water content sharply decreases with depth.

5. Wetting front. This is the zone of the steep water content gradient and forms a sharp boundary between the wet and dry soil. Beyond this zone, there is no visible penetration of water.

Consider the water distribution patterns during infiltration under shallow ponding of constant depth in three uniform soil profiles of sand, loam and clay (Fig. 2.25). As can be observed, the advance of the wetting front is the most rapid in sand, thereby requiring more water per unit volume to saturate it than the other two. Also, the wetting front is the sharpest in sand and the least so in clay, obviously, due to hydraulic characteristics at that suction value. Eventhough the depth of water standing on the surface (ponding) had been the same the sand profile was wetted throughout its entire 1.16 m depth within 4 hours, whereas loam and clay required 12 and 20 hours, respectively.

Cumulative infiltration as a function of the square root of time shows that the curves are almost straight in the beginning as the pressure gradients predominate over the gravitational force (Fig. 2.26). However, as the profile gets wetted more deeply, the gravity force dominates. As this occurs, the infiltration rate approaches a constant value approximately equal to the hydraulic conductivity at saturation. Figure 2.26 also shows that at a $t^{1/2}$ value of 150 hours, the sand, loam and clay profiles absorb nearly 65 cm, 25 cm and 7 cm water, respectively.

In the profiles of sand, loam and clay mentioned above if we consider the processes of infiltration and redistribution-cumevaporation occurring simultaneously following a rainstorm, the water distribution patterns would be considerably different. It was seen that during a rainstorm, the upper horizons of sand remained unsaturated (rainfall rate < saturated hydraulic conductivity), while those of the finer textured soil profile approached saturation as the maximal rainfall rate nearly equalled or exceeded the hydraulic

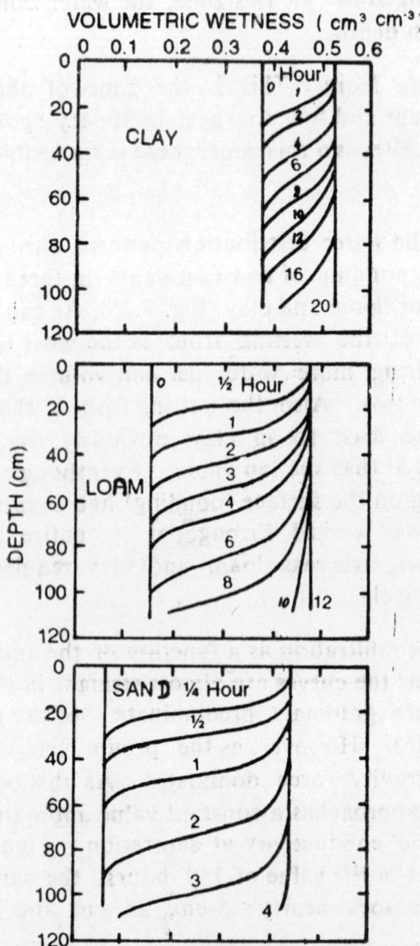

Fig. 2.25. **Successive distribution of soil moisture during infiltration into uniform profiles of sand, loam and clay, all at an initial matric suction of 1/3 bar. Numbers on the curve indicate time.**

conductivity of these soils. Total penetration of weter during the first rainstorm duration (one-fourth of a day) was the deepest in the sand but did not exceed more than 0.5 m. During the 8-day redistribution-cum-evaporation phase the penetration of water was the shallowest in clay and the deepest in sand. Therefore, the upward moisture content gradients

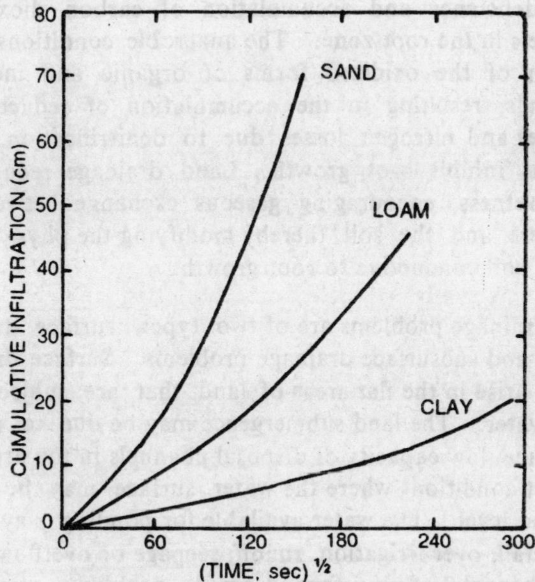

Fig. 2.26. Cumulative infiltration into uniform profiles of sand, loam, and clay (all at initial matric suction of 1/3 bar) as a function of the square root of time.

due to evaporation during the evaporation phase are steeper in clay than in coarser textured soil.

DRAINAGE

Drainage means removal of excess surface and subsurface water by means of some water-conveying devices. In agriculture we are more concerned with lowering of the water table and the relationship between the water table depth and the crop performance. Drainage is essential for crop production. The growth of crops is adversely affected by continuous soil saturation or by ponded water on the surface. Since wet soil requires more heat to warm up than dry soil, due to the high specific heat of water, the growing season for winter crops is shortened in poorly drained soils. Saturation of soil may encourage certain plant diseases and parasites. Poor drainage also leads to the development of salt-affected soils. As shown in Fig 2.27, the high water table may limit root penetration. Soil saturation stops the gaseous exchange causing

oxygen deficiency and accumulation of carbon dioxide to toxic levels in the root zone. The anaerobic conditions cause reduction of the oxidized forms of organic and inorganic compounds, resulting in the accumulation of reduced toxic substances and nitrogen losses due to denitrification. Such conditions inhibit root growth. Land drainage reduces excessive wetness, encouraging gaseous exchange between the atmosphere and the soil thereby modifying the physical properties of soil conductive to root growth.

The drainage problems are of two types : surface drainage problems and subsurface drainage problems. Surface drainage problems arise in the flat areas of land that are subjected to ponded water. The land submergence may be due to uneven land surface, low capacity of disposal channels in the area and the outlet conditions where the water surface may be above the ground level. The water available for ponding may come from rainfall, over irrigation, runoff, seepage or overflow from stream channels. Subsurface drainage problems may arise due to low soil permeability, humid elimate and rise of ground water causing development of a shallow water table.

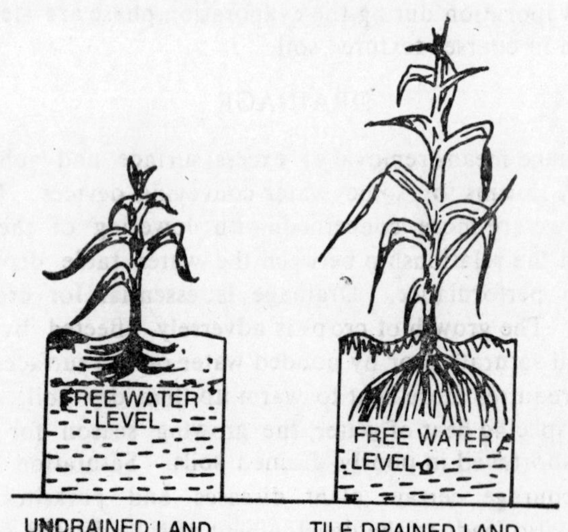

Fig. 2.27. Root and shoot development of crops grown on drained and undrained land.

The methods used for land drainage can, therefore, be classified into two broad categories, *viz.*, surface drainage and subsurface drainage. In the case of surface drainage, water is removed directly from the land by land smoothing, land grading, bunding or ditching. Land surfaces are also reshaped to eliminate ponding and to create slopes so as to induce gravitational flow over land and through channels to an outlet. The excess water can also be diverted from the land by diversion ditches, dykes, etc.

In subsurface drainage, surface and subsurface drains are established in the soil profile to collect and convey excess ground water to gravity or pumped outlets The discharge by permitting the pressure to drop induces the flow of excess ground water through the soil into the drains. The ranges of depth and spacing generally used for installing tile drains in the field are given in Table 2.5.

Table 2.5. Average depth and spacing of tile drains in various soil types.

Soil type	*Hydraulic conductivity, cm/day*	*Spacing of drains, m*	*Depth of drains, m*
Clay	0.15	10-20	1-1.5
Clay loam	0.15-0.5	15-25	1-1.5
Loam	0.5-2.0	20-35	1-1.5
Fine sandy loam	2.0-6.5	30-40	1-1.5
Sandy loam	6.5-12.5	30-70	1-1.5
Peat	12.5-25	30-100	1-2

The drainage systems can be designed using empirical criteria from data collected through evaluation of existing drainage systems or by theoretical analysis of the problem based on the physical laws of water movement and testing the theory through the evaluation of existing systems.

EVAPORATION FROM THE FREE WATER SURFACE

Evaporation is the process by which water is changed from the liquid state to the vapour state. A water body is made up of H_2O molecules, each of which is in the state of motion in different directions, and attracted to each other by a force that is inversely proportional to the square of the distance between them and directly proportional to the product of their masses. Evaporation takes place when some molecules have attained enough kinetic energy to eject themselves from the water mass. Since the escaping molecules are attracted by other molecules underneath them, only those molecules which possess greater kinetic energy are able to escape from the water surface. The kinetic energy of the molecules escaping from the water surface decreases because of negative work done on it by the intermolecular forces. Therefore, some water molecules continuously escape from the surface (Fig. 2.28) while others return and penetrate into the mass of water. The rate of evaporation is determined by the excess rate of those molecules leaving the water mass over those molecules returning. At a high temperature, the water molecules attain greater kinetic velocities and, therefore, tend to fly off into the lower layers of the atmosphere resulting in greater evaporation.

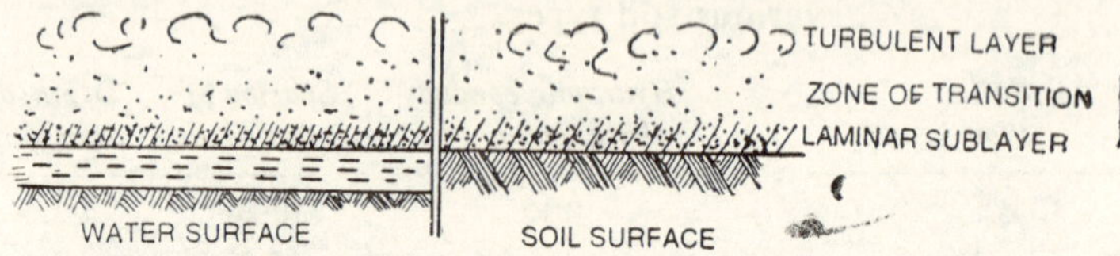

Fig. 2.28. Air structure near a flat exposed surfac (schematic).

Analogous to heat, water and electricity flow, the process of evaporation is dependent on a driving force and a resistance. The motion of the water molecules escaping from the water surface produces a pressure which is called vapour pressure. It is actually the partial pressure of the water vapour in the atmosphere, which is a mixture of gases and water vapour. According to the ideal gas law, the partial gas pressure is directly proportional to the number of gas molecules present at constant temperature and volume of the mixture. The difference in the vapour pressure between the layer from which water is evaporating and that of the turbulent atmosphere constitutes the driving force. The resistance is dependent on the nature of the vapour flow medium. It can be diffusion impedance to liquid and vapour flow in the system, the thickness of the laminar layer adjacent to the evaporating surface or the resistance caused by the turbulent air. Under the conditions of no turbulence or calm atmosphere, continuous evaporation causes an increase in the vapour pressure, eventually leading to the saturation of the atmospheric layers near the water surface and cessation of evaporation. Under such conditions condensation takes place and the saturation deficit is zero, *i.e.*,

$$(e^0 - e) = 0$$

where e^0 = saturation vapour pressure of water and e = actual vapour pressure of water.

Thus, evaporation and condensation are continuous processes. It is obvious that evaporation will be faster than condensation because the overlying air layers are often not saturated. Assuming that a layer of a thin film of vapour-saturated air is formed adjacent to the water surface, as shown in Fig. 2.28, under isothermal conditions, the evaporation function analogous to Darcy's equation can be expressed as

$$E = -K' \frac{de}{dz}$$

where E = evaporation rate ; z = thickness of the layer : and K' = transfer coefficient.

It is seen that there exists a gradient in the vapour pressure in the air layer immediately above the film saturated with water vapour. The evaporation is a function of this vapour pressure gradient. Under absolutely calm atmospheric conditions, a situation may occur where the gradient is practically zero. Under normal conditions, the air movement causes mixing of air layers encouraging turbulence and convection, thereby reducing the vapour pressure and permitting further evaporation. The stronger the wind, the more effective is the turbulent action in the air, causing greater difference between the surface and overlying air temperatures, and the greater is the effect of convection.

In 1802, Dalton considered all these factors and first enunciated the fundamental law of evaporation from the free water surface. This law states that evaporation will occur only if the actual vapour pressure of the air above the water surface is less than that at the water surface. Thus, the rate of evaporation is expressed as

$$E=(e_0-e_d)f(U)$$

where E = rate of evaporation ; e_0 = mean vapour pressure at the water surface ; e_d = mean vapour pressure in the air at some observational height above the water surface ; and $f(U)$ = a function whose value depends on the horizontal wind velocity.

From Equation it is clear that if other factors remain constant evaporation is proportional to the wind speed and the vapour pressure deficit (e_0-e_d).

EVAPORATION FROM THE SOIL SURFACE

Evaporation from the soil surface is essentially the evaporation of water surrounding the soil particles as thin films and filling the pore spaces between them. Therefore, the atmospheric conditions that govern the evaporation from a free water surface will also govern the rate of evaporation from the soil surface. However, in the case of evaporation from the free water in bulk, the supply of water is not limited while the evaporation from the soil will frequently be affected due to insufficient supply of water in the soil. Furthermore, the water molecules escaping from the soil will have to overcome greater resistance due to the attraction of soil particles than while escaping from a free water surface. When the water content of the surface soil is reduced below a threshold value, evaporation practically ceases.

The foregoing discussion reveals that evaporation of water from the soil demands three basic physical requirements: First, a continuous supply of heat to change the state of water from liquid to vapour should be available. Secondly, a vapour pressure gradient between the soil surface and the surrounding atmosphere should be maintained. Thirdly, there must be a continuous supply of water from or through the soil profile at the surface. The first two conditions, namely, supply of energy and removal of vapour are external to the evaporating body and are influenced by the meteorological factors such as air temperature, humidity, wind velocity and radiation, which together determine the evaporative demand of the atmosphere or evaporation potential. The evaporation potential is defined as the maximum evaporation rate from the surface of a free water in bulk under the given meteorological conditions. The third condition determines the maximum rate at which soil can transmit water to the plane of evaporation. The evaporation rate is thus determined either by the evaporation potential or by the soil's own ability to transmit water to the plane of evaporation under the given condition, whichever is lesser.

Physical properties of water

Water differs markedly from substances having closely related electronic structures, a point which we can illustrate by considering the series CH_4, NH_3, H_2O, HF, and Ne. Each member contains 10 protons and 10 electrons, but the number of hydrogen atoms decreases along the series from methane to neon. For this sequence, the melting points are —184°, —78°, 0°, —92°, and —249°, respectively. A similar striking variation in boiling points occurs, *viz.*, —161°, —33°, 100°, 19°, and —246°. The relatively high melting and boiling points for water, compared with substances having a similar electronic structure, are indicative of its strong intermolecular forces. In other words, thermal agitation does not easily disrupt the water-water bonding. This attraction between molecules is responsible for many characteristic features of water.

Hydrogen bonding. The strong intermolecular forces in water result from the structure of the H_2O molecule (Fig. 2.29). The internuclear distances between the oxygen and each of the two hydrogens are approximately 0.99Å, and the H—O—H angle is about 105°. The oxygen atom is strongly electronegative and tends to draw electrons away from the hydrogen atoms. This leaves the oxygen with a partial negative charge (δ in Fig. 2.29), while the two hydrogens become positively charged (δ^+). The positively charged hydrogens of a water molecule are electrostatically attracted to the negatively charged oxygens in two neighbouring water molecules. This leads to "hydrogen bonding" between water molecules, with an energy of about 4.8 kilocalories (kcal)/mole of hydrogen

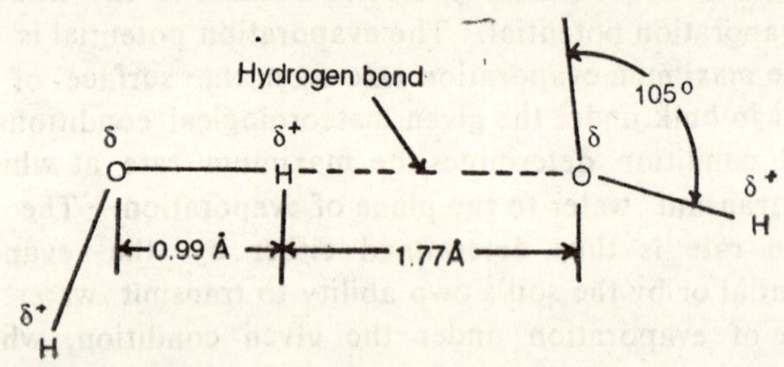

Fig. 2.29. Schematic structure of water molecules, indicating the hydrogen bonding resulting from the electrostatic attraction between the net positive charge on a hydrogen (δ^+) in one molecule and the net negative charge on an oxygen ($\delta-$) in a neighbouring water molecule.

bonds. Such bonding of water molecules to each other leads to increased order in aqueous solutions. In fact, liquid water becomes nearly crystalline in local regions, a condition which is extremely important for determining the molecular interactions and orientations that occur in aqueous solutions.

Ice is a coordinated crystalline structure in which essentially all the water molecules are joined by hydrogen bonds. When heat is added so that the ice melts, some of these intermolecular hydrogen bonds are broken. The heat of fusion of ice at 0° is 80 gal/g or 1.44 kcal/mole. The total rupture of the intermolecular hydrogen bonds involving each of its hydrogens would require 9.6 kcal/mole of water. The heat of fusion thus indicates that (100) (1.44)/(9.6), or at most 15% of the hydrogen bonds are broken when ice melts (some energy is needed to overcome van der Waals attractions, so that less than 15% of the hydrogen bonds are actually broken upon melting). Conversely, over 85% of the hydrogen bonds remain intact for liquid water at 0°. Since 1.00 cal is needed to heat 1 g of water 1°, (1.00 cal/g—°) (25°) (18 g/mole) ($^1/_{1000}$ kcal/cal), or 0.45 kcal/mole is required to heat water from 0° to 25°. If all of this energy were used to break hydrogen bonds, over

80% of the bonds would still remain intact at 25°. Such bonding leads to the semicrystalline order found in aqueous solutions. The extensive amount of intermolecular hydrogen bonds present in the liquid state contributes to the unique and biologically important properties of water that we will discuss throughout this chapter.

The energy required to separate molecules from a liquid and move them into an adjacent vapour phase without a change of temperature is called the *heat of vapourization.* For water, the heat of vapourization at 100° is 539 cal/g or 9.71 kcal/mole. Per gram, this is the highest heat of vapourization of any known liquid, and it reflects the large amount of energy required to disrupt the extensive hydrogen bonding in aqueous solutions. More pertinent to plant physiology is the heat of vapourization of water at temperatures encountered by plants. At 25° each mole of water evaporated takes 10.5 kcal, which means that a substantial heat loss by the plant accompanies the evaporation of water in transpiration. Most of this vapourization energy is needed to break hydrogen bonds so that the water molecules can become separated in the gaseous phase. For example, if 80% of the hydrogen bonds remained at 25°, then (0.80) (9.6), or 7.7 kcal/mole would be needed just to rupture them—other energy is needed to overcome the attractive van der Waals forces and for the expansion in going from a liquid to a gas. The heat loss upon evapouration of water is one of the primary means of temperature regulation in land plants. It dissipates much of the energy gain resulting from the net radiation balance.

Surface tension. Water is further characterized by having an extremely high surface tension, which is evident at an interface between water and air. Surface tension can be defined as the force transmitted across a line in the surface, and its value at an air-water interface is 72.8 dynes/cm at 20°. Perhaps a better way of defining surface tension is to regard it as the amount of energy required to expand the surface by unit area. (Since a dyne is an erg/cm. note that surface tension has not only the dimensions of force per unit length but also the units of ergs/cm^2, or energy per unit area.) To see why energy is required to expand the water surface, let us consider water molecules which are brought from the interior of an aqueous phase to an air-water interface. If this involves only an expansion of the surface area—*i.e.*, if there is no accompanying movement of other water molecules from the surface to the interior—then a loss in the water-water attraction of intermolecular hydrogen bonds occurs with no compensating air-water attraction. Energy (72.8 ergs to increase the area by 1 cm^2) is thus needed to break the hydrogen bonds involved in moving water molecules from the interior of the solution to the air-water interface. In fact, the term "surface free energy" is more appropriate from a thermodynamic point of view than is the conventional term "surface tension."

The surface tension of an aqueous solution is ordinarily only slightly influenced by the composition of the adjacent gas phase, but it can be markedly affected by certain solutes. Molecules are far apart in a gas—the density of dry air at 0° and 760 mm Hg is only 1.293 mg/cm^3—so that the frequency of collisions or other interactions between molecules in the gas phase and those in the liquid phase is relatively low. Certain solutes, such as sucrose or KCl, do not preferentially collect at the air-liquid interface and consequently have little effect on the surface tension of an aqueous solution. On the other hand, fatty acids and certain lipids may become concentrated at interfaces and greatly reduce surface tensions. For example, 10 mM caproic acid (a 6-carbon fatty acid) lowers the surface tension about 15 dynes/cm below the value for pure water, while only 0.05 mM capric acid (a 10-C fatty acid) lowers it by about 25 dynes/cm. Substances such as soaps (salts of fatty acids) or denatured proteins with large hydrophobic side chains may collect nearly exclusively at the interface and reduce the surface tension of aqueous solutions to as low as 20 dynes/cm. A common feature of such "surfactant" molecules is that they have both polar and nonpolar regions.

Capillary rise. The intermolecular attraction occurring between like molecules in the liquid state, such as the water-water attraction based on hydrogen bonds, is referred to as ***cohesion***. The attractive interaction between a liquid and a solid phase, such as between water and the walls of a small tube or capillary, is called *adhesion*. When the water-wall attraction is appreciable compared with the water-water cohesion, the walls are said to be *wettable*, and water then rises in the capillary. At the opposite extreme, where the intermolecular cohesive forces within the liquid are substantially greater than the adhesion between the liquid and the wall material, the upper level of the liquid in a capillary is lower than the surface of the free solution. Such capillary depression occurs for liquid mercury in glass capillaries. For water in glass capillaries or in xylem vessels, the attraction between the water molecules and the walls is great, therefore the liquid rises. Since capillary rise has important implications in plant physiology, we will discuss its characteristics in a quantitative manner.

As an example appropriate to the evaluation of water rise in a plant, let us consider a wettable capillary of inner radius r dipping into some aqueous solution (Fig. 2.30). The strong adhesion of water molecules to the wettable wall leads to a rise of the fluid in the capillary. Since there is also a strong water-water cohesion in the bulk solution, water, is concomitantly pulled up into the lumen of the capillary as water rises along the walls. Viewed another way, the air-water surface greatly resists being stretched, a property reflected in the high surface tension of water at gas-liquid interfaces. This resistance tends to minimize the area of the air-water interface, a condition achieved if water also moves up in the lumen as well as along the sides of the capillary. The tendency for water to rise along the walls of the capillary is thus transmitted to a volume of fluid. Let the height that the liquid rises in the capillary be h and the contact angle that the liquid makes with the capillary wall be α (Fig. 2.30). The extent of the rise depends on α, so the properties of the contact angle will now be examined a little more closely.

The size of the contact angle depends on the magnitude of the liquid-solid adhesive force compared with that of the liquid-liquid cohesion. Specifically, Young's equation indicates that the adhesion equals $(1+\cos\alpha)/2$ times the cohesion. (Sometimes this relation is called the Young and Dupre equation—see Adamson, Bikerman, and/or Davies and Rideal for an authoritative treatment.) Consequently, when the adhesive force equals (or exceeds) the cohesive force in the liquid, cos α must be unity, therefore the contact angle α equals zero (cos $0° = 1$). This is the case for water in capillaries made of clear smooth glass or having walls with polar groups on the exposed surface. When the adhesive force equals half of the value of the cohesive force, cos α is then zero and the contact angle is then 90° (cos $90° = 0$). In this case, the level of the fluid in the capillary is the same as that in the bulk of the solution. This latter condition is closely approached in water-polyethylene adhesion, where α equals 94°. As the liquid-solid adhesive force becomes relatively less compared with the intermolecular cohesion in the liquid phase, the contact angle (indicated in Fig. 2.30) increases toward 180° and capillary depression occurs, *e.g.*, water has an α of about 110° for paraffin, while the contact angle for mercury interacting with a glass surface is near 150°.

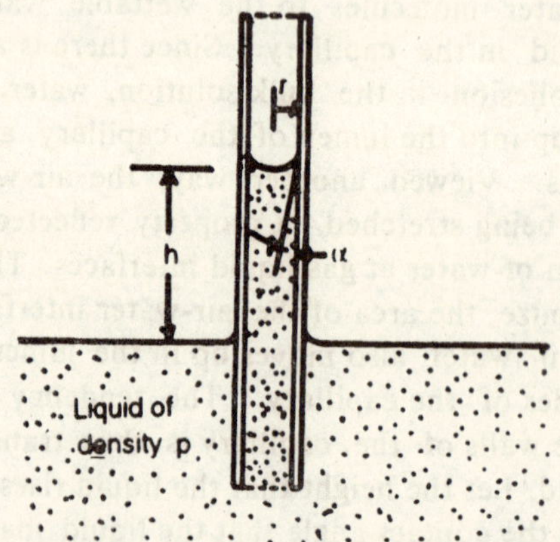

Fig. 2.30. Quantities involved in the capillary rise of a liquid.

We can calculate the extent of the capillary rise by considering the balance of two forces : (1) gravity acting downward and (2) surface tension, which leads to a force in the upward direction in the case of wettable walls in a vertical tube (cf. Fig. 2.30). The force pulling upward acts along the inside perimeter of the capillary, a distance of $2\pi r$, with a force per unit length equal to σ, the surface tension. However, the component of the force actually acting vertically upward is $2\pi r\sigma \cos \alpha$ (where α is the contact angle illustrated in Fig. 2.30). This upward force is balanced by the gravitational force acting on the liquid of density ρ and volume essentially $\pi r^2 h$.* The gravitational force is simply the mass involved times the gravitational acceleration g(F$=ma$, Newton's second law of motion). In the present case gravity acts on a mass of fluid of approximately $\pi r^2 h\rho$, so the gravitational force is $\pi r^2 h\rho g$ acting downward. Since this force is balanced by $2\pi r\sigma \cos \alpha$ pulling in the opposite direction, the extent of rise, h, is given by equating the two forces ($\pi r^2 h\rho g = 2\pi r\sigma \cos \alpha$), which leads to

$$h = \frac{2\sigma \cos \alpha}{r\rho g}$$

Equation 23 indicates the readily demonstrated property that the extent of liquid rise in a capillary in inversely proportional to the radius of the tube. For water in glass capillaries as well as in many of the fine channels encountered in plants, where the cell walls have a large number of exposed polar groups, the contact angle can be near zero, in which case $\cos \alpha$ appearing in Equation can be set equal to 1. The density of water is 0.998 g/cm³ at 20° (actually, ρ in Equation is the difference between the liquid density and that of the displaced air, the latter being about 0.001 g/cm³), and the acceleration due to gravity, g, is about 980 cm/sec². Using a surface tension for water at 20° of 72.8 dynes/cm (App. II), we obtain the following relation between the height of the rise and the capillary radius :

$$h_{(cm)} = \frac{0.149 \text{ cm}^2}{r_{(cm)}}$$

or

$$h_{(m)} = \frac{1.49 \times 10^{-5} \text{m}^2}{r_{(m)}}$$

where the subscripts (cm) or (m) in Equation mean that the dimensions involved are expressed in centimeters or meters, respectively.

Capillary rise in the xylem. **Although Equations refer only to the height of capillary rise in a static sense, they**

they still have important implications concerning the movement of water in plants. To be specific, let us consider a xylem vessel having a radius of 20 μm. From Equation water will rise in it to a height of $(1.49\times10^{-5}\ m^2)/(20\times10^{-6} m)$, or 0.75 m. This capillary rise would be sufficient to account for the extent of the upward movement of water in small plants, although it says nothing about the rate of such movement. For water to reach the top of a 30 meter tree by capillary action, however, the vessel would have to be $(1.49\times10^{-5} m^2)/(30\ m)$ or 5×10^{-7} m, which is 0.5 μm in radius. This is much smaller than that observed for xylem vessels, indicating that capillary rise in channels of the size of the xylem cells cannot account for the extent of the water rise in tall trees. Furthermore, the vascular system is not open to the air at the upper end, and thus xylem vessels are not really analogous to the capillary depicted in Fig. 2.30.

The numerous interstices in the cell wall of xylem vessels form a mesh-work of many small tortuous capillaries, which can lead to an extensive capillary rise of water in a tree. A representative "radius" for these channels in the cell wall might be 5 nm (50 Å). Based on Equation a capillary of 5 nm radius could support a water column of $(1.49\times10^{-5})/(5\times10^{-9})$, or 3×10^{3} m—which is 3 kilometers (km)—far in excess of the needs of any plant. In other words, the cell wall could act as a very effective wick for water rise in its numerous small interstices, although the actual rate of such movement up a tree would generally be far from sufficient to replace the water lost by transpiration.

Because of the appreciable water-wall attraction which can develop both at the top of a xylem vessel and in the numerous interstices of its cell walls, water that is already present in the lumen of a xylem vessel can be substained or supported at great heights. The upward force, transmitted to the rest of the solution in the xylem vessel by water-water cohesion, overcomes the gravitational pull downward. Thus, the key feature for holding up water that is already present in the xylem vessel against the pull of gravity is the very potent attractive interaction (adhesion) of water to the cell wall surfaces at the top of the vessel. Two aspects of this warrant further consideration. First, what happens if the lumen of the xylem vessel becomes filled with air ? Will water then refill it ? The capillary rise of water is not sufficient to refill most xylem vessels greater than about 1 m in length, so that, in general, air-filled vessels are lost for conduction, such as the loss which occurs in the inner annual rings of most trees. The other aspect is more subtle, and we will consider it a little more extensively.

Tensile strength. The pulling on water columns which occurs in capillary rise and in the sustaining of water in the xylem requires that water be put under tension or negative pressure. In other words, water must have a definite tensile strength, the maximum tension (force per unit area) that it can withstand before breaking. In fact, the intermolecular hydrogen bonds resist the pulling apart of water in a column. If 80% of the hydrogen bonds are intact in water at 25°, their energy will be (0.80) (9.6) or 7.7 kcal/mole, which is (7.7 kcal/mole)/(18 g/mole) or 0.43 kcal/g of water. For a density of 1.00 g/cm^3, and replacing kcal by 4.184×10^{10} ergs, we calculate that the energy of the hydrogen bonds is 1.8×10^{10} ergs/cm^3. The tension that is applied to a water column acts against this attractive energy of the hydrogen bonds. When the fracture is just about to occur at each hydrogen bond, the maximum possible tensile strength is developed. Thus, the maximum tensile strength would represent an input of energy of 1.8×10^{10} ergs/cm^3. Since an erg is a dyne-cm and a bar is 10^6 dynes/cm^2, the maximum tensile strength of water corresponds to 18,000 bars. We might also attempt to calculate the maximum tensile strength of water by considering the electrostatic force of attraction between the δ^+ on the hydrogen of one water molecule and the δ^- on the oxygen of a neighbouring one (cf. the hydrogen bond in Fig. 2.29). Dividing this force by the area of the water molecules involved, we could then calculate the tension developed—but such an approach involves formidable geometrical considerations.

Experimentally, water withstands negative pressures (tensions) only up to approximately 300 bars at 20° without breaking. (Observed tensile strengths depend on the wall material, the diameter of the vessel in which the determinations are made, and any solutes present in the water sec Hayward.) Any local imperfections in the semicrystalline structure of water—such as those caused by H^+ and OH^-, which are always present, even in pure water—markedly reduce the observed tensile strength from the maximum value predicted based on hydrogen bond strengths. Nevertheless, the measurable tensile strength for water provided by the intermolecular hydrogen bonds (up to 300 bars) is nearly 10% of that for copper or aluminium, and is sufficiently high to meet the demands encountered for water movement in plants. In summary, the great cohesive forces between water molecules allow an appreciable tension to exist in an uninterrupted water column in a wettable capillary or tube such as a xylem vessel. This property is important for the continuous movement of water from the root through the plant to the surrounding atmosphere during transpiration.

In contrast to the stable situation for metals, water under tension is actually in a metastable state. If gas bubbles form in the water under tension in the xylem vessels, the column of liquid can be ruptured. Thus the introduction of another phase can disrupt the metastable state describing water under tension. Minute air bubbles sometimes spontaneously from in the xylem as the external temperature rises. These usually adhere to the walls of the xylem vessel (or the pores of the perforation plate), and the gas in them slowly redissolves. However, if they grow large enough or a number of small bubbles coalesce, that xylem vessel could cease functioning. Moreover, freezing of the solution in xylem vessels can lead to bubbles in the ice (the solubility of air is quite low in ice), and these bubbles could then interrupt the water columns upon thawing. Some plants are in fact damaged by freezing and thawing of xylem sap and the consequent loss of water continuity in the xylem vessels.

Electrical properties. Another important physical characteristic of water is its extremely high dielectric constant, again a consequence of its molecular structure. A high dielectric constant for a solvent lowers the electrical forces between charged solutes. To quantify the magnitude of electrical effects occurring in a fluid, let us consider two ions having charges Q_1 and Q_2 and separated from each other by a distance r. The electrical force exerted by one ion on the other is expressed in Coulomb's law,

$$\text{electrical force} = \frac{Q_1 Q_2}{4\pi\epsilon_o D r^2}$$

where ϵ_o is a proportionality constant known as the permittivity of a vacuum. In Equation. D is a dimensionless quantity called the dielectric constant. It equals unity in a vacuum and is 1.00058 in air at 0° at a pressure of 760 mm Hg.

Any substance composed of highly polar molecules, such as water, generally has a high dielectric constant. Specifically, D for water is 80.2 at 20° and 78.4 at 25°, which are extremely high values for dielectric constants of liquids. By contrast, the dielectric constant of the nonpolar liquid hexane is only 1.87—such a low value is typical of many organic solvents. Based on these two vastly different dielectric constants and Equation we can calculate that the attractive electrical force between ions such as Na^+ and Cl^- is (80.2/1.87), or 43 times greater in hexane than in water. The much stronger attraction between Na^+ and Cl^- in hexane reduces the amount of NaCl that will dissociate, compared with the dissociation of this salt in aqueous solutions. Stated another way, the much weaker electrical forces in water, compared with those in organic solvents, permit a larger concentration of ions to be in solution. Water is thus a good solvent for charged particles.

The electrostatic interaction between ions and water partially cancels or screens out the local electrical fields of the ions. Cations attract the negatively charged oxygens of the water molecule, while anions reciprocally attract its positively charged hydrogen atoms. The oriented water molecules around the charged particles produce local electrical fields opposing the fields of the ions. The resultant screening diminishes the electrical interaction between the ions and allows more of them to remain in solution. This is the molecular basis for the high dielectric constant of water.

The energy for the interaction between water and nonpolar molecules is usually less than the energy required to break water-water hydrogen bonds. Nonpolar compounds are therefore not very soluble in water. Certain substances such as detergents, phospholipids, and proteins can have both a nonpolar and also a polar region in the same molecule. In aqueous solutions these compounds can form aggregates, called *micelles*, in which the nonpolar groups of the molecules are in the center, while the charged or polar groups are external and interact with water. The lack of appreciable attractive electrostatic interaction between polar or charged species and the nonpolar (hydrophobic) regions of membrane lipids underlies the ability of biological membranes to limit the passage of such solutes into and out of cells and organelles.

Chemical potential

The chemical potential of species *j* is a thermodynamic way of quantitatively describing the free energy associated with it and available for performing work. Since there are many concepts to be mastered before understanding such a statement, we will first briefly explore the concept of free energy.

Free energy and chemical potential. Plants and animals require a continual input of free energy. If we were to remove the sources of free energy, organisms would drift toward equilibrium and consequent cessation of life. The ultimate source of free energy is the sun. Photosynthesis converts its plentiful radient energy into free energy, which is stored first in intermediate energy "currencies" like ATP and then in the altered chemical bonds which result when CO_2 and H_2O react to form a carbohydrate and O_2. When the overall photosynthetic reactions are reserved during respiration, the available free energy is reconverted into suitable energy currencies like ATP. In turn, such free energy is used to perform biological work such as transporting amino acids into cells, pumping blood, powering the electrical machinery of the brain, or lifting weights. For every such process on both a molecular and a global scale the free energy decreases. In fact, the structure of cells, as well as that of ecosystems, is governed by the initial supply of free energy and by the inexorable laws of thermodynamics which describe the expenditure of that free energy.

Instead of viewing the whole universe, the thermodynamicist focuses his attention on some small part of it referred to as a "system." Such a system can be potassium ions dissolved in water, a plant cell, a leaf, or even a whole plant. Something happens to the system ; the potassium concentration is increased, some protein is made in a cell, a leaf abscises, or water moves up a tree. We then say that the system changes from state A to state B (the change may of course be brought about in many different ways). The minimum amount of work needed to cause the change from A to B can be defined as the free energy increase associated with it. Alternatively, the free energy decrease in going from state B back to state A represents the maximum amount of work which can be derived from the transition. (If there were no frictional or other irreversible losses, we could do without the words "minimum" and "maximum.") Thus, the limits to the work done on or by a system when it undergoes a transition from one state to another concern changes in free energy. Knowledge of the free energy under one condition compared with another allows us to predict the direction of spontaneous change or movement—a spontaneous change in a system at constant temperature and pressure always proceeds in the direction of decreasing free energy.

Most systems of interest in biology are subject to a constant pressure (atmospheric) and remain at constant temperature, at least for short periods. In discussing the energetics of processes at constant temperature and pressure, the appropriate quantity is known as the *Gibbs free energy.* The Gibbs free energy has a very useful property in that it always decreases for a spontaneous process at constant temperature and pressure. Under such conditions, the decrease in Gibbs free energy equals the maximum amount of energy available for work, while if it increases for some transition, the change in Gibbs free energy represents the minimum amount of work required. For our present purposes, we will simply note that the Gibbs free energy of an entire system is made up of additive contributions from each of the species present. We will therefore shift our emphasis to the individual components of the system and the manner in which their free energy can be described.

It turns out that we can assign to every chemical component in a system a free energy per mole of that species. The quantity is called the *chemical potential* of species *j* and is given the symbol μ_j. We can view μ_j as a property of species *j* that tells how that species will react to a given change *e.g.*, a transition of the system from state A to state B. During the transition the chemical potential of species *j* changes from μ_j^A to μ_j^B. If μ_j^B is less than μ_j^A, then the free energy per mole of species *j* decreases. Such a process can take place spontaneously as far as species *j* is concerned—water flowing downhill is an example of such a spontaneous process. (We are restricting our attention to changes that occur when the system is at a constant temperature and subjected to a constant pressure.) In principle, we can harness the spontaneous change to do work. The maximum amount of work that can be done per mole of species *j* is μ_j^A— μ_j^B (Fig. 2.31). Suppose now that μ_j^B equals μ_j^A. As far as species *j* is concerned, no work is involved in the transition from state A to state B—species *j* is then at equilibrium (cf. Fig. 2.31). (We will show later that not only are living systems as a whole far from equilibrium, but also many chemical species are not in equilibrium across cellular membranes). Finally, let us consider the case where μ_j^B is greater than μ_j^A. A change which increases the chemical potential of a substance can only occur if some other change occurs in the system—this other change supplying the free energy required. The pumping of blood along arteries using ATP to cause contraction of the heart muscles is an instance where the increase in chemical potential of a mole of water in the blood is accompanied by an even larger decrease in the free energy associated with ATP. The minimum amount of work that must be done per mole of species j to cause the energetically uphill transition is $\mu_j{}^B - \mu_j{}^A$ (Fig. 2.31).

Initial chemical potential	Final chemical potential	Comment
μ_j^A	μ_j^B	Transition can occur spontaneously as far as spacies j is concerned: maximum work available per mole of species j equals $\mu_j^A - \mu_j^B$
μ_j^A	μ_j^B	Equilibrium
μ_j^A	μ_j^B	Change will not occur unless free energy at least equal to $\mu_j^B - \mu_j^A$ is supplied per mole of species j

Fig. 2.31. Possible changes in the chemical potential of species j which can accompany a transition from initial state A to final state B—as might occur in a chemical reaction or in crossing a membrane. The heights of the bars representing μ_j correspond to the relative values of the chemical potential.

The condition for a spontaneous change, namely a decrease in chemical potential, has important implications for discussing fluxes from one region to another. In particular, we can use the chemical potential difference between two locations as a measure of the "driving force" for the movement of that component. Thus, the bigger the chemical potential difference. $\mu_j{}^A - \mu_j{}^B$, the faster the spontaneous change takes place—in this case, the larger the flux of species j from region A to region B.

The chemical potential of a substance clearly depends on its chemical composition, but it is also influenced by other factors. Here we shall argue in an intuitive way about what environmental factors might influence μ_j. To begin with, we note that chemical potential depends on the "randomness" (entropy) of the system, and concentration is a quantitative description of this randomness. For example, the passive process of diffusion describes neutral molecules spontaneously moving from some region to another where their concentration is lower. In more formal terms, diffusion of species j actually proceeds in the direction of decreasing μ_j, which in this case is the same as that of decreasing concentration. The importance of concentration gradients, or differences, in leading to the movement of a substance was also established in the discussion of diffusion. Here we will present the effect of concentration on chemical potential in a somewhat more sophisticated manner, using that part of concentration which is thermodynamically active.

Chemical potential also depends on pressure, which for situations of interest in plant physiology means hydrostatic pressure. The existence of pressure gradients can cause movements of fluids—the flow of crude oil in long-distance pipelines, blood in arteries, and sap in xylem vessels—which is another way of saying that pressure differences can affect the direction for spontaneous changes.

Electrical potential also affects the chemical potential of charged particles, hence it must be considered when predicting the direction of their movement. It is a matter of definition that work must be done to move a positively charged particle to a higher electrical potential. Accordingly, if an electrical potential difference is imposed across an electrolyte solution containing a uniform concentration of ions throughout, then cations will spontaneously move in one direction (toward the cathode, *i.e.*, to regions of lower electrical potential), while anions will move in the opposite direction (toward the anode). It is clear, then, that in order to describe the chemical potential of a charged species we must include an electrical term in μ_j.

Another contributor to chemical potential is gravity. We

can readily appreciate that position in a gravitational field affects μ_j, inasmuch as work must be done to move the substance upward. Although the gravitational term can be neglected for ion and water movements across plant cells and membranes, it is important for water movement in a tall tree.

In sum, the chemical potential of some species should depend on its concentration, the pressure, the electrical potential, and gravity. We can compare the chemical potentials of a species on two sides of a barrier in order to decide whether or not the species is in equilibrium. If μ_j is the same on the two sides, we would not expect a net movement of species *j* to occur spontaneously across the barrier. We will use the relative values of the chemical potential of species *j* at various locations to predict the directions for spontaneous movement of that chemical substance (*viz.*, toward lower μ_j), just as comparison of temperatures can be used to predict the direction for heat flow (toward lower T). Finally, we will find that $\Delta\mu_j$ from one region to another gives a convenient measure of the driving force on species *j*.

Analysis of chemical potential. For convenience and also because it has proved experimentally valid, we will represent the chemical potential of any species *j* by the following sum of the various components into which it can be analyzed :

$$\mu_j = \mu_j^* + RT \ln a_j + \overline{V}_j P + z_j FE + m_j gh$$

One measure of the elegance of a mathematical expression is the amount of usefulness or information which it contains. Based on such a criterion. Equation truly represents one of the most elegant relations in all of biology. After defining and describing the various contributions to μ_j indicated in Equation we will consider in greater detail the various terms for the important special case where species *j* is water.

Chemical potential, like electrical or gravitational potentials is a relative quantity. By this we mean that it must be expressed relative to some arbitrary level. Thus, an unknown additive constant, or reference level, μ_j^*, is included in Equation. Since it contains an unknown constant, the actual value of chemical potential is not determinable. For most applications of chemical potential in physiology, we are interested in the difference in the chemical potential between two particular locations, in which case only relative values of the chemical potential are important anyway. The arbitrariness of μ_j^* is of little consequence because it is added to each of the chemical potentials under comparison. Therefore, μ_j^* cancels out when the chemical potential in one location is subtracted from that in another in order to obtain the chemical potential difference between the two locations. Figure 2.31 further illustrates the usefulness of considering differences in chemical potential instead of the actual value of μ_j. (We will return to μ_j^* when we consider μ_j in the standard state). Finally, we note that the units of μ_j and of μ_j^* are energy per mole of the substance, *e.g.*, joules/mole or cal/mole.

In certain cases we can adequately describe the chemical properties of species j by using the concentration of that solute, c_j. Owing to molecular interactions, however, this usually requires that the total solute concentration below. Molecules of solute species j interact with each other as well as with other solutes in the solution, and this influences the behaviour of species j. Such intermolecular interactions increase as the solution becomes more concentrated. The use of concentrations for describing the thermodynamic properties of some solute thus indicates an approximation, except in the limiting case of infinite dilution where interactions between solute molecules are negligible. Where exactness is required, *activities*—which may be regarded as "corrected" concentrations—are employed. Consequently, for general thermodynamic considerations, such as Equation the influence of the amount of a particular species j on its chemical potential is not handled by its concentration but by the activity of that species, a_j. The activity of solute j is related to its concentration by means of an activity coefficient, γ_j :

$$a_j = \gamma_j c_j$$

The activity coefficient is usually less than unity, since the thermodynamically *effective* concentration, or *activity* of a species is generally less than is its actual concentration.

For an ideal solute, γ_j is one and the activity of species j

then equals its concentration. This condition can be approached for real solutes in certain dilute aqueous solutions, especially for neutral species. Activity coefficients for charged species can be appreciably less than unity because of the importance of electrical interactions.

The activity of a solvent is defined in a manner different from that for a solute. The a_j of a solvent is defined as $\gamma_j N_j$, where N_j is the mole fraction of the solvent. An ideal solvent has $\gamma_{solvent}$ equal to 1, while an ideal solution has all activity coefficients equal to unity.

In the expression for chemical potential, Equation, the term involving the activity appears as $+RT \ln a_j$. Therefore, the greater the activity of species *j*—loosely speaking, the higher its concentration—the larger will be its chemical potential. The appearance of the logarithmic form can be "justified" in a number of different ways, all of which are based on agreement with empirical observations. The factor RT multiplying $\ln a_j$ in Equation gives the activity term the units of energy per mole.

The term $\bar{V}_j P$ in Equation represents the effect of pressure on chemical potential. Since essentially all measurements in plant physiology are made on systems subjected to atmospheric pressure, it is convenient to define P as the pressure in excess of this, and we will adopt such a convention here. $\bar{V}_j$ is the differential increase in volume of a system when a differential amount of species *j* is added, other species, temperature, pressure, electrical potential, and gravitational position remaining constant:

$$\bar{V}_j = \left(\frac{\partial V}{\partial n_j}\right)_{n_i T, P, E, h}$$

The subscript n_i on the partial derivative in Equation means that the numbers of moles of all species present, other than species *j*, are held constant when the derivative is taken—the four other subscripts are included to remind us of the additional parameters which must be held constant.

$\bar{V}_j$ is called the *partial molal volume* of species *j*. (μ_j is actually the partial molal Gibbs free energy, $(\partial G/\partial n_j)_{n_i, T, P, E, h}$. The partial molal volume of a substance is often nearly equal to the volume of a mole of that species, but because there is in general a slight change in total volume when substances are mixed; $\overline{V}_j$ is not exactly equal to it. We will indicate that the value of $\overline{V}_j$ for water is about 18 cm²/mole when we specifically consider the chemical potential of water.

Since work is often expressed as pressure times change in volume, note that $\overline{V}_j P$ has the correct units for μ_j of energy per mole ($\overline{V}_j$ is the volume per mole). To help justify the form of the pressure term, let us imagine that the solution containing species *j* is built up by adding small volumes of species *j* while the system is maintained at a constant pressure P. The work done to add a mole of species *j* would then be the existing pressure times some volume change of the system which characterizes a mole of that species, *viz*., $\overline{V}_j$.

The influence of electrical potential on the μ_j of an ion is handled by the term z_jFE in Equation where z_j is an integer representing the charge number of species *j*, F is a constant known as the Faraday and E is electrical potential. Because water is uncharged ($z_w = 0$), the electrical term does not contribute to its chemical potential. However, electrical potential is of central importance when discussing ions and also the origin of membrane potentials. The full expressed for chemical potential, μ_j, given by Equation including z_jFE, is often referred to as *electrochemical potential* when discussing the properties of changed particles.

For completeness, the gravitational term $m_j gh$ is included in Equation. It is apparent that to raise an object of mass m_j per mole to a vertical height *h* requires an amount of work $m_j gh$, where *g* is the gravitational acceleration (close to 980 cm/sec². In the case of water, which is the primary concern of this chapter, m_w is 18.016 g/mole. The gravitational term in the chemical potential is not only important for the fall of rain, snow or hail, but also it affects the percolation of water downward through porous soil and the upward movement of water in a tree. The gravitational term $m_j gh$ can have units of (g/mole)(cm/sec^2)(cm) or (g-cm^2/sec^2)/(mole), which is ergs/mole.

Standard state. The additive constant term $\mu_j{}^*$ in Equation is actually the chemical potential of species j for a suitable reference state. From the preceding definitions of the various quantities involved, this reference state is attained when the activity of species j is unity (RT ln $a_j=0$), the hydrostatic pressure equals atmospheric pressure ($\overline{V}_jP=0$), the species is uncharged or the electrical potential is zero ($z_jFE=0$), we are at the zero level for the gravitational term ($m_jgh=0$), and the temperature equals that for the system under consideration. Under these conditions μ_j equals $\mu_j{}^*$.

An activity of unity is defined in different ways for the solute and the solvent. To describe liquid properties, such as the dielectric constant, the heat of vaporization, and the boiling point, the most convenient standard state is that of the pure solvent. For a solvent, $a_{solvent}$ equals $\gamma_{solvent}$ times $N_{solvent}$, hence the activity is unity when the mole fraction $N_{solvent}$ is unity ($\gamma_{solvent}=1$ for pure solvent). For example, the standard reference state for water is pure water at atmospheric pressure and at the temperature and gravitational level of the system under consideration. Water has an activity of unity when N_w as given by Equation is one—the concentration of water is then (1000 g of solvent)/(18.016 g/mole), or 55.5 molal. The accepted convention for a solute, on the other hand, is that a_j is unity when γ_jc_j equals 1 molal. For example, if γ_j equals one, a solution that has a 1 molal concentration of solute j has an activity of 1 molal for that solute. Thus, the standard state for an ideal solute is when its concentration is 1 molal, in which case RT ln a_j is zero.

The value of the chemical potential for a solute in the standard state, $\mu_j{}^*$, actually depends on the solvent. In order to be explicit, let us consider a two-phase system consisting of water with an overlying layer of olive oil. It turns out that $\mu_j{}^*$ for a polar molecule is smaller for an aqueous solution (where the species readily dissolves) than it is for an organic phase such as olive oil (where it is not as soluble). If we put the polar solute into our system, shake the water and olive oil together, and wait for the two phases to separate out again, the solute concentration will be higher in the bottom or water phase. Let us next analyze this event using symbols, where phase A represents water and phase B is olive oil. Waiting for the phase separation is equivalent to waiting for the attainment of equilibrium, and so $\mu_j{}^A$ is then equal to $\mu_j{}^B$. In the present case this means that $\mu_j{}^{*,A}+RT \ln a_j{}^A$ equals $\mu_j{}^{*,B}+RT \ln a_j{}^B$ by Equation. But the polar solute is more soluble in water than it is in olive oil, hence $RT \ln a_j{}^A$ is greater than $RT \ln a_j{}^B$. Because we are at equilibrium, this means that $\mu_j{}^{*,A}$ must be less than $\mu_j{}^{*,B}$, as we already indicated. Such a two-phase separation procedure is actually followed in determining the olive oil-water partition coefficient, $a_j{}^B/a_j{}^A$ (which is essentially the same as $c_j{}^B/c_j{}^A$ for dilute solutions). Moreover, since the diffusion of many substances across membranes is correlated with their lipid solubility, such partition coefficients are of great utility in discussing the movement of solutes into and out of cells or organelles.

Hydrostatic pressure. Because of the rigid cell walls, fairly large hydrostatic pressures can exist in plant cells, while such pressures in animal cells generally are relatively small. In addition to being involved in the support of the plant, hydrostatic pressures are important for the movement of water and solutes in the xylem and probably also in the phloem. The term expressing the effect of pressure on the chemical potential of water is $\bar{V}_w P$ Equation, where $\bar{V}_w$ is the partial molal volume of water and P is the hydrostatic pressure in the aqueous solution in excess of the ambient atmospheric pressure. The approximate value of $\bar{V}_w$ can be estimated using the following argument. When 1 mole (18.0 g) of water is added to water, the volume increases by 18.0 cm^3, since the density of water at 20° is 0.998 g/cm^3. In other words, $\bar{V}_w$ is 18.0 cm^3/mole in the case of pure water. Although $\bar{V}_w$ (defined according to the partial derivative given in Equation can be influenced by the various solutes actually present, it generally is close to 18.0 cm^3/mole for a dilute solution, a value that we will use for calculations in this book.

Various units are used for expressing pressures, atmospheres and bars being commonly employed in plant physiology. A pressure of one atmosphere is defined as that pressure which can support either a column of mercury 76.0 cm high or a column of water 10.3 m high. However, most current research concerning the water relations of plants uses pressures in bars. A bar is 10^6 dynes/cm^2, which equals 0.987 atmosphere. Pressure is force per unit area, which is dimensionally the same as energy per unit volume, *e.g.*, one bar is 0.100 joule/cm^3 or 0.0239 cal/cm^3. Since $\overline{V}_w$ has the units of cm^3/mole, the quantity $\overline{V}_w P$ and hence μ_w itself can be expressed in joules/mole or cal/mole.

Water activity and osmotic pressure. The presence of solutes in an aqueous solution tends to decrease the activity of water (a_w). As a first approximation, we can view the decrease in a_w as more and more solutes are added as simply a dilution effect. In other words, the concentration (or mole fraction) of water becomes less when the water molecules are displaced by those of the solute. As its activity thus decreases, we would expect a lowering of the chemical potential of water, as does indeed occur (cf. the RT ln a_j term in μ_j). We also know that the presence of solutes can lead to an osmotic pressure (π) in the solution. An increase in the concentration of solutes raises the osmotic pressure, indicating that π and a_w change in opposite directions. In fact, the osmotic pressure and water activity are related, a fundamental definition of π being.

$$RT \ln a_w = -\overline{V}_w \pi$$

where the subscript w refers to water. As solutes are added, a_w decreases from its value of unity for pure water, ln a_w is therefore negative, and π (defined by Eq.) positive, which is consistent with the arguments presented above. Equation also indicates that RT ln a_w, a term appearing in the chemical potential of water, can be replaced by $-\overline{V}_w \pi$.

Unfortunately there is much variation on the use of the terms "osmotic pressure" and "osmotic potential", as well as the algebraic sign of the quantity. Osmotic pressures were originally measured using an osmometer, a device employing a membrane that in the ideal situation was permeable to water but not to any of the solutes present. When pure water was placed on one side of the membrane and some solution on the other, there was a net diffusion of water toward the side with the solutes. To counteract this tendency and establish equilibrium, a hydrostatic pressure had to be applied to the solution side. This pressure was called the *osmotic pressure*—it depended of course on the presence of the solutes. Does an isolated solution have an osmotic pressure ? In the sense of requiring an applied hydrostatic pressure to maintain equilibrium, the answer is no. Yet the same solution when placed in an osmometer can manifest an osmotic pressure. Thus, some physical chemistry texts have found it convenient to say that a solution has an osmotic *potential*—*i.e.*, it could show an osmotic pressure if placed in an osmometer—and that osmotic potentials are a property characteristic of solutions. Many plant physiology books have come to a similar conclusion, *e.g.*, Salisbury and Ross, but have defined the *negative* of the same quantity as the osmotic potential, in order to have the water potential (a quantity we will introduce shortly) depend directly on the osmotic potential and not on its negative. We will not try to resolve this dilemma. Instead, we will consistently refer to π defined by Equation as osmotic pressure. In principle, it is always possible to measure π as a pressure by using an osmometer.

Using another "colligative" property is a direct and convenient way of evaluating osmotic pressure and water activity, as defined by Equation. The four colligative properties of a solution—those that depend on the concentration of solutes regardless of their nature, at least in a dilute enough solution—are the freezing point depression, the boiling point elevation, the lowering of the vapour pressure of the solvent, and osmotic pressure. Thus, if the freezing point of cell sap is measured, $\pi^{cell\ sap}$ can be calculated using the freezing point depression of 1.86° for a 1 molal solution and the *Van't Hoff relation.*

The Van't Hoff relation. For many purposes in plant physiology it is convenient to relate osmotic pressure directly to the concentration of solutes instead of expressing π in terms of the water activity, a_w, as is done in Equation. In general, the greater the concentration of solutes, the more negative ln a_w becomes and the larger is the osmotic pressure. Thus, some way of expressing a_w in terms of the properties of the solutes would seem to be an appropriate approach for expressing π as a function of the various solute concentrations. The ensuring derivation not only will show how a_w can be so expressed, but will also serve to indicate the many approximations necessary in order to achieve a rather simple expression for π.

The activity of pure water is unity, while in general a_w equals $\gamma_w N_w$, where γ_w is the activity coefficient of water and N_w is its mole fraction. The mole fraction of a species is simply the ratio of the number of moles of that species to the total number of moles of all species in the system. Thus, N_w is given as :

$$N_w = \frac{n_w}{n_w + \sum_j n_j} = 1 - \frac{\sum_j n_j}{n_w + \sum_j n_j}$$

where n_w is the number of moles of water, n_j is the number of moles of solute j, and the summation, $\sum_j$, is over all solutes in the system being considered. Equation also expresses the familiar relation that N_w equals 1 minus the mole fraction of solutes. For an ideal solution, γ_w equals 1. This condition of unit activity coefficient is approached for dilute solutions in which case n_w is much greater than $\sum_j n_j$. (The expression $n_w \gg \sum_j n_j$ can be taken as one way of defining a dilute solution, where the symbol $\gg$ means "is much greater than".) Using Equation and assuming a dilute solution, we obtain the following relations for ln a_w :

$$\ln a_w \cong \ln N_w = \ln\left\{1 - \frac{\sum_j n_j}{n_w + \sum_j n_j}\right\} \cong -\frac{\sum_j n_j}{n_w + \sum_j n_j}$$

$$\cong -\frac{\sum_j n_j}{n_w}$$

The penultimate step in Equation is based on the series expansion of a logarithm—$\ln (1-x) = -x - x^2/2 - x^3/3 - \ldots$, a series which converges rapidly for $|x| \ll 1$--and the last two steps employ the approximation ($n_w \gg \sum_j n_j$) relevant to a dilute solution. Equation is thus restricted to dilute ideal solutions. Nevertheless, it is a useful expression which clearly indicates that in the absence of solute ($\sum_j n_j = 0$), $\ln a_w$ is zero and a_w is unity while the presence of solutes decreases the activity of water from the value of one for pure water.

To obtain a rather familiar form for expressing the osmotic pressure, we can incorporate the approximation for $\ln a_w$ given by Equation into the definition of osmotic pressure ($RT \ln a_w = -\bar{V}_w \pi$, Equation which yields

$$\pi_s \cong -\frac{RT}{\bar{V}_w}\left(-\frac{\sum_j n_j}{n_w}\right) = RT \sum_j \frac{n_j}{\bar{V}_w n_w} = RT \sum_j c_j$$

where $\bar{V}_w n_w$ is the total volume of water in the system (essentially the total volume of the system for a dilute aqueous solution), $n_j / \bar{V}_w n_w$ is the number of moles of species j in that volume and is therefore the concentration of species $j (c_j)$, and the summations in Equation are over all solutes.

Osmotic pressure is often expressed by Equation, known as the Van't Hoff relation, but this is justified only in the limit of *dilute ideal* solutions. An ideal solution has ideal solutes dissolved in anideal solvent. Since in writing the string of equalities in Equation we assumed that γ_w was essentially unity, our subsequently derived expression (Equation) strictly applies only when water acts as an ideal solvent ($\gamma_w = 1.00$). To emphasize that we are neglecting any factors which cause γ_w to deviate from unity and thereby affect the measured osmotic pressure (such as the interaction between water and colloids, which we will shortly discuss), π_s instead of π has been used in Equation, and we will follow this convention throughout the text. We must also consider the nonideality of solutes. We can handle departures of the solutes from ideality by replacing c_j in Equation with a_j, which is equal to $\gamma_j c_j$. Thus, the Van't Hoff relation for a dilute solution of *real* solutes can be represented by $\pi_s = RT\sum_j \gamma_j c_j$. The effect of solute concentration on osmotic pressure as described by Equation was first clearly shown by the botanist W. Pfeffer in 1877. Both the measurement of osmotic pressures and recognition of their effects have been crucial for an understanding of the water relations of plants.

Relatively dilute solutions have fairly large osmotic pressures. The cellular fluid expressed from plants like pea or spinach often contains about 0.3 mole of osmotically active particles per kg or liter of water. This fluid is referred to as being 0.3 osmolal by analogy with molality, which refers to the total concentration, *e.g.*, 0.1 molal $CaCl_2$ is nearly 0.3 osmolal since most of the $CaCl_2$ is dissociated. Note that molality (moles of solute/100 g solvent) is the correct unit for the Van't Hoff relation in its exact form, as well as for the other colligative properties. However, molarity (moles of solute/liter of solution, symbolized by M) is often a more convenient unit. Below 0.1 M, molarity and molality are nearly the same for a small solute in an aqueous solution, while at high concentrations the molarity can be considerably less than the molality. At 20°, RT has the value of 24.37 liter-bars/mole. Using Equation, the osmotic pressure of an 0.3 osmolal solution is (24.37 liter-bars/mole) (0.3 mole/liter), or 7.3 bars (essentially 7 atmospheres). Thus, an 0.3 osmolal solution has a substantial osmotic pressure.

Matric potential. For certain applications in plant physiology, another term is frequently included in the chemical potential of water, *viz.*, $\bar{V}_w\tau$, where τ is the matric potential, or matric pressure. The use of this term is sometimes convenient for dealing explicitly with interactions occurring at interfaces, although these interfacial forces can also be adequately represented by their contributions to π or P. In other words, the matric potential does not represent a new or different contribution to μ_w—it is only a book-keeping device for handling interfacial interactions. To help make this statement more meaningful, we will briefly consider the influence on the chemical potential of water of liquid-solid interfaces at the surfaces of colloids.

When water molecules are associated with interfaces such as those provided by the surfaces of colloidal particles suspended in an aqueous solution, they have less of a tendency either to react chemically in the bulk solution or to escape into a surrounding vapour phase. The presence of interfaces thus lowers the thermodynamic activity of the water (a_w), especially near the surfaces of the colloids. But the presence of solutes also lowers water activity an effect that can be expressed by Equation. As a useful first approximation, we can consider that these two different effects lowering water activity are additive in a solution which contains both ordinary solutes and colloids. Osmotic pressure (π) depends on the actual activity of water regardless of the reason for a_w being different from unity, *i.e.*, π still equals $-(RT/\bar{V}_w)\ln a_w$, as it must by Equation. Combining these various ideas (and recalling that $a_w = \gamma_w N_w$) we can write the following relation :

$$\pi = -\frac{RT}{\bar{V}_w}\ln \gamma_w N_w = -\frac{RT}{\bar{V}_w}\ln \gamma_w - \frac{RT}{\bar{V}_w}\ln N_w = \tau + \pi_s$$

where τ is a matric pressure resulting from the water-solid interactions at the surfaces of the colloids.

Note that π_s in Equation is to be identified with the osmotic pressure of all solutes, including the *concentration* of the colloids, as represented by Equation ($\pi_s \cong RT\sum_j c_j$). In other words, we were actually dealing only with—$(RT/\bar{V}_w)\ln N_w$

in the derivation of π_s, since γ_w was then set equal to unity. For an exact treatment when many interfaces are present—*e g.*, in the cytoplasm of a typical cell—we cannot set γ_w equal to unity, since the activity of water, and hence the osmotic pressure (π), is strongly affected by the proteins and other colloids present. In such a case, Equation suggests a simple way in which matric pressure may be related to the reduction of the activity coefficient of water caused by the interactions at interfaces. (Note that if we let $\tau = -(RT/\bar{V}_w) \ln \gamma_w$, then γ_w less than unity leads to a positive τ). Equation should not be viewed as a relation defining τ, rather, it is one example where it might be useful to represent interfacial interactions by a separate term which can be added to π_s, the effect of the solutes on π.

Although π and a_w may be the same throughout some system, both π_s and τ in Equation may vary. For example, water activity in the bulk of the solution may be predominantly lowered by the presence of the solutes, while at or near the surface of colloids the main factor decreasing a_w from one could be the interfacial attraction and binding of water. This latter effect does not change the mole fraction of water but does reduce its activity coefficient, γ_w.

Other areas of plant physiology where matric potentials have been invoked are in descriptions of the chemical potential of water in soil or in cell walls. The interfacial interactions between water and the many surfaces lining the interstices of the cell wall lead to a tension in the cell wall water. Such a tension is simply a negative hydrostatic pressure, *i.e.*, P is less than zero. Although it is not necessary, P is sometimes used to refer only to positive pressures, while negative P in cell wall water is often defined as a positive matric potential. Some recent books define this same negative hydrostatic pressure in water in cell wall interstices or between soil particles as a negative matric potential. It is far more straightforward to refer to the quantity which reduces μ_w in such pores as a negative P.

Water potential. From the definition of chemical potential (Equation) and the formal expression for osmotic pressure

(Equation), we can write the chemical potential of water (μw) as

$$\mu w = \mu w^* - \bar{V} w \pi + \bar{V} w P + m w g h$$

where the electrical term ($z_j FE$) is not included, since water carries no charge ($zw=0$). The quantity $\mu w - \mu w^*$ has proved to be of considerable importance for discussing the water relations of plants. It represents the work involved in moving one mole of water from some point in a system (at constant pressure and temperature) to a pool of pure water at atmospheric pressure, at the same temperature as the system under consideration, and at the zero level for the gravitational term. More to the point, a difference between two locations in the value of $\mu_w - \mu_w^*$ indicates that water is not in equilibrium, hence there will be a tendency for water to flow toward the region where $\mu w - \mu w^*$ is lower.

A quantity proportional to $\mu w - \mu w^*$ and increasingly used in studies of plant water relations is the so-called *water potential*, ψ. This is conveniently defined as

$$\psi = \frac{\mu_w - \mu_w}{V_w} = P - \pi + \rho_w g h$$

which follows directly from Equation plus the identification of $m_w / \bar{V}_w$ with the density of water, ρ_w. Equation indicates that an increase in hydrostatic pressure raise the water potential, while an increase in osmotic pressure lowers it.

Some older terminology describing plant water relations can be identified from Equation. For example,$-\psi$ is the quantity called "diffusion pressure deficit" or "suction pressure" (actually, the term $\rho w g h$ was nearly at ways neglected). Although these expressions are quite colorful, the terms *water potential* (ψ) and *chemical potential of water* (μw) are increasingly being used in biology and other disciplines, and we will use them in this text. Hydrostatic pressure (P) is often referred to as "turgor pressure" in considerations of the protoplast, or, from a complementary point of view, as "wall pressure" in considerations of the pushing of the cell wall back against the protoplast.

We have already mentioned that work must be performed

to raise an object in the gravitational field of the earth. Thus, vertical position also affects the chemical potential, consequently, the term $\rho_w g h$ must be included in the water potential given by Equation at least when water moves an appreciable distance vertically in the gravitational field. In plant physiology the magnitude of $\rho_w g$ is conveniently expressed as 0.098 bar/m—if water moves 10 m vertically upward in a tree, the gravitational contribution to the water potential is increased by about 1 bar. For most applications in this text, such as consideration of chemical reactions or the crossing of membranes, there is no change in vertical position and the gravitational term will generally be omitted from μ_j and ψ.

Central vacuole and chloroplasts

The water relations of both the large central vacuole and the chloroplasts will be considered next, using the water potential as defined above. We will focus our attention on the situation where there is no net flow of water across the limiting membranes surrounding these subcellular compartments and for the special case of nonpenetrating solutes.

The central vacuole occupies up to 90% of the volume of a mature plant cell, which means that under such circumstances most of the cellular water is in the vacuole. The vacuolar volume is generally from 10^4 to 10^5 μm^3 for the mesophyll cells in a leaf, but can be much larger in certain algal cells. Because its size is similar to that of the cell, the vacuole cannot be easily removed from the plant cell without rupturing its surrounding membrane, the tonoplast. In other words, procedures which remove the cell wall and open up the plasmalemma usually also break the tonoplast, especially for small cells. Chloroplasts are much smaller than the central vacuole, often having volumes near 30μm^3 *in vivo* (although the sizes vary considerably with the plant species). When chloroplasts are carefully isolated, suitable precautions will ensure that their limiting membranes remain intact. Such intact chloroplasts can be placed in solutions having various osmotic pressures and the resulting movement of water into or out of the isolated plastids can be precisely measured.

Water relations of the central vacuole. In order to predict whether and in what direction water will move, we need to know the value of the water potential in the various compartments under consideration. At equilibrium the water potential is the same in all communicating phases, such as those separated by membranes. For example, when water is in epuilibrium across the tonoplast, the value of the water potential is the same in the vacuole as it is in the cytoplasm. No force then drives water across this membrane, and no net flow of water occurs into or out of the vacuole.

The tonoplast most likely does not have an appreciable difference in hydrostatic pressure across it. A higher internal hydrostatic pressure would cause a typical slack (folded) tonoplast to be mechanically pushed outward. The observed lack of motion thus indicates that ΔP is zero across a slack tonoplast. If the tonoplast were taut, ΔP would cause a stress in the membrane, analogous to the cell wall stresses discussed in the last chapter. For a tonoplast 10 nm thick, surrounding a spherical vacuole 10 μm in radius, the stress would be $r\Delta P/2t$, which is $(10\times 10^{-6}\text{ m})\Delta P/[(2)(10-10^{-9}\text{ m})]$, or $500\Delta P$. If the hydrostatic pressure were only 0.01 bar higher in the vacuole compared with the cytoplasm, in a taut tonoplast there would then be a stress of (500)(0.01), or 5 bars. This exceeds the tensile strength of biological membranes—most membranes rupture before a stress of even 1 bar develops in them. Thus, P may well have essentially the same value in the cytoplasm and the vacuole. With this simplifying assumption, and for the equilibrium situation where ψ is the same in the two phases, Equation ($\psi=P-\pi+\rho_w gh$ where $\Delta h=0$ across a membrane) gives

$$\pi^{cytoplasm}=\pi^{vacuole}$$

where $\pi^{cytoplasm}$ is the osmotic pressure in the cytoplasm and $\pi^{vacuole}$ that in the central vacuole.

The vacuole appears to be a relatively homogeneous aqueous phase, while the cytoplasm is a more complex phase containing many colloids and membrane-bounded organelles. Since the vacuole contains few colloidal or other interfaces, any matric potential in it is probably negligible compared with the osmotic pressure resulting from the vacuolar solutes. Expressing osmotic pressure by Equation ($\pi=\pi_s+\tau$) and assuming that $\tau^{vacuole}$ is negligible, we can replace $\pi^{vacuole}$ in Equation by $\pi_s{}^{vacuole}$, *i.e.*, the decrease in vacuolar water activity is due almost solely to the presence of solutes. On the other hand, water activity in the cytoplasm is presumably considerably lowered both by solutes and by the presence of interfaces. In other words, individual water molecules in the cytoplasm are never very far away from proteins, organelles, or order colloids, so that interfacial interactions can appreciably affect $a_w{}^{cytoplasm}$. In terms of Equation this means that both $\pi_s{}^{cytoplasm}$ and $\tau^{cytoplasm}$ contribute to $\pi^{cytoplasm}$. Hence, Equation indicates that $\pi_s{}^{cytoplasm}+\tau^{cytoplasm}$ equals $\pi_s{}^{vacuole}$. Since $\tau^{cytoplasm}$ is a positive quantity, we conclude that at equilibrium, the osmotic pressure in the vacuole due to solutes, $\pi_s{}^{vacuole}$, must be larger than $\pi_s{}^{cytoplasm}$. Equation leads then to the prediction that the vacuole has a higher concentration of osmotically active solutes than does the cytoplasm.

There are various speculations concerning the role of the large central vacuole in plant cells. The vacuole might act as a storage site for metabolites, ions, or toxic products. Because of the large central vacuole, the cytoplasm occupies a thin layer around the periphery of the cell. Therefore, for its volume, the cytoplasm has a relatively large surface area (across which diffusion can occur). Vacuoles also provide large, relatively simple compartments in which hydrostatic pressures lead to the cellular turgidity necessary for support of the plant. The actual function of the central vacuole most likely entails a combination of these various possibilities.

Boyle-Van't Hoff relation. The volume of a chloroplast or other membrane-bounded body changes in response to variations in the osmotic pressure of the external solution, π°. This is a consequence of the properties of membranes, which generally allow water to move readily across them, at the same time restricting the passage of certain solutes, such as sucrose. For example, if π° outside a membrane-bounded aqueous compartment were raised, we would expect that water would flow out across the membrane to the region of lower chemical potential and thereby decrease the compartment volume, while in general there would be very little movement of solutes in such a case. This differential permeability leads to the "osmometric behaviour" characteristic of many cells and organelles. The conventional expression quantifying this volume response to changes in the external osmotic pressure is the *Boyle-Van't Hoff relation* :

$$(\pi^o V - b) = RT \sum_j \phi_j n_j = RTn$$

where π^o is the osmotic pressure of the external solution, V is the volume of the cell or organelle, φ is the so-called non-osmotic volume (osmotically inactive volume) and is frequently considered to be the volume of a solid phase within volume V which is not penetrated by water, n_j is the number of moles of species j within $V-b$, the osmotic coefficient (ϕ_j) is generally construed as a correction factor, and $n = \sum_j \phi_j n_j$ is the apparent number of osmotically active moles in $V-b$.

In this chapter we will derive the Boyle-Van't Hoff relation using the concept of chemical potential, and we will extend the treatment by employing irreversible thermodynamics. Although the derived Boyle-Van't Hoff expression will be used only to interpret the osmotic responses of chloroplasts, the equations that will be developed are quite general and can be applied equally well to data obtained with mitochondria, whole cells, or other membrane-surrounded bodies.

The Boyle-Van't Hoff relation applies to the equilibrium situation when equality of the water potential has been achieved across the membranes surrounding chloroplasts exposed to a particular external osmotic pressure. When ψ^i equals ψ^o, net water movement across the membrane ceases, and the volume of a chloroplast has a steady value. (The superscript i refers to the inside of the cell or organelle, the superscript o designates the outside.) If we were to measure the chloroplast volume under such conditions, the external solution would generally be at atmospheric pressure ($P^o = 0$ bars). From Equation ($\psi = P - \pi$), the water potential in the external solution is then

$$\psi^o = -\pi^o$$

As is the case for ψ^o, the water potential inside the chloroplasts depends on osmotic pressure. In addition, the internal hydrostatic pressure (P^i) may be different from atmospheric pressure and should be included in the expression for ψ^i. Macromolecules and solid-liquid interfaces inside the chloroplasts can lead to a matric potential, τ^i. To allow for this possibility, the internal osmotic pressure (π^i) will be considered as the sum of the solute and the interfacial contributions, in the manner expressed by Equation. Using Equation ψ^i is therefore

$$\psi^i = P^i - \pi^i = P^i - \pi_s{}^i - \tau^i$$

We will now re-express $\pi_s{}^i$ and then equate ψ^o to ψ^i. For a dilute ideal solution inside the chloroplasts, $\pi_s{}^i$ equals $RT \sum_j n_j{}^i / (\bar{V}_w n_w{}^i)$ by Equation. The solutions of real solutes necessitate the introduction of an activity coefficient for solute j, γ_j, which approaches one as the solution becomes more dilute. To handle deviations from ideal behaviour for the solutes, $n_j{}^i$ can thus be replaced by $\gamma_j{}^i n_j{}^i$. Making these changes and equating the water potential outside the chloroplast to that inside the condition for water equilibrium across the limiting membranes of the chloroplast becomes

$$\pi^o = RT \frac{\sum_j \gamma_j{}^i n_j{}^i}{\bar{V}_w n_w{}^i} + \tau^i - P^i$$

Although this expression may look a little frightening, it states only that at equilibrium the water potential is the same on the two sides of the membranes and that we can explicitly recognize various contributors to the two ψs involved.

To appreciate the refinements this thermodynamic treatment introduces into the customary expression describing the osmotic responses of cells and organelles. Equation should be compared with the conventional Boyle-Van't Hoff relation The volume of water inside the chloroplast is $\bar{V}_w n_w{}^i$, since $n_w{}^i$ is the number of moles of internal water and $\bar{V}_w$ is the volume of one mole of water. This factor in Equation can be identified with $V-b$ in Equation. Instead of being designated the "non-osmotic volume," b is more appropriately called the "nonwater volume," for it includes the volume of the internal solutes including that of colloids. In other words, the total volume (V) minus the nonwater volume (b) equals the volume of internal water ($\bar{V}_w n_w{}^i$). The osmotic coefficient, ϕ_j in Equation is generally less than l, as is the activity coefficient of an internal solute, $\gamma_j{}^i$, in Equation. But $\gamma_j{}^i$ cannot simply be equated to ϕ_j since other factors are also involved. Finally the possible hydrostatic and matric contributions included in Equation are neglected in the usual Boyle-Van't Hoff relation. In short, although certain approximations and assumptions have crept into Equation (*e.g.*, that solutes do not cross the limiting membranes, and using dilute solution considerations), it is still a far more satisfactory expression for describing osmotic responses than is the conventional Boyle-Van't Hoff relation.

Osmotic responses of chloroplasts. To illustrate the use of Equation in interpreting osmotic data, we will consider osmotic responses of pea chloroplasts suspended in external solutions of various osmotic pressures. Since it is customary to plot the volume V versus the reciprocal of the external osmotic pressure, $1/\pi^{\circ}$, certain algebraic manipulations are needed in order to put Equation into a more convenient form for discussing such osmometric studies. By transferring $\tau^i - P^i$ to the left side of Equation and then multiplying both sides by $\bar{V}_w n_w{}^i/(\pi^{\circ} - \tau^i + P^i)$. $\bar{V}_w n_w{}^i$ can be shown to equal $RT\sum_j \gamma_j{}^i n_j{}^i/(\pi^{\circ} - \tau^i + P^i)$. The measured chloroplast volume (V) can be represented by $\bar{V}_w n_w{}^i + b$, *i.e.*, as the sum of the aqueous and nonaqueous contributions. Hence, V should be equal to $RT\sum_j \gamma_j{}^i n_j{}^i/(\pi^{\circ} - \tau^i = P^i) + b$, if the modified form of the Boyle-Van't Hoff relation is obeyed. Figure 2.32 indicates that the volume of pea chloroplasts varies linearly with $1/\pi^{\circ}$ over a considerable range of external osmotic pressures. Consequently, $\tau^i - P^i$ occurring in both Equation and the above expression must either be negligibly small for pea chloroplasts, compared with the various values of π° employed, or

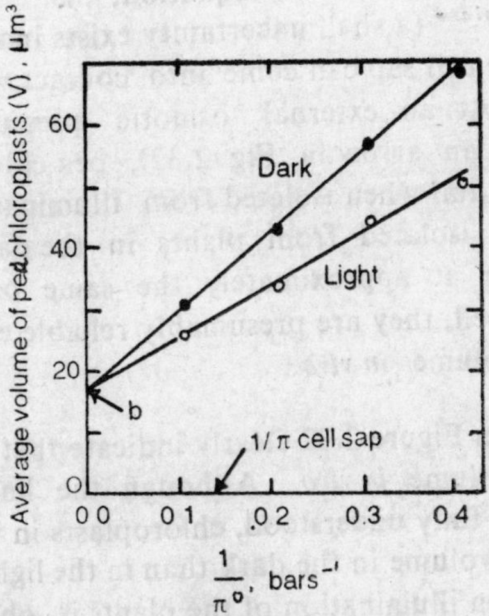

Fig. 2.32. Volumes of chloroplasts at various external osmotic pressures π^o. Pea chloroplasts were isolated from plants in the light or the dark as indicated. (Source P.S. Nobel, *Biochimica et Biophysica Acta* 172 : 134-143 (1969). Data reprinted by permission.)

perhaps proportional to π^o. Actually, the situation is rather complicated, since both τ^i and γ_j^i depend on the chloroplast volume. For simplicity, we will ignore these effects and consider only the observed proportionality between V and $1/\pi^o$. In other words, we will assume that $V-b$ (or, $\overline{V}_w n_w^i$) equals $RT\sum_j \gamma_j^i n_j^i/\pi^o$ for pea chloroplasts. We will return to such considerations, where we will further refine the Boyle-Van't Hoff relation to include the more realistic case in which solutes can cross the limiting membranes.

The relatively simple measurement of the volumes of pea chloroplasts for various external osmotic pressures can yield a considerable amount of information about the organelles. If we measure the volume of the isolated chloroplasts at the same osmotic pressure they encounter in the cytoplasm, we could then determine the volume the organelles have when in the plant cell. Cell sap expressed from pea leaves was found to have an osmotic pressure of 7.0 bars. Such "sap" comes mainly from the central vacuole, but since we expect $\pi^{cytoplasm}$ to be essentially equal to $\pi^{vacuole}$ Equation, $\pi^{cell\ sap}$ is about the same as $\pi^{cytoplasm}$ (a shall uncertainty exists inasmuch as during extraction the cell sap can come into contact with water in the cell walls). At an external osmotic pressure of 7.0 bars (indicated by an arrow in Fig. 2.32), pea chloroplasts have a volume of 29 μm^3 when isolated from illuminated plants, and 35 μm^3 when isolated from plants in the dark. Since these volumes occur at approximately the same osmotic pressure found in the cell, they are presumably reliable estimates of pea chloroplast volumes *in vivo*.

The data in Figure 2.32 clearly indicate that light affects the chloroplast volume *in vivo*. Although the basis for the size changes is not fully understood, chloroplasts in many plants do have a larger volume in the dark than in the light. The decrease in volume upon illumination of the plants is observed by both phase contrast and electron microscopy as a flattening or decrease in thickness of the chloroplasts. This flattening, which amounts to about 20% of the thickness for pea chloroplasts *in vivo*, is in the vertical direction for the chloroplast. We also note that the slopes of the osmotic response curves in Fig. 2.32 are equal to $RT\sum_j \gamma_j{}^i n_j{}^i$. The slope is 50% greater for pea chloroplasts from plants in the dark than from those in the light, suggesting that chloroplasts in the dark contain more osmotically active particles. In fact, illumination causes a marked efflux of K^+ and Cl^- from the chloroplasts. Also, some Mg^{++} enters the chloroplasts in the light, perhaps by an exchange of one Mg^{++} for two K^+. Such a process would also reduce $\sum_j \gamma_j{}^i n_j{}^i$, since only one Mg^{++} would enter for every two K^+ that leave. The entry of Mg^{++} apparently causes its concentration to rise by about 10 mM in the chloroplasts, which may act as a control mechanism for photophosphorylation and CO_2 fixation, both of which depend on magnesium. Although such speculation takes us far away from our present consideration of water, it indicates the kind of insights that can be obtained from osmometric studies.

The intercept on the ordinate in Figure 2.32 is the chloroplast volume theoretically attained in an external solution of infinite osmotic pressure—$1/\pi^\circ$ equaling zero is the same as a π° of infinity. For such an infinite π°, all of the internal water would be removed ($n_w{}^i=0$ moles), and the volume, which is actually obtained by extrapolation, would be that of the non-aqueous components of thc chloroplasts. Thus, the intercept on the ordinate of a V versus $1/\pi^\circ$ plot corresponds to b in the conventional Boyle-Van't Hoff relation (Equation). This intercept (indicated by an arrow in Fig. 2.32 equals 17 μm^3 for chloroplasts both in the light and dark. (The extra solutes in a chloroplast in the dark correspond to less than 0.1 μm^3 of solids.) Using the chloroplast volumes obtained for a π° of 7.0 bars and 17μm^3 for the nonwater volume b, we find that the water content of pea chloroplasts in the dark is $[(35\ \mu m^3 - 17\ \mu m^3)/(35\ \mu m^3)](100)$ or 51% by volume, and in the light $[(29-17)/29](100)$ or 41%. These relatively low water contents in the organelles are consistent with the extensive internal lamellar system and the abundance of CO_2 fixation enzymes in chloroplasts. Thus, osmometric responses of certain cells and organelles can be used to provide information on their fractional water content.

Water potential and plant cells

In this section we will shift our emphasis from a consideration of the water relations of subcellular bodies to those of whole cells, and extend the development to include the case of water fluxes. Whether water enters or leaves a plant cell, and how much, depends on the water potential outside compared with that inside. The external water potential (ψ°) can often be varied experimentally, and the direction as well as the magnitude of the resulting water movement will give information about ψ^i. Moreover, the equilibrium value of ψ° can be used to estimate the internal osmotic pressure, π^i.

A loss of water from plants—indeed, sometimes also an uptake—occurs at cell-air interfaces. As we would expect, the chemical potential of water in cells compared with that in the adjacent air determines the direction for net water movement at such locations. Thus, we must obtain an expression for the water potential in a vapour phase, and then relate this ψ_{wv} to ψ for the liquid phases in a cell. We will specifically consider the factors influencing the water potential in the component at the plant cell-air interface, *viz.*, the cell wall. We will find that $\psi^{cell\ wall}$ is dominated by a negative hydrostatic pressure resulting from surface tension effects in the cell wall pores.

Incipient plasmolysis. For usual physiological conditions, a positive hydrostatic pressure exists inside a plant cell, *i.e.*, it is under turgor. By suitably adjusting the solutes in an external solution that bathes such a turgid cell, it is possible to reduce P^i to zero and thereby to obtain an estimate of π^i, as the following argument will indicate.

Let us place the cell in pure water ($\pi^\circ=0$ bars) at atmospheric pressure ($P^\circ=0$ bars). ψ° is than zero (ignoring the gravitational term, $\psi=P-\pi$, Equation while ψ^i is $P^i-\pi^i$ (the inner phase can be the cytoplasm). If the cell can be assumed to be in equilibrium with this external solution $\psi^\circ=\psi^i$), P^i must then equal π^i. Suppose that π° is now gradually raised from its initial zero value. The osmotic pressure outside the cell can be increased for example by adding solute to the external solution. If the cell remains in equilibrium, then π° equals π^i-P^i ($P^\circ=0$ bars, since the external solution is at atmospheric pressure). As π° is increased, P^i will tend to decrease, while π^i usually does not change very much. More precisely, because the cell wall is quite rigid, a plant cell will not change its volume appreciably in response to changes in P^i. (Since the cell wall has elastic properties some water flows out as P^i decreases and the cell shrinks, a point which we will return to shortly.) Therefore, if the cell volume can be assumed not to change in response to changes in P^i and no internal solutes leak out, π^i will remain constant. As the external osmotic pressure is raised, π° will eventually become equal to π^i. In such a plant cell at equilibrium, P^i is zero, which means that no internal hydrostatic pressure is exerted against the cell wall when π° equals π^i. The cell would thus lose its turgidity. If π° were increased even further, an appreciable amount of water would flow out of the cell and plasmolysis would occur as the plasmalemma pulls away from the rigid cell wall. Ignoring for the moment any overall volume changes of the cell, we find that the condition under which water just begins to move out of the cell—referred to as the point of incipient plasmolysis,

$$\pi^{o}_{plasmolysis} = \pi^{i}$$

Equation suggests that only a relatively simple measurement ($\pi^{o}_{plasmolysis}$) is needed in order to estimate the osmotic pressure (π^{i}) occurring inside an individual plant cell.

The existence of an internal hydrostatic pressure within a plant cell leads to stresses in its cell wall and a resulting elastic deformation or strain. The decrease of P^{i} to 0 bars—which occurs in taking a turgid plant cell to the point of incipient plasmolysis—must therefore be accompanied by a contraction of the cell as the wall stresses are removed. This decrease in volume means that some water will actually flow out of the cell before the point of incipient plasmolysis is reached. If no internal solute leaves as the cell shrinks, then the osmotic pressure inside will increase (the same amount of solute in a smaller volume). As a useful first approximation, we can assume that the increase in osmotic pressure reciprocally follows the change in volume, *i.e.*, the product of π^{i} and the cellular volume is approximately constant. With this assumption, the osmotic pressures in cells—described by Equation, and determined by using the technique of measuring the point of incipient plasmolysis—can be corrected to its original value by using the ratio of the initial volume to the final volume of the cell. The change in volume is only a few percentage points for some plant cells, in which case fairly accurate estimates of π^{i} can be obtained from plasmolytic studies alone.

Measurements of $\pi^{o}_{plasmolysis}$ often give values near 7 bars for cells in storage tissues like onion bulbs or carrot roots and in leaves like those of pea or spinach. These values of the external osmotic pressure provide information on various contributors to the water potential inside the cell, as Equation indicates. For the case in which water is in equilibrium within the plant cell at the point of incipient plasmolysis, the term π^{i} in Equation is the osmotic pressure both in the cytoplasm and in the vacuole. Under certain circumstances it is convenient to replace π^i by $\pi_s^i + \tau^i$ (Equation), where π_s^i is the osmotic pressure contributed by the internal solutes and τ^i is the matric potential. This relation was invoked for example when the implications of Equation ($\pi^{cytoplasm} = \pi^{vacuole}$) were discussed, and the various arguments presented at that time extend to the present case. Specifically, since the possible matric potential in the vacuole is probably negligible, $\pi^{\circ}_{plasmolysis}$ should be a rather good estimate of $\pi_s^{vacuole}$. Moreover, $\pi^{cytoplasm}$ is most probably significant, as the water activity in the cytoplasm can be lowered by the many interfaces present. Thus, $\pi^{\circ}_{plasmolysis}$ is an upper limit for $\pi_s^{cytoplasm}$, We see, therefore, that determination of the external osmotic pressure at the point of incipient plasmolysis provides information on the osmotic pressure existing in different compartments within the plant cell.

Activity and water potential of water vapour. Water molecules in an aqueous solution continually escape into a surrounding gas phase. At the same time, water molecules also condense back into the liquid. The rates of escape and condensation depend *inter alia* on the chemical activities of water in the gas and the liquid phases the two flows becoming equal at equilibrium. The gas phase adjacent to the given solution then contains Y, much water as it can hold at that temperature and still remain in equilibrium with the liquid. It is thus *saturated* with water vapor for the particular solution under consideration. The partial pressure exerted by the water vapor in equilibrium with pure water is known as the *saturation vapour pressure*.

The vapour pressure at equilibrium depends on the temperature and the solution but is independent of the relative or absolute amounts of liquid and vapor. When air adjacent to pure water is saturated with water vapour (100% relative humidity), the gas phase then has the maximum water vapour pressure possible at that temperature—unless it is supersaturated, a non-equilibrium situation. This saturation vapour pressure of pure water increases markedly with temperature, *e.g.*, it is equivalent to 4.6 mm of Hg at 0° and 55.3 mm at 40°. Thus, air of 100% relative humidity at 0° would be only (100)(4.6)/(55.3), or 8% relative humidity when heated at constant pressure to 40°. As solutes are added to the liquid phase and the activity of its water is thereby lowered, water molecules have less of the tendency to leave the solution, so that the water vapor pressure in the gas phase at equilibrium becomes less—this is one of the colligative properties of solutions that we mentioned earlier. In fact, for dilute solutions the actual partial pressure of water vapour (P_{wv}) at equilibrium depends linearly on the mole fraction of water (N_w) in the liquid phase—this is *Raoult's law*. For pure water, N_w equals one and P_{wv} has its maximum value, *viz.*, P_{wv}^*, the saturation vapour pressure.

The partial pressure of the water vapour determines the activity of water in a gas phase, a_{wv}. If water vapour behaved as an ideal gas, then

$$a_{wv} = \frac{P_{wv}^*}{P_{wv}} = \frac{\%\ \text{relative humidity}}{100}$$

where P_{wv} is the partial pressure of water vapour in the air under consideration, and P_{wv}^* is the saturation vapour pressure in equilibrium with pure liquid water at the same temperature. To handle deviations from ideality. P_{wv} is replaced by $\gamma_{wv}P_{wv}$, where γ_{wv} is the activity coefficient (more properly, the "fugacity" coefficient) of water vapour. If water vapour obeyed the ideal gas law—$P_jV = n_jRT$, where P_j is the partial pressure of species j in a volume V that contains n_j moles of that gas—γ_{wv} would equal unity. This is an extremely good approximation for most situations of interest in plant physiology. The relative humidity defined by Equation is a readily measured quantity reflecting the activity of water in a particular gas phase, and it has been extensively used in studying the water relations of plants.

For water existing as water vapour in air, the chemical potential of the water as defined by Equation is simply $\mu_{wv}^* + RT \ln a_{wv} + m_{wv}gh$. (We will also discuss chemical potential in a vapour phase in App. VII.) Using the expression for the activity of water vapour given by Equation, the water potential of water vapour in a gas phase such as air, ψ_{wv}, is then

$$\psi_{wv} = \frac{\mu_{wv} - \mu_{wv}^*}{\bar{V}_w} = \frac{RT}{\bar{V}_w} \ln \left(\frac{\%\ \text{relative humidity}}{100} \right) + \rho_w gh$$

were a definition of water potential analogous to that given in Equation [$\psi=(\mu_w-\mu_w^*)/\bar{V}_w$] has been used, and $m_{wv}/\bar{V}_w$ (the same as $m_w/\bar{V}_w$) has been replaced by the density of water, ρ_w. Note that $\bar{V}_w$, not $\bar{V}_{wv}$) is used in the definition of ψ_{wv} in Equation. This is necessary since the fundamental term representing free energy per mole is the chemical potential—we want to compare $\mu_j-\mu_j^*$ for the two phases when predicting changes at an interface—and thus the proportionality factor between $\mu_w-\mu_w^*$ or $\mu_{wv}-\mu_{wv}^*$ and the more convenient terms, ψ or ψ_{wx}, must be the same in each case, *viz.*, $\bar{V}_w$.

Let us next cosider the various parameters at the surface of pure water ($a_w=1$) at atmospheric pressure (P=0 bars) in equilibrium with its vapour. In such a case, P_{wv} equals P_{wv}^*, and we can take the zero level of the gravitational term as the surface of the liquid. Hence, μ_w equals μ_w^*, and μ_{wv} is μ_{wv}^*. Since there is no tendency for a net gain or loss of water molecules from the liquid phase at equilibrium ($\mu_w=\mu_{wv}$), we conclude that the constant μ_w^* must equal the constant μ_{wv}^* when the two phases are at the same temperature.

What happens to P_{wv} as we move vertically upward from pure water at atmospheric pressure in equilibrium with water vapour in the air ? If we let h be zero at the surface of the water, ψ equals zero ($\psi=P-\pi+\rho_w gh$. Equation, and since water vapour is by supposition in equilibrium with the liquid phase, ψ_{wv} is also zero. Equation indicates that as we go vertically upward in the gas phase, $\rho_w gh$ must make an increasingly positive contribution to ψ_{wv}. Since we are at equilibrium, the other term in ψ_{wv}, which equals $(RT/\bar{V}_w)\ln(P_{wv}/P_{wv}^*)$ by Equation must make a compensating negative contribution. Thus, at equilibrium P_{wv} must decrease with altitude. We can further appreciate this conclusion simply by noting that gravity attracts the molecules of water vapour and other gases towards the earth, *i.e.*, air pressure is higher at sea level than it is on the top of a mountain. This effect on P_{wv} is generally not very important in plant physiology, since rather great heights are needed before the partial pressure of water vapour is reduced significantly. In particular, let us consider a 7% reduction in P_{wv} from P_{wv}^* (its value at the water surface at equilibrium) to $0.93\ P_{wv}^*$).

Since $RT/\bar{V}_w$ is 1350 bars at 20° $(RT/\bar{V}_w) \ln (P_{wv}/P_{wv}^*)$ would change by (1350 bars) ln (0.93), or −99 bars $\rho_w g$ equals 0.098 bar/m, hence a height of (99 bars)/(0.098 bar/m), or 1010 m would be required for the gravitation term to increase by the same amount. In other words, not until we reached a height of 1 km would the partial pressure of water vapour at equilibrium decrease by 7% from its value at sea level, indicating that the gravitational term generally has relatively little influence on ψ_{wv} over the distances involved for a particular plant.

Plant-air interface. Water equilibrium across the plant-air interface occurs when the water potential in the leaf cells equals that of the surrounding atmosphere. (This presupposes that the leaf and the air are at the same temperature. To measure ψ^{leaf}, the leaf can be placed in a closed chamber and the relative humidity adjusted so that there is no net water gain or loss by the leaf. Such a determination is rather demanding, since small changes in relative humidity have large effects on ψ_{wv}, as we will now show.

Extremely large negative values are possible for ψ_{wv}^{air}. In particular, $RT/\bar{V}_w$ at 20° is 1350 bars. In the expression for ψ_{wv} (Equation this relatively large factor multiplies ln (% relative humidity/100), and a wide range of relative humidities can occur in the air. By Equation and ignoring $\rho_w gh$, a relative humidity of 100% corresponds to a water potential in the vapour phase of (1350 bars) [ln (100/100)], which equals 0 bars (ln 1 = 0). This ψ_{wv}^{air} would be in equilibrium with pure water at atmospheric pressure, which also has a water potential of 0 bars. For a relative humidity of 99%, ψ_{wv}^{air} given by Equation equals (1350)[ln (99/100)], which is −13.5 bars—remember that ln $(1-x)$ is about $-x$ for $|x| \ll 1$, such as is the value 0.01 in the present case. Hence, going from 100% to 99% relative humidity corresponds to a decrease in water potential of 13.5 bars. Small changes in relative humidity indeed reflect marked differences in the water potential in air ! Finally, we note that a relative humidity of 50% at 20° leads to a ψ_{wv}^{air} of (1350) ln (50/100), or −940 bars.

One important consequence of the large negative values of ψ_{wv}^{air} is easily appreciated by considering values of the water potential that may occur in plant leaves. The actual value of ψ^{leaf} depends on the ambient conditions as well as the plant type and its physiological status. The range of values for most mesophytes is -3 to -30 bars, with -5 bars being typical for leaves of a garden vegetable such as lettuce. From Equation, the relative humidity corresponding to a water potential of -5 bars is 99.6%. Even during a rainstorm the relative humidity of the air rarely exceeds 99%. Since relative humidity is lower than 99.6% under most natural conditions, water is continually being lost from a leaf having an internal water potential of -5 bars. (Again, it should be emphasized that throughout this discussion the leaf is assumed to be at the same temperature as the air. A few desert plants have a ψ^{leaf} as low as -50 bars. Even in this case of adaptation to arid climates, however, water still tends to leave the plant unless the relative humidity is above 96%.

Pressure in the cell wall water. Plant cells come into contact with air at the cell walls bounding the intercellular air spaces. Thus, the water potential in the cell walls must be considered with respent to ψ_{wv} in the adjacent gas phases. The main contributor to ψ in cell wall water is often the negative hydrostatic pressure arising from surface tension effects at the numerous air liquid interfaces of the cell wall interstices. In turn, $P^{cell\ wall}$ can be related to the geometry of the cell wall pores and the contact angles that occur.

The magnitude of the negative hydrostatic pressure that develops in cell wall water can be analyzed by considering the pressure that occurs in a liquid within a cylindrical pore—the argument being basically the same as the one presented earlier in this chapter in discussing capillary rise. One of the forces acting on the fluid in a narrow pore of radius r is the result of surface tension, σ. This force equals $2\pi r\sigma \cos \alpha$, where α is the contact angle, a quantity which can be essentially zero for wettable walls. Since the adhesive forces at the wall are transmitted to the body of the fluid by means of its cohesion, a tension or negative hydrostatic pressure develops in the fluid.

The total force resisting the surface adhesion can be regarded as this tension times the area over which it acts, πr^2, hence the force is $\pi r^2 \times (-P)$. Equating the two forces gives the following expression for the pressure developed in the fluid contained within a cylindrical pore :

$$P = -\frac{2\sigma \cos \alpha}{r}$$

Adhesion of water at interfaces generally creates negative hydrostatic pressures (such as that described by Equation in the bulk of the fluid. Such a negative P arising from interfacial interactions has sometimes been treated in plant physiology as a positive matric pressure, a convention that we mentioned earlier.

The strong water-wall adhesive forces, which are transmitted throughout the interfibrillar spaces by water-water hydrogen bonding, conceivably can greatly reduce the water potential in the cell wall, as the following calculation indicates. At 20° the surface tension of water is 72.8 dynes/cm or 7.28×10^{-5} bar-cm. The voids between the fibers in the cell wall are often about 10^{-6} cm (10 nm or 100 Å) across ; for wettable walls, cos α can equal one. For water in such cylindrical pores, Equation indicates that P would be $-(2)(7.28 \times 10^{-5}$ bar-cm$)(1)/(0.5 \times 10^{-6}$ cm$)$, or −290 bars. This is an estimate of the negative hydrostatic pressure or tension that *could* develop in the aqueous solution within cell wall interstices of typical dimensions, supporting the contention that $\psi^{cell\ wall}$ can be markedly less than zero. Moreover, the previous discussion of the tensile strength of water indicates that the hydrogen bonding in water can withstand such a tension. We shall now consider the actual pressures that develop in the cell wall water.

The relative value of the water potential in the cell wall compared with that in the adjacent vapour phase is of considerable importance in transpiration and for water loss from plant material in general. From Equation $\psi_{wv}{}^{air}$ corresponding to a relative humidity of 95% at 20° is (1350) ln (0.95), or −69 bars. According to Equation a P of −69 bars would occur in a water-containing cylindrical pore with wettable walls having a radius of $-(2)(7.28\times10^{-5})(1)/(-69)$, or 2.1×10^{-6} cm, which is 21 nm or 210 Å. In other words, water in 21 nm radius interstices in the cell walls of leaf mesophyll cells would be in equilibrium with air of 95% relative humidity in the intercellular spaces. If the relative humidity near the cell wall surface were decreased, water in such interstices could be lost, but it would still remain in the finer pores where a more negative P or larger tension can be created. Hence, if the plant material is exposed to air of moderate to low relative humidity—such as can happen for an herbarium specimen—then larger and larger tensions resulting from interfacial interactions develop in the cell walls as the plant material dries out. Water is then retained only in finer and finer pores of the wall, as it must, according to Equation.

We shall now show the effect of the availability of water adjacent to that in wettable cell walls on the contact angle and on $P^{cell\ wall}$. Suppose that pure water in mesophyll cell wall interstices 10 nm across is in equilibrium with water vapour in the intercellular air spaces where the relative humidity is 99%. ψ_w, for this relative humidity is -13.5 bars at 20°. Hence, at equilibrium the (pure) water in the cell wall interstices must be under a hydrostatic pressure of -13.5 bars ($\psi = P - \pi + \rho wgh$, Equation. From Equation ($P = -2\ \sigma \cos \alpha/r$), and letting σ be 7.28×10^{-5} bar-cm at 20°, we calculate that cos α must be $-(-13.5$ bars) $(0.5\times10^{-6}$ cm)/[(2)$(7.28\times10^{-5}$ bar-cm)], or 0.0464 for the cell wall pores which are 5 nm in radius. When cos α is 0.0464, α is 87°. For the wettable cell walls, α *can* be 0°, in which case cos α is unity and the maximum negative pressure does develop, *viz.*, -290 bars for the pores 5 nm in radius. On the other hand, when water is available, it can be pulled into the interstices by such possible tensions and there by cause α to increase toward 90°, which leads to a decrease in the cos α and consequently a decrease in the absolute value of the pressure developed. In the present example, α is 87°, P is -13.5 bars, and the pore 5 nm in radius is nearly filled. Thus, the water in the interstices can be nearly flush with the cell wall surface. In fact, the large tensions that could be present in the cell wall generally do not occur in living cells, since water is usually available within the plant and is "pulled" into the interstices, thus nearly filling them, depending of course on their dimensions Equation.

The only contribution to the water potential of the cell wall that we have been considering is the negative pressure that develops because of air-liquid interfaces, which can be the case for pure water, When the gravitational term in Equation 35 is ignored, $\psi^{cell\ wall}$ actually equals $P^{cell\ wall} - \pi^{cell\ wall}$. The bulk solution in the cell wall pores generally has an average osmotic pressure of 3 to 15 bars. The local $\pi^{cell\ wall}$ next to the Donnan phase is considerably greater because of the many ions present there ; by way of compensation, $P^{cell\ wall}$ is also higher near the Donnan phase than in the bulk of the cell wall water.

Water flux. When the water potential inside a cell differs from that outside, the water is no longer in equilibrium and we can expect a water movement toward the region of lower water potential. Our attention here will be specifically focused on water flow into and out of plant cells. This volume flux of water, J_{Vw}, is assumed to be proportional to the difference in water potential ($\Delta\psi$) across the membrane or membranes restricting the flow. The proportionality factor indicating the permeability to water flow at the cellular level is expressed by a water conductivity coefficient, L_w :

$$J_{Vw} = L_w \Delta\psi = L_w(\psi^o - \psi^i)$$

In Equation J_{Vw} is the volume flow of water per unit area of the barrier per unit time. It can have units of $cm^3/(cm^2\text{-sec})$ or cm/sec, which is a velocity. In fact, J_{Vw} is the average velocity of water moving across the barrier being considered. L_w can be given in cm/(sec-bar), in which case the water potentials would be expressed in bars. (L_w is usually experimentally the same as L_P, a coefficient also describing water conductivity.

When Equation is applied to cells, ψ^o is the water potential in the external solution, and ψ^i usually refers to the water potential in the vacuole. L_w then reflects the conductivity for water flow across the cell wall, the plasmalemma, and the tonoplast, all in series. For a group of barriers in series with an overall water conductivity coefficient L_w, $1/L_w$ equals $\sum_j 1/L_{wj}$, where L_{wj} is the water conductivity coefficient of barrier j. To see why this is so, let us go back to Equation, $J_{Vw}=L_w\Delta\psi$. When the barriers are in series, J_{Vw} is the same across each one, and so J_{Vw} equals $L_{wj}\Delta\psi_j$ (also $\Delta\psi_j=J_{Vw}/L_{wj}$), where $\Delta\psi_j$ is the drop in water potential across barrier j, *i.e.*, $\sum_j \Delta\psi_j=\Delta\psi$. We therefore obtain the following string of equalities : $J_{Vw}/L_w=\Delta\psi=\sum_j \Delta\psi_j=\sum_j J_{Vw}/L_{wj}=J_{Vw}\sum_j 1/L_{wj}$; hence, $1/L_w$ must equal $\sum_j 1/L_{wj}$. (We also note that $1/L_w$ corresponds to a resistance and the resistance of a group of resistors in series is the sum of the resistances, $\sum_j 1/L_{wj}$.)

We shall now consider a membrane separating two solutions that differ only in osmotic pressure. This will help us to relate L_w to the permeability coefficient of water, P_w, and also to view Fick's first law $[J_j=P_j(c_j^\circ-c_j^\circ)$, in a slightly different way. The proper form for Fick's first law when describing the diffusion of the solvent water is $J_{Vw}=P_w(N_w^\circ-N_w^i)$. By Equation, J_{Vw} also equals L_w $(\psi^\circ-\psi^i)$, which becomes L_w $(\pi^i-\pi^\circ)$ in our case (recall that $\psi=P-\pi+\rho_w gh$, Equation. Thus, a flux of water toward a region of higher osmotic pressure (Equation) is equivalent to J_{Vw} toward a region where water is less concentrated in the sense of having a lower mole fraction (Fick's first law applied to water). Let us pursue this one step further. Using Equations through 40 we can obtain the following string of equalities : $N_w\cong 1-\sum_j n_j/n_w\cong 1-\overline{V}_w\pi_j/RT$.

Therefore, $N_w^\circ-N_w^i$ equals $(1-\overline{V}_w\pi_s^\circ/RT)-(1-\overline{V}_w\pi_s^i/RT)$, or $\overline{V}_w(\pi_s^i-\pi_s^\circ)/RT$. When we incorporate this last relation into our two expressions for the volume flux of water, we obtain $J_{Vw}=L_w(\pi^i-\pi^\circ)=P^w(N_w^\circ-N_w^i)=P_w\overline{V}_w(\pi_s^i-\pi_s^\circ)/RT(\pi^i$ is here the same as $\pi_s{}^i$). Hence, L_w equals $P_w\bar{V}_w/RT$, which states that the water conductivity coefficient (L_w) is proportional to the water permeability coefficient (P_w). Next, we will estimate a possible value for L_w. P_w is generally at most 10^{-2} cm/sec for plasmalemmas, RT is 24,370 cm^3·bars/mole at 20° and $\bar{V}_w$ is 18 cm^3/mole. Hence, L_w might be (10^{-2} cm/sec)(18 cm^2/mole)/(24,370 cm^3-bars/mole), or 0.74×10^{-5} cm/sec-bar). When P_w is less than 10^{-2} cm/sec, L_w will be proportionally less than 0.74×10^{-5} cm/sec-bar).

When the value of the water conductivity coefficient is known, the water potential difference necessary to give an observed water flux can be calculated by using Equation. For the internodal cells of *Nitella* and *Chara*, L_w for water entry can be about 10^{-5} cm/(sec-bar)—a large value for L_w, see Dainty. For convenience of calculation, we will consider cylindrical cells 10 cm long and 1 mm in diameter as an approximate model for such algal cells. The surface area across which the water flux occurs is $(2\pi r)(l)$, where r is the cell radius (0.05 cm) and l is the cell length (10 cm). Thus, the area equals (2π)(0.05 cm)(10 cm), or π cm^2. (The area of each end of the cylinder (πr^2) is much less than $2\pi rl$; in any case, a water flux from the external solution across them would not be expected, since they are in contact with other cells, not the bathing solution. The volume of a cylinder is $\pi r^2 l$, which equals $(\pi)(0.05\text{ cm})^2(10\text{ cm})$, or $\pi/40$ cm^3 in the present case. Internodal cells of *Nitella* and *Chara* grow relatively slowly—a change in volume of about 1% per day being a possible growth rate for the fairly mature cells which were used in determining L_w. This growth rate means that the water content increases by about 1% of the volume per day. The total volume flux of water into the above cell is therefore (0.01) π/40) cm^3/day. Such an inflow of water per day is equivalent to [(π/4000) cm^3/day]($^1/_{24}$ day/hr)($^1/_{3600}$ hr/sec). or $3\pi \times 10^{-9}$ cm^3 of water moving into a cell/sec. This water influx occurs across a surface area of π cm^2 per cell, as we just calculated, hence the rateof volume flow of water per cm^2 is ($3\pi \times 10^{-9}$ cm^3/sec)/(π cm^2), or 3×10^{-9} cm^3/(cm^2—sec). But such a quantity is precisely what is meant by J_{Vw} in Equation. Using the appropriate value of L_w for these internodal cells, 10^{-5} cm/sec-bar, Equation indicates that $\Delta\psi$ is (3×10^{-9} cm/sec)/(10^{-5} cm/sec-bar) or 3×10^{-4} bar. In order words, the internal water potential (ψ^i) needed to sustain the water influx accompanying a growth of about 1% per day is 3×10^{-4} bar less than the outside water potential (ψ^o). *Nitella* and *Chara* can grow in pond water, which is a dilute aqueous solution often having a water potential near -0.07 bar. Thus, ψ^i need be only slightly more negative than -0.07 bar to account for the influx of water accompanying a growth of 1% per day.

3

Photophysiology

The radiation from the sun produces a direct heating effect, and also produces photochemical transformations. After these transformations have been completed, the energy appears in the form of heat. Accordingly, all the sun's energy eventually ends up as heat, but certain portions of the radiation from the sun take part in vital photochemical processes before becoming heat.

Light as an ecological factor is generally highly directional. It differs in this respect from the temperature factor since heat often reaches the living organism from many different directions at the same time. Light is extremely variable. It changes over a tremendous range, often very rapidly. Many organisms can respond to a value of light that is only one ten-billionth of full sunlight. Although we often think of temperature as varying widely on the earth's surface, the magnitude of its variation is extremely small compared to that of light. Light can change from a value near zero to its maximum within a few hours.

Light is essential for most plants and animals, though some can do without it. For the continued existence of organisms two requirements must be met. First, light must not be so strong as to cause serious harm at any stage in the life history. We shall see that a considerable variation exists in the upper limits of tolerance to the light factor. Second, for those animals and plants that require light it must be sufficient in intensity and in duration. The intensity of the light must be above the threshold for the organism concerned, and the total amount of light received during the period when it is needed must be adequate.

DISTRIBUTION OF LIGHT

In discussing the distribution of light it is convenient to start with sunlight as the chief source and trace the radiation as it passes through the air and water environments. The magnitude of the solar radiation as it reaches the outer atmosphere referred to as the *solar constant* has the value of about 1.9 g-cal/cm²/min. At sea level the intensity of solar radiation averages about 1.5 g-cal/cm²/min. If the radiation received from the sun were evenly distributed over the surface of the globe, it would be sufficient to melt a layer of ice 35 m thick during the course of a year. At a latitude of 44° N the energy received on the earth's surface from the sun is equivalent to the light that would be produced by hanging a 250-watt lamp over each square meter of the ground.

We know very well, however, that the radiation from the sun is not evenly distributed either in time or in space. We wish to inquire what its variations both in quality and in quantity are. Light changes in spectral distribution and in angular distribution. The light factor also varies in intensity and in duration, resulting in differences in the total amount of light falling on each unit of surface for each unit of time, such as a day or a month. The changes in the light factor in these respects will be examined as they affect the terrestrial and the aquatic environments.

Light on Land

Spectral Composition. The spectral distribution of light as it reaches the earth's surface differ as to the exact wavelength limits to be assigned to the different portions of the spectrum. Roughly speaking radiation of wavelength longer than 7600 Angstrom units is considered to be infrared, and that of wavelength shorter than about 3600 A is designated as ultraviolet. Almost one-half of the total emission of the sun is infrared radiation, and almost one-half is visible light. These proportions remain approximately the same at the earth's surface, regardless of the total intensity of sunlight. The ultraviolet component, however, is always only a small percentage of the total radiation, and it may be reduced to immeasurably small quantities under certain circumstances.

Intensity of Light. The intensity of light reaching the earth's surface varies with the angle of incidence and with the amount of absorption by the atmosphere and by obscuring features. The lower the altitude of the sun, the smaller is the angle of incidence and the longer is the path of the light through the atmosphere with corresponding reduction in intensity. Changes in the sun's altitude result from differences in latitude as well as from changes in the season and in the time of day. The greatest intensity of sunlight occurs at positions on the earth's surface and at times at which the sun is most nearly overhead. At higher latitudes, the intensity of light is correspondingly reduced. At 50° N latitude, for example, during the period of the equinox in March and in September when the day is everywhere 12 hours long, the intensity of sunlight is only about one-half of what it is at the equator (Fig. 3.1).

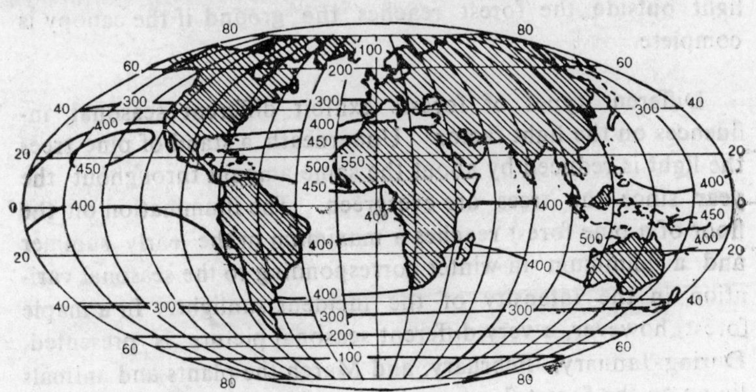

Fig. 3.1. Total solar radiation (g-cal/day/cm² of horizontal surface) on March 21 with average cloudiness.

Latitude thus has a definite effect; but other factors may have much greater influence upon the light factor. Moisture, clouds, and dust in the atmosphere have a profound and irregular effect in reducing illumination. Living organisms may also act to diminish the intensity of daylight, as is clearly shown by the forest vegetation, and this represents a reciprocal action in which the inhabitants modify the light factor in their own environment.

Different forest communities vary widely in the degree to which they diminish the sun's radiation. Cottonwood (popular) trees tend to grow rather widely spaced, and the relatively sparse foliage allows many patches of sunlight to reach the ground. The canopies of pines and oaks usually have fewer gaps. Measurements made in Illinois revealed that the portion of the forest floor exposed to direct sunlight was 84 per cent for popular, 77 per cent for pine, and 35 per cent for oak. In elm-maple forests and in tropical rain forests the canopy typically is complete, with the result that none of the direct sunlight reaches the forest floor. Other measurements showed that the light on the forest floor which had passed through the leaves of the canopy was reduced in intensity to 3.5 per cent of its value above the tree tops in the oak forest ; corresponding figures for elm-maple and rain forests were 0.4 per cent and 0.2 per cent, respectively. Thus less than 1 per cent of the light outside the forest reaches the ground if the canopy is complete.

Different types of forests exhibit different seasonal influences on the light factor. Underneath a stand of pine trees the light is reduced by about the same amount throughout the year since the trees are evergreen. The illumination on the floor of a pine forest reaches a maximum in the early summer and a minimum in winter corresponding to the seasonal variation in the intensity of the incident sunlight. In a maple forest, however, a very different seasonal picture is presented. During January, February, and March the plants and animals living on the forest floor receive increasing amounts of light, but in April the leaves begin to appear and the intensity of light drops rapidly reaching a minimum in the middle of the summer. Then, as the leaves of the maple trees begin to curl and drop off a larger amount of light is allowed to filter through the forest canopy. The striking seasonal variation in the amount of ultraviolet radiation penetrating different forest types is shown in Fig. 3.2. The changes in the total visible illumination are of a corresponding nature. Accordingly, if you were a beetle, or, if you prefer, a lily growing in a maple forest, the brightest month of the year for you would be April and the darkest would be July. In this situation living

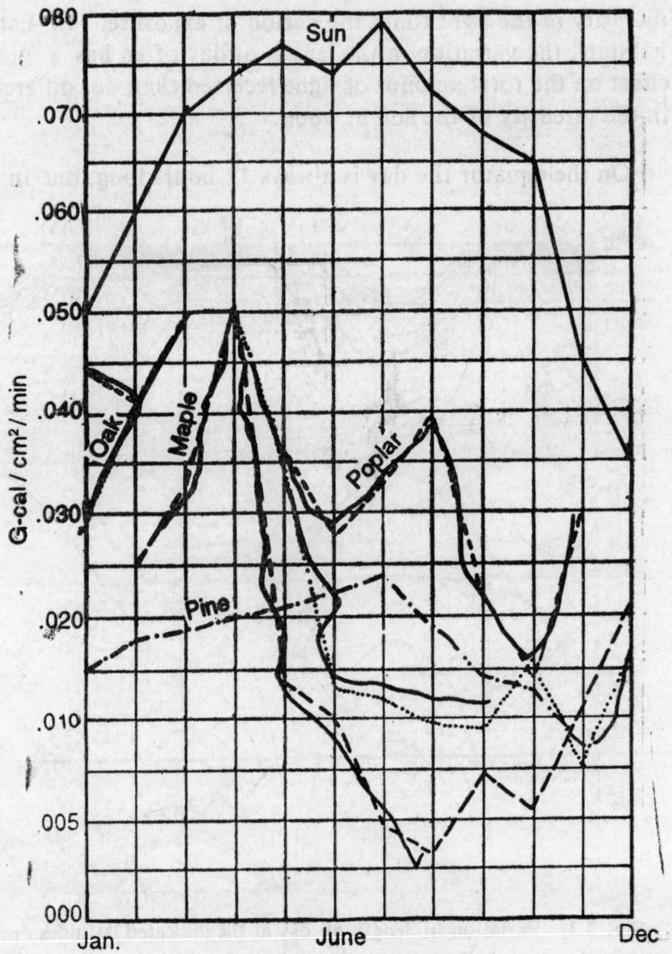

Fig. 3.2. Mean monthly maximum intensities of solar ultraviolet radiation (2900–4000A) in forest communities of northern Indiana.

organisms have modified the environment in respect to the light factor so that the seasons have been practically reversed.

Duration and Amount of Light. The total amount of light received by an organism is determined both by the intensity of the light and the duration of the period of irradiation. The situation is similar to the exposure of a photographic plate. The amount of blackening of the plate is determined by the intensity of the light times the period of exposure. In natural habitats, the variation in the length of day often has a greater effect on the total amount of light received than do differences in the intensity of the sun at noon.

On the equator the day is always 12 hours long, but in the

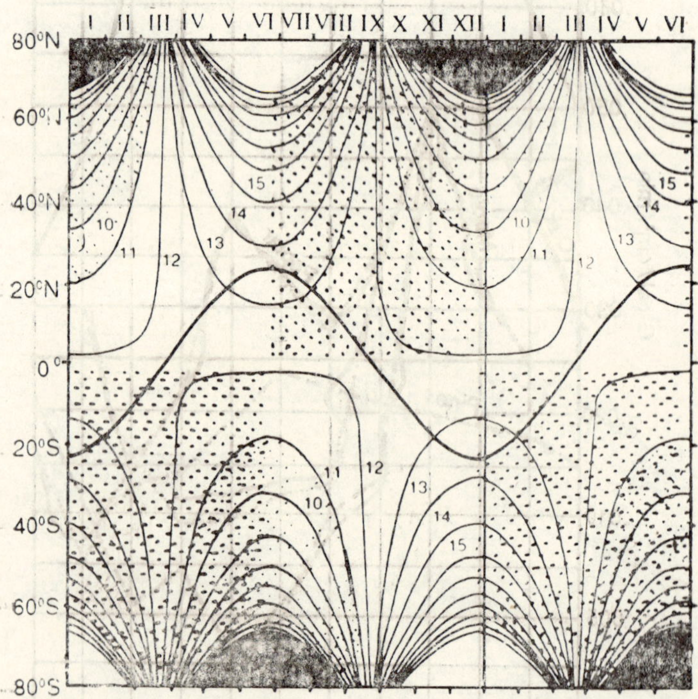

Fig. 3.3. Variation in length of day at the indicated latitudes during the indicated months. Arrowheads indicate periods of decreasing daylength. The heavy line shows the course of zenith sun.

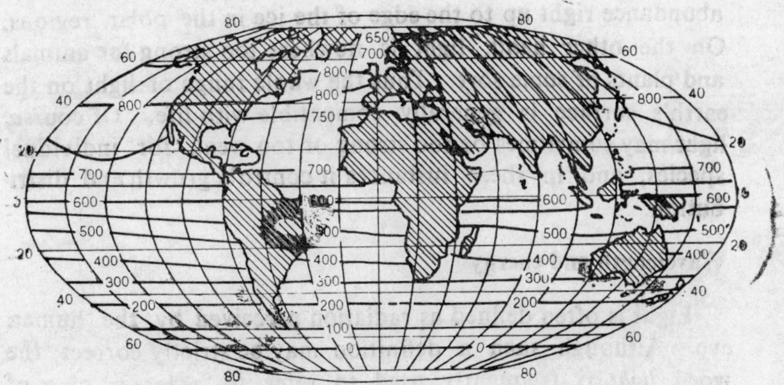

Fig. 3.4. Total solar radiation (g-cal/day/cm² of horizontal surface) on June 21 with cloudless sky.

temperate regions the day grows longer as spring progresses. This effect is accelerated at higher latitudes, and the day becomes 24 hours long during the summer in the polar regions. The day becomes correspondingly shorter after the summer solstice (Fig 3.3). Up to moderately high latitudes the increase in length of day during summer has more effect on the total amount of light received per day than the reductions in solar intensity due to the greater angle of incidence. The greatest northward development of this influence is illustrated in Fig. 3.4 in which the total solar radiation, determined by the combined action of intensity and duration, is plotted for June 21 under conditions of cloudless sky. At that period the amount of solar radiation received each day at 50° N latitude is considerably greater than that received at the equator, and the daily amount of light received in Scandinavia is roughly the same as that falling on tropical Africa. If you are amazed to find potatoes growing in Norway, or wheat being harvested in Alaska, don't forget that once the plants have begun to develop in the late spring, sufficient light exists in these latitudes for the plants to continue growing all day and a good part of the "night."

The foregoing review of the circumstances of light on the earth's surface has shown that light is everywhere sufficient for life of some type at least for part of the year. Plants and animals exist from pole to pole. Mosses and lichens grow in abundance right up to the edge of the ice in the polar regions. On the other hand, light is nowhere too strong for animals and plants of some sort. Thus the whole range of light on the earth's surface is generally compatible with life. Of course, light may sometimes be too strong or too weak for individual species, and in these instances it controls growth and distribution.

Wavelength and energy

Light is often defined as radiation perceived by the human eye. Although such a definition may be strictly correct, the word *light* is frequently used to refer to a larger class of electromagnetic waves. In this section we will discuss the range of electromagnetic radiation that is important in plant and animal physiology, including the subdivisions into various wavelength intervals. The wavelength of light will be related to its energy. After noting the units for radiant flux and

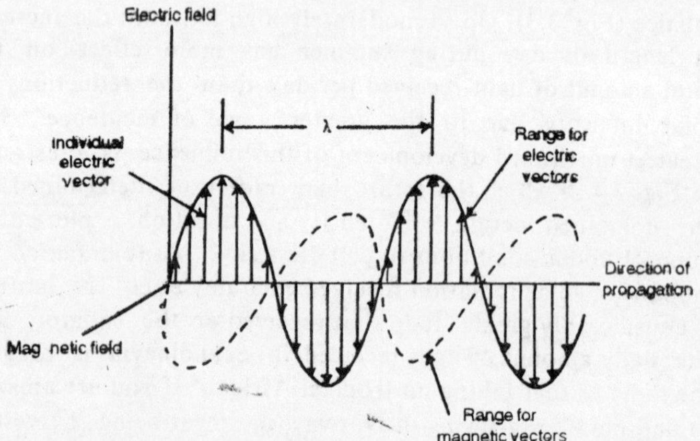

Fig. 3.5. Light can be represented by an electromagnetic wave corresponding to oscillations of the local electric and magnetic fields. The oscillating electric vectors at a porticular instant in time are indicated by arrows in the diagram. A moment later, the entire pattern of electric and magnetic fields will shift in the direction of propagation of the wave. A wavelength (λ) is the distance between two successive points of the same phase along the wave.

illumination, we will briefly consider some of the characteristics of solar radiation reaching the earth.

Light waves. Our concern here is with the regular and repetitive changes in the intensity of the minute electric and magnetic fields which indicate the passage of light wave (Fig. 3.5). Light can travel in a solid (glass), in a liquid (water), in a gas (air), and even in a vacuum (between the sun and the earth's atmosphere). One way to characterize light is by its wavelength—the separation in space between successive points of the same phase, such as between two successive peaks of a wave train (Fig. 3.5). A wavelength is the distance per *cycle* of the wave. One unit for wavelengths of light is the Angstrom (Å), which equals 10^{-8} cm or 10^{-10} m. Another is the millimicron. However, the presently preferred unit for wavelengths of light is the nanometer (nm)—like the millimicron, it is 10^{-9} m—and it will be used in this text (1 nm=10 Å).

Table 3.1. Definitions and characteristics of the various wavelength regions of light. The ranges of wavelengths leading to the sensation of a particular colour are somewhat arbitrary and vary with individuals. Both frequencies and energies in the table refer to the particular wavelength indicated in column 3 for each wavelength interval. Wavelength magnitudes are those in a vacuum.

Colour	*Approximate wavelength range (nm)*	*Representative wavelength (nm)*	*Frequency (cycles/sec) (Hertz)*	*Energy* (*ev per photon*)	*Energy* (*kcal/"mole" of photons*)
Ultraviolet	below 400	254	11.30×10^{14}	4.88	112.5
Violet	400—425	410	7.31×10^{14}	3.02	69.7
Blue	425—490	460	6.52×10^{14}	2.70	62.2
Green	490—560	520	5.77×10^{14}	2.39	55.0
Yellow	560—585	580	5.17×10^{14}	2.14	49.3
Orange	585—640	620	4.84×10^{14}	2.00	46.2
Red	640—740	680	4.41×10^{14}	1.82	42.1
Infrared	above 740	1400	2.14×10^{14}	0.88	20.4

The wavelength regions of major interest in biology are the ultraviolet, the visible, and the infrared (Table 3.1). Wavelengths below about 400 nm are referred to as ultraviolet (UV)—meaning on the other side of, or beyond violet, in the sense of having a shorter wavelength. The visible region extends from approximately 400 nm to 740 nm and is subdivided into various bands such as blue, green, or red (Table 3.1). These divisions are based on the subjective colour experienced by humans. The infrared (IR) region has wavelengths longer than those of the red end of the visible spectrum, up to approximately 40 μm.

Besides wavelength, we can also characterise a light wave by its frequency of oscillation, ν, and by a velocity of propagation that has the magnitude v. These three quantities are related as follows :

$$\lambda\nu = v \qquad \text{...(1)}$$

where λ is the wavelength. For all wavelengths the speed of light in a vacuum is a constant, generally designated as c, which experimentally equals 299,792$\pm$1 km/sec, or 3.00×10^8 m/sec. Light passing through a medium other than a vaccum has a speed less than c. For example, light having a wavelength of 589 nm in a vacuum is decreased in speed 0.03% by air, 25% by water, and 40% by dense flint glass. The wavelength undergoes a decrease in magnitude equal to the decrease in the speed of light in these various media, while the unchanging property of a wave propagating through different media is its frequency. Note that ν is the frequency of the oscillations of the local electric and magnetic fields illustrated in Figure 3.5.

For light, $\lambda_{vacuum}\nu$ equals v_{vacuum} by Equation 1, where v_{vacuum} is the constant c. Therefore, if we know the wavelength in a vacuum, we can calculate the frequency. In fact, the wavelengths given for light generally refer to values in a vacuum, as is the case for columns 2 and 3 in Table 3.1 (λs in air differ only slightly from the magnitudes listed). As a specific example, let us select a wavelength in the blue region of the spectrum, *e.g.*, 460 nm. By equation 1, the corresponding frequency equals c/λ_{vacuum}, or $(3.00\times10^8$ m/sec$)/(460\times10^{-9}$m/cycle), which is 6.52×10^{14} cycles/sec for such blue light. Since ν does not change from medium to medium, it is often desirable to describe light by its frequency, as has been done in column 4 of Table 3.1.

Light in Water

The sunlight available for plants and animals in the aquatic environment has entered the water from the air and hence has first been subjected to all the changes imposed upon it by the conditions above the surface. Ten per cent or more of the light is lost by reflection at the surface or in the special conditions just beneath the surface. In addition, in passing downward, the light is further modified by the water medium in respect to intensity, spectral composition, angular distribution, and time distribution.

Extinction and Modification of Light. The light factor in the water environment is subject to a number of variable influences that modify it profoundly. In order to consider the situation first in its simplest terms, imagine a lake from which we have pumped out all the natural water and refilled with distilled water. Our lake then contains water of uniform and maximum transparency from top to bottom. Light is reduced in intensity both by absorption and by scattering, and the rate of reduction is measured as the extinction rate, although the term absorption rate is sometimes used loosely for the combined effect. Pure water causes the extinction of light at different rates in different parts of the spectrum. Measurements of illumination made with a photometer placed in a watertight case and lowered into our imaginary lake would reveal the changes presented in Fig. 3.6. At a depth of 70 m the blue component of sunlight has been reduced to about 70 per cent of its intensity at the surface. At the same depth the yellow component of sunlight has been reduced to 6 per cent of its incident value. The orange and red components have been extinguished very much more rapidly. At a depth of 4 meters, red light has already been diminished to about 1 per cent of its surface intensity.

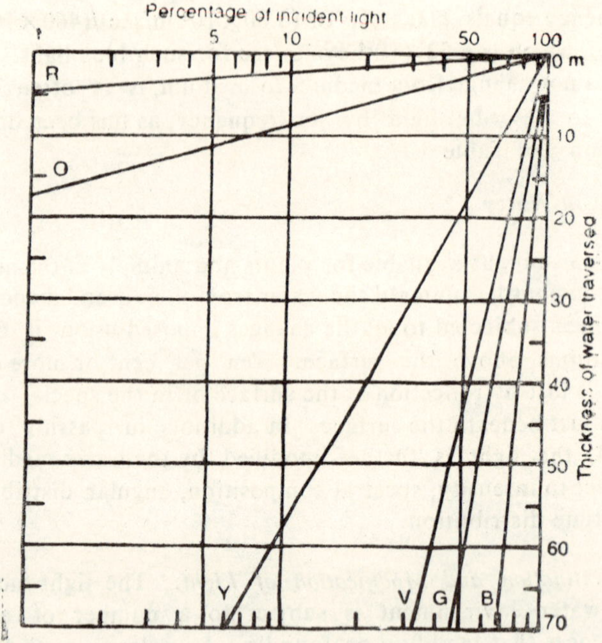

Fig. 3.6. Reduction in intensity (logarithmic scale) of the colour components of sunlight (indicated by initial letters) at increasing depths (linear scale) in a lake of optically pure water.

Since we know that the daylight incident upon the surface of our lake differs markedly in intensity in different parts of the spectrum, we are dealing with the unequal extinction of an unequal spectrum. As shown in Fig. 3.7 the extinction by pure water of light at the two ends of the spectrum is much more rapid than in the middle of the spectrum. The blue component of light is also the most penetrating in the clearest ocean and lake waters. At depths of 100 m or more the blue light becomes completely predominating; this was observed directly by Beebe in his bathysphere dive off Bermuda. An observer looking straight down into the water from the deck of a ship in the clear tropical parts of the ocean sees the blue colour resulting from this selective absorption. Since sunlight has been shorn progressively of its longer and shorter wavelengths, the only component remaining to be scattered upward again to the eye is the blue. Anyone who has seen the intense indigo of the tropical oceans will never forget it.

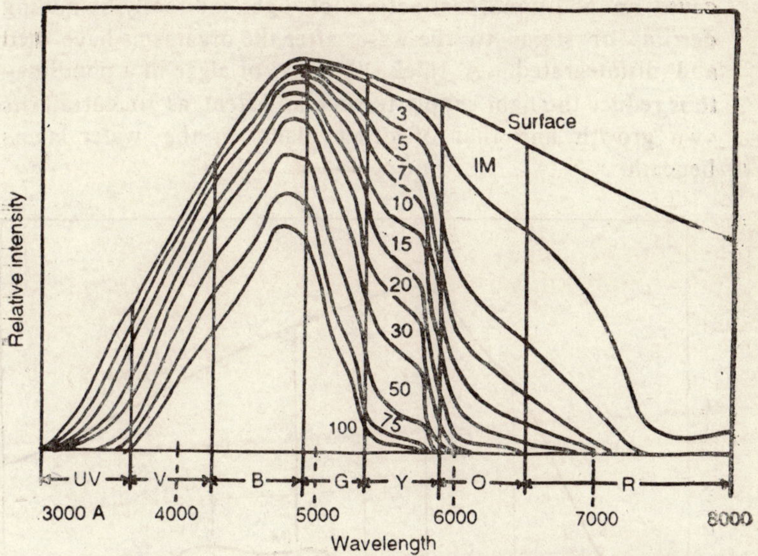

Fig. 3.7. The spectral distribution of solar energy at the earth's surface and after modification by passage through the indicated meters of pure water. Similar light conditions would be found in the clearest ocean and lake waters.

The foregoing has shown that even pure water absorbs light at a very rapid rate compared to air and causes a profound-change in spectral distribution. In the clearest parts of the ocean and in exceptionally clear lakes the optical properties of the water are closely similar to those of pure water. Other natural waters contain suspended particles and dissolved material in sufficient quantities to cause a further reduction in transparency and a further alteration in spectral composition.

Suspended material in the water includes living organisms that increase the extinction of the light and thus modify their own environment in this respect. Illumination is reduced by beds of kelp along the seacoast and by submerged or floating vascular plants along the shores of inland waters in much the same way that light is cut down by the larger vegetation on land. In the free water of ponds, lakes, and the oceans the phytoplankton is sometimes sufficiently abundant to produce a noticeable reduction of light. Plankton populations may cause an additional extinction of light indirectly by adding detritus or stains to the water after the organisms have died and disintegrated. A thick "bloom" of algae in a pond may thus reduce the light supply to such an extent as to curtail its own growth and that of other plants in the water layers beneath.

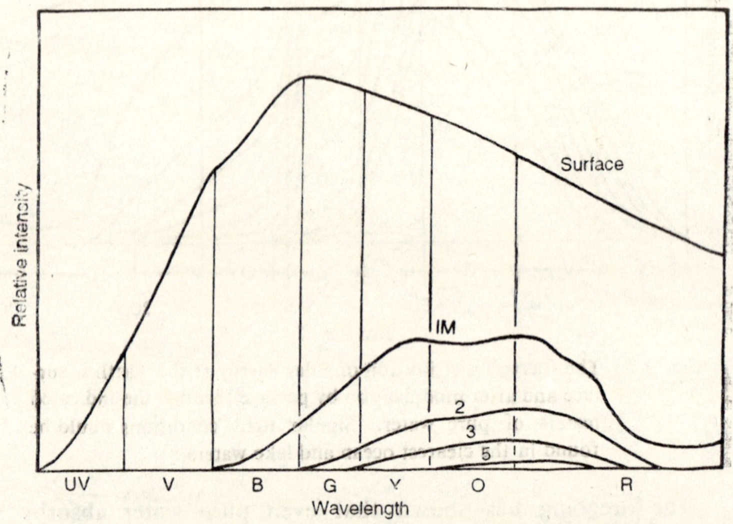

Fig. 3.8. Reduction in intensity and shift in spectral composition of light in heavily stained Rudolph Lake, Wisconsin.

In temperate and coastal seas and in the majority of clear inland lakes fine particles or stains are present that tend to absorb or scatter the blue component of light more strongly than occurs in pure water. As a result the green component of sunlight is usually the most penetrating in these situations and gives the water its characteristic emerald colour. The organic stains occurring in some ponds and rivers absorb the shorter wavelengths so strongly that the red or orange components of sunlight become the most penetrating. In Rudolph Lake, Wisconsin, for example, the combined absorption of the water and of stains in the water causes not only a very rapid reduction in the light with depth but also a shift in the position of the moximum light to the red region of the spectrum (Fig. 3.8).

Since natural waters differ very greatly in transparency and in selective absorption, a study of the penetration of each part of the spectrum in each body of water would have to be made for a complete description of the light conditions. An approximate comparison of the illumination in various natural waters can be made on the basis of the relative transparency to the central part of the spectrum. Sample values are given in Table 3.2 for representative bodies of water. Since the great

Table 3.2. Transparency of water to central part of visible spectrum (50000—60000 A)

	Extinction per Meter (%)	*k**	*Depth for 1% of Surface Light*	*Secchi Disc Depth†*
Distilled water	3.8	0.039	118 m	44 m
Caribbean Sea	4.1	0 041	110	41
Continental slope	7.2	0.072	60	24
Gulf of Maine	10.0	0.10	42	17
Woods Hole Harbor	26.0	0.30	16	6
Crater Lake, Oregon	6	0.06	77	28
Crystal Lake, Wis.	15	0.16	28	11
Trout Lake. Wis.	33	0.40	11	4
Midge Lake, Wis.	78	1.5	3	1

*Extinction coefficient, *k*.

$$\frac{I}{I_0} = e^{-kL}$$

where I_0=initial intensity.
I=intensity at depth, L, in meters.
e=2.7.

† Depth at which a 20-cm white disc disappears when lowered from the surface.

differences in the extinction of light per meter have a cumulative effect, an even greater contrast is presented in the depths at which a given fraction of the light is found. The depth at which the light intensity is reduced to 1 per cent of its surface value is of particular importance because it represents the approximate lower limit for plants as will be discussed later. In general the table and the representative curves in Fig. 3.9 show the profound variation in the intensity of light available for organisms at increasing depths in different natural waters.

Another way in which light is changed by water is in its angular distribution. The direct beam of light from the sun is bent by refraction toward the perpendicular at the water surface. Since waves or ripples almost always exist in natural situations, the light from the sun tends to be broken up to a considerable extent in passing through the surface. Within the water itself further diffusion is produced by scattering.

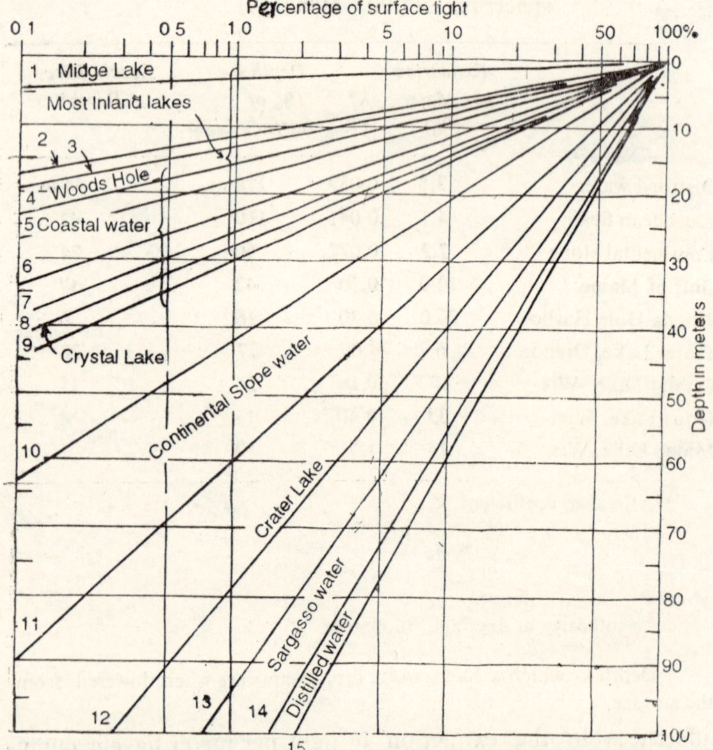

Fig 3.9. Comparison of the transparencies of representative natural waters.

When larger amounts of suspended materials are present, scattering proceeds at a very rapid rate, with the result that measurable light is passing in every direction. In the upper strata the light passing in the direction of the refracted rays from the sun is the strongest. With increasing depth the direction of strongest light intensity tends to move toward the vertical, and a limiting pattern of diffusion is established. Thus, although a variable degree of diffusion occurs, the directional character of the light in the water is never entirely lost.

Changes in Transparency. Parts of the ocean and of lakes that are well stirred exhibit a uniform transparency from the surface downward, but, when strong density gradients exist, considerable changes in transparency may occur with depth. Furthermore, the transparency of coastal marine areas and of inland water bodies often changes profoundly from season to season, owing to differences in amount of stirring, in the discharge of muddy rivers, and in the growth of plankton. Such variations in transparency are added to the seasonal changes in the intensity of the light received at the surface and in the length of the day. The combined effect of these influences results in surprisingly great fluctuations in the light factor at subsurface levels.

Observations made off the coast of Massachusetts illustrate the magnitude of the foregoing seasonal influences. In this region the maximum hourly intensity of light received at the surface during the summer averaged about twice that received during the winter period. The average total solar radiation received per day in summer was about four times as great as that received during the winter. Significant departures from the average for periods up to several weeks occurred, and such variations are undoubtedly important in causing changes in the response to the light factor from year to year. Values for individual days may also depart markedly from the average. During the investigation referred to above the maximum light received on the brightest day of summer was forty times greater than that received during the dullest day in winter. In the course of the same year the extinction coefficient of the water varied fourfold. Since the effect of extinction is cumulative, the illumination in the subsurface layers varies more because of changes in transparency than because of seasonal changes in radiation reaching the surface. By combining the two effects it was found that for an organism living at a depth of only 80 m the maximum hourly light intensity was 7000 times greater and total daily radiation was 10,000 times greater in May when high incident light was combined with high transparency, than in December when both incident illumination and transparency were low. The magnitude of seasonal changes in illumination in the aquatic environment are therefore frequently very much greater than in the terrestrial environment. Any water body that displays a much higher transparency in winter than in summer will have less light available during the summer in its deeper layers and thus suffer a reversal of the usual seasonal change in the light factor. The foregoing instances are sufficient to show that in the aquatic environment light becomes profoundly altered quantitatively and qualitatively and that its changes may go far beyond those experienced by plants and animals on land.

Orientation

Light often plays a significant role in orienting the growth or locomotion of plants and animals. Since orientation to light is often associated with reactions to other factors in the environment, such as gravity, we shall first consider the subject in general terms. Orientation is brought about either by the differential growth or movement of parts of the organism or by a change in the direction of locomotion of the whole organism. The question of why plants and animals grow or move in the directions they do has occupied the attention of investigators for a long time. Many of the theories and terms used have been in conflict, and for a more extended discussion reference should be made to Fraenkel and Gunn (1940) and to Griffin (1953).

The term *tropism* is best used for orientation by growth or turgor movements as exhibited by sessile forms. These forms are usually plants, but essentially the same type of orientation is exhibited by many sessile animals such as the hydroids. If the orientation is to gravity, the term *geotropism* is used. If it is to light, the growth movement is referred to as *phototropism*, and other prefixes are used for other orienting factors. On the other hand, the orientation of the locomotion of motile organisms is best referred to as a *taxis*, although it is sometimes also called a tropism. Here the forms involved are usually animals, although motile plants such as the green flagellates and motile Plant gametes or zoospores are included. Applying suitable prefixes, we obtain the terms *geotaxis*, *phototaxis*, etc., for orientation by the corresponding factors. Orientation in the direction of an orienting force is referred to as a positive tropism or taxis ; orientation in the opposition direction is a negative tropism or taxis. Occasionally a transverse tropism or taxis is displayed.

Control of speed of locomotion by the intensity of the factor is termed a *kinesis*. Gravity never changes significantly in intensity, but the speed of swimming, flying or creeping is influenced by alterations in the strength of other factors ; the common occurrence of photokinesis has already been mentioned.

The position of the main axis of the body is the primary orientation of the plant or animal. This is usually determined by gravity, as is seen in the upward growth (negative geotropism) of the shoot and the downward growth (positive geotropism) of the root of a plant seedling (Fig. 3.10). In the aquatic environment the buoyant action of water often reduces the effect of gravity so that primary orientation is to light or to current. When the source of illumination is from the side the primary orientation of, a land plant will be a compromise between the influence of gravity and that of light.

The primary orientation of motile organisms is also usually to gravity, and a walking or a flying animal continually adjusts its position to it.

Superimposed upon the primary orientation, many secondary responses to orienting influences take place. The tips of green plants grow toward the light and hence exhibit positive

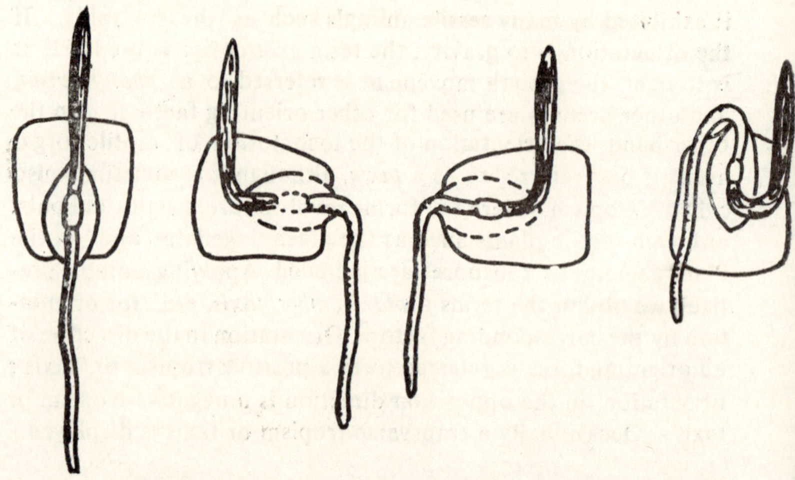

Fig. 3.10. Negative geotropism of shoots and positive geotropism of roots of four kernels of maize that have germinated in different positions.

phototropism. The leaves are oriented generally at right angles to the incident radiation, thus receiving maximum illumination, and they sometimes form a symmetrical pattern (Fig. 3.11) and sometimes a mosaic, as is displayed by ivy leaves on a wall.

In such species as the sunflower the top portion of the plant or its leaves are turned by turgor changes during the course of the day, keeping always in the direction of the sun. In regions where light and heat from the sun are excessive the leaves of some plants. Such as the compass plant *silphium laciniatum*, are oriented so as to present their edges towards the sun. These examples will suffice to illustrate the turning reaction of phototropism by actual growth or by turgor movements.

Ecological Aspects of Photosynthesis

Light is fundamentally important as the essential direct source of energy for the growth of all green plants and the purple bacteria. The photosynthetic plants from the first step in the ecological cycle in every natural situation. They are the

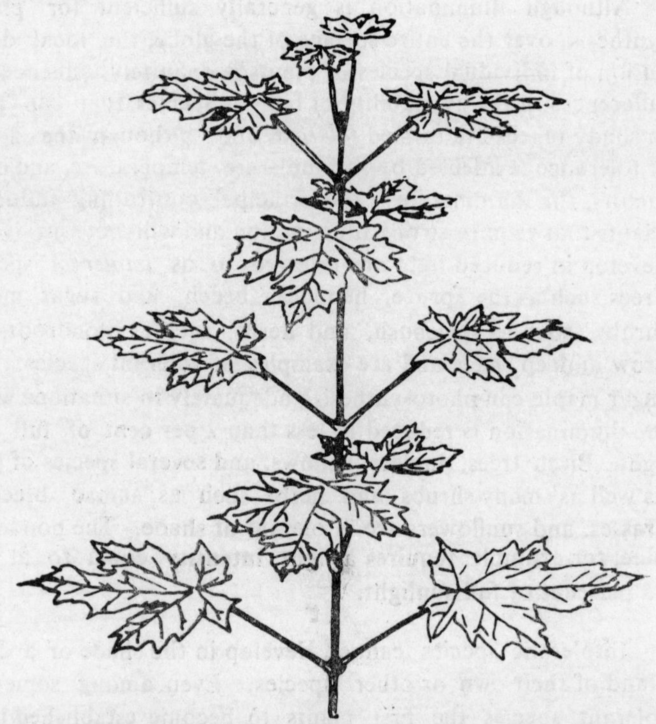

Fig. 3.11. Alternating position and horizontal orientation of leaves on shoot of Norway maple, resulting in maximum exposure to sunlight.

first link in the food chain and hence are the base of the production pyramid.

Land Plants. On the surface of the land enough light is received everywhere for the growth of plants of some sort. Within the soil and in the interior of caves light is insufficient to allow photosynthesis. A striking illustration of the limiting action of the light factor is found in caves to the walls of which electric lamps have been secured for the benefit of tourists. Within the cone of light immediately around each lamp mosses have grown from spores brought in by air currents. These "islands" of plant growth present a sharp contrast to the complete lack of vegetation in the remainder of the cave.

Although illumination is generally sufficient for photosynthesis over the entire surface of the globe, the local distribution of individual species of plants is definitely influenced by differences in the availability of light. Species that can grow in shady places are termed *tolerant*, and, although the degree of tolerance is affected by soil moisture, temperature, and other factors, the illumination is the principal controlling influence. Plants that require strong illumination and will not survive or develop in reduced light are referred to as *intolerant* species. Trees such as the spruce, hemlock, beech, and sugar maple, shrubs, such as spicebush, and herbs, such as bloodroot, can grow in deep shade and are examples of tolerant species. The sugar maple can photosynthesize adequately in situations where the illumination is reduced to less than 2 per cent of full sunlight. Birch trees, poplars, willows, and several species of pine, as well as many shrubs and herbs such as sumac, bluestem grasses, and sunflowers are intolerant of shade. The ponderosa pine, for example, requires a light intensity equal to at least 25 per cent of full sunlight.

Intolerant species cannot develop in the shade of a dense stand of their own or other species. Even among somewhat tolerant species the first plants to become established in an open area will grow to better advantage than individuals subsequently developing around them because of the competition for light and other needs. Differential growth often results in "pyramiding" in the development of such a group of plants. As we have seen in an earlier section, the light on the floor of the temperate deciduous forest become seriously reduced when the trees leaf out. Many of the smaller plants are adapted to a period of rapid growth in the early spring while light on the forest floor is sufficiently intense. In the marginal rainforest of Panama a dense undergrowth is possible because during the dry season the leaves fall from the trees and allow light to penetrate. In the equatorial rainforest of Columbia, on the other hand, trees shed their leaves individually and a continuous canopy is maintained, with the result that little vegetation can grow on the forest floor.

In such rainforests many smaller plants have gained access

to sufficient illumination by the evolution of the epiphytic habit, that is, by growing in the crotches or on the horizontal branches of the trees. The vertical distribution of the many species of bromeliads in the forest trees of Trinidad reveals their varying degrees of tolerance (Fig. 3.12). One group of species grows only near the tree tops in situations fully exposed to the sun, a second group is found at intermediate levels in partial sunlight, and a shade-tolerant third group inhabits the lower branches of the forest where the illumination is quite inadequate for the other groups.

Aquatic Plants. Reduction of light presents even more serious problems in the aquatic environment. As we have seen illumination diminishes rapidly with depth even in clear water and becomes changed in spectral composition and in other respects. Plants attached to the bottom in the marine environment consist principally of algae with a few species of vascular plants such as the eel grass. In fresh water vascular plants as well as algae are well represented in the submerged vegetation.

Early investigators reported an apparent colour zonation in the depth of occurrence of attached plants. In Puget Sound, for example, the green algae are generally the most abundant in shallow water, brown algae dominate the zone from 5 to 20 m, and red algae are usually most numerous in depths of 10 to 30 m. These colour types of algae often occur at depths where the complementary colour of the penetrating daylight predominates—red algae, for example, tend to be abundant in deep water where the blue or green component of daylight is the strongest. It was formerly believed that the predominating colour of the light controlled the depth distribution of the colour types of algae because the plants would absorb light of complementary colour more efficiently. This generality was referred to as *chromatic adaptation.* Many exceptions occur, however, and we now know that the wavelength of light does not control in any precise way the depth of distribution of algae on the basis of their colour. Furthermore, the pigment of algae living in weak light deep in the water is sufficiently thick to absorb all the light incident upon it, regardless of the part of the spectrum in which it occurs. Evidence exists that some

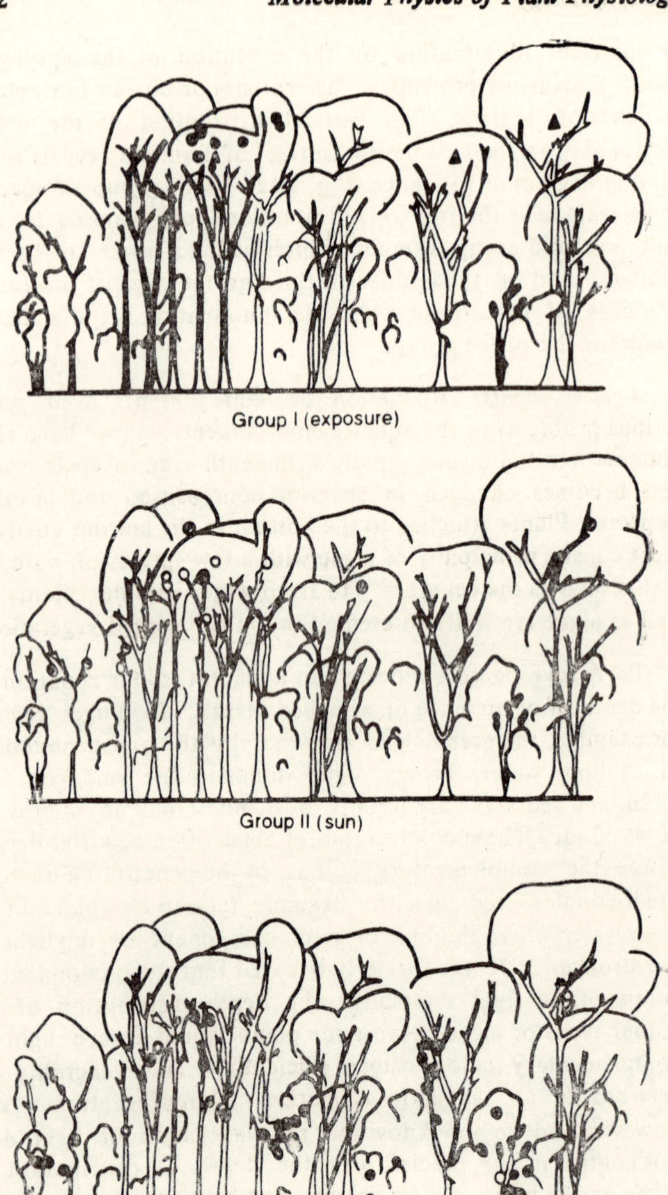

Fig. 3·12. **Vertical distribution of various species of bromeliads in the forest trees of Trinidad.**

of the light energy absorbed by pigments other than chlorophyll can be transferred to the chlorophyll present and thus enhance the ability of certain algae to live in weak blue-green light.

When algae are grown in the laboratory in light of different colours, they often tend to change colour, and this response has also been referred to as chromatic adaptation. Experiments have shown that individual plants photosynthesize most efficiently and grow best in that colour and in that intensity in which they have been living. In the course of time fixed algae can adapt themselves to habitats with illumination of widely different conditions of colour and intensity. In view of the great range in the transparency of natural waters, it is not surprising that the maximum depth at which attached plants can live varies widely in different situations (Table 3.3). In general

Table 3.3. Maximum depth in meters for Growth of attached plants.

Crater Lake, Ore.	120
Crystal Lake, Wis.	20
Trout Lake, Wis.	12
Mediterranean Sea	160
Challenger Bank	100
Of Iceland	50
Puget Sound	30
Baltic Sea	20
Off Cape Cod	10

benthic plants will not grow at depths at which the light intensity is less than 0.3 per cent of the surface value.

A glance at a chart of the oceans or of a typical lake reveals that only in a very narrow fringe around the margin is the water sufficiently shallow for enough light to reach the bottom for the growth of plants. We are forced to the realization that in any deep body of water far more organic matter is synthesized by the phytoplankton present everywhere in the surface layers than by the benthic plants limited to the littoral zone. Photosynthesis, the first step in the ecological cycle of the sea upon which all marine life ultimately depends, is carried forward primarily by the minute diatoms and other microscopic planktonic plants. The growth of the phytoplankton is limited not only by the rapid diminution of light with depth but also by the fact that vertical movements of the water subject the plants to continually changing light conditions.

The greatest depth at which phytoplankton can grow successfully in any body of water may be ascertained by suspending bottles containing samples of the plant population at various levels beneath the surface. Photosynthesis going on within the bottle adds oxygen to the water, and respiration taking place simultaneously removes oxygen from the water. The rate of respiration alone may be measured by placing at each level a blackened bottle from which all light is excluded. The sum of the gain in oxygen in the clear bottles and the loss in the dark bottles gives a measure of total photosynthesis. The increment of oxygen in the clear bottles is the amount produced by photosynthesis minus the amount consumed in respiration and represents the actual increase in energy content or growth of the plants.

With the diminution of light in any environment photosynthesis is reduced but respiration remains approximately the same, provided, of course, that temperature and other factors are essentially unchanged. For each species a point is reached in the reduction of the illumination at which the rate of photosynthesis is just equal to the rate of respiration ; this is known as the compensation point or better as the *compensation intensity*. At light intensities below this value photosynthesis may still go on but the plant is fighting a losing battle : it cannot survive indefinitely under these conditions because the loss of energy due to the catabolic processes represented by respiration exceed the gain in energy brought about by the anabolic process of photosythesis. With further light reduction a minimum intensity is reached below which no photosynthesis at all can take place.

In the aquatic environment the level in the water at which the compensation intensity is found is called the *compensation depth*. Sample values for phytoplankton based on measurements in the middle of the day are presented in Table 3.4. For phytoplankton in general the compensation intensity, has been

Table 3.4. Compensation depths in meters for phytoplankton. (for midday periods)

Sargasso Sea	100
English Channel	35
Gulf of Maine	30
Off British Columbia	19
Trout Lake, Wis.	16
Woods Hole harbor	7

found to be about 1 per cent of the value of full sunlight at the surface. In Fig. 3.13 the change in the amount of photosynthesis at various depths during the course of the day is plotted for a type of diatom representative of the marine plankton. It will be observed that during the noon hours the illumination close to the surface was excessive and depressed photosynthesis. The highest rate of oxygen production occurred at 5 m in the middle of the day. No significant amount of

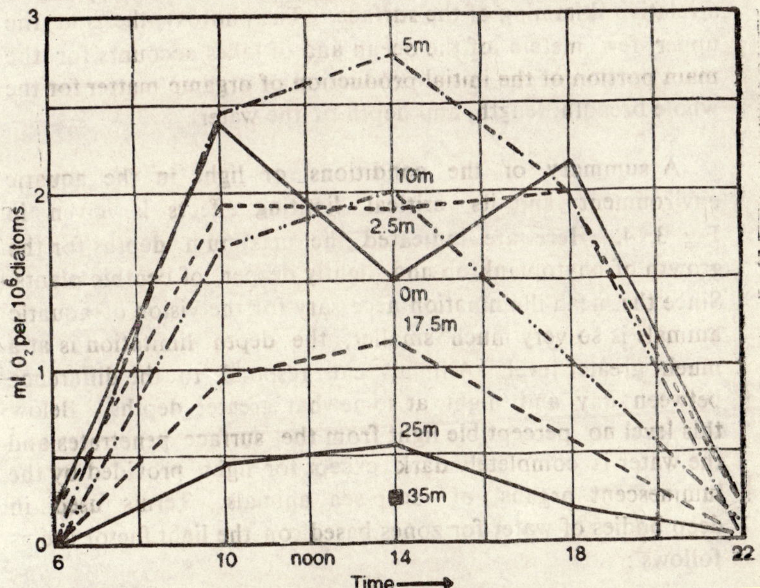

Fig. 3.13. Photosynthesis of a species of phytoplankton (*Coscinodiscus*) at the indicated depths during the course of the day off Stoke Point, England.

photosynthesis occurred at any depth before 0600 in the morning or after 2200 and none below 35 m at 1400.

If a plant is to grow, its photosynthesis during the day must build up enough organic matter to more than make up for the material lost by respiration not only during the day but also during the night. In other words, the crucial value for the continued existence of the plant is the compensation depth for the 24-hour period. This will obviously occur at a shallower depth than the values reported for experiments limited to the middle of the day. The compensation depth for the complete day ranges between 20 and 30 m during the summer in clear coastal water of the temperate oceans. In the winter and in less transparent water the compensation depth occurs correspondingly nearer the surface. No constructive growth is possible for diatoms or other pelagic plants below these levels. When we recall that the average depth of the ocean is more than 4000 m, it is apparent that the zone in which productive growth of the crucially important phytoplankton can take place is a relative skimming of the surface. The photosynthesis in the upper few meters of the ocean and of lakes accounts for the main portion of the initial production of organic matter for the whole breadth, length, and depth of the water.

A summary of the conditions of light in the aquatic environment and its critical limiting effects is given in Fig. 3.14. Here are indicated the maximum depths for the growth of phytoplankton and slightly deeper, of benthic plants. Since the mean illumination necessary for the vision of aquatic animals is so very much smaller, the depth limitation is at a much greater level. Animals can respond to the difference between day and night at somewhat greater depths. Below this level no perceptible light from the surface penetrates and the water is completely dark except for light provided by the luminescent organs of deep-sea animals. Terms used in deep bodies of water for zones based on the light factor are as follows :

Euphotic Zone : Sufficient light for photosynthesis.

Disphotic Zone : Insufficient light for photosynthesis but sufficient light for animal responses.

Aphotic Zone : No light of biological significance from the surface.

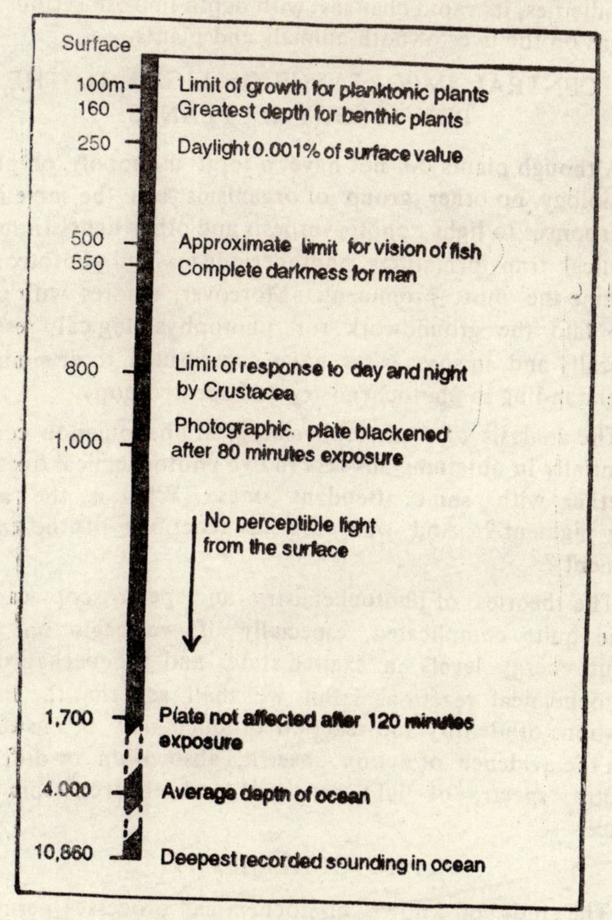

Fig. 3.14. Summary of limitations by light factor in aquatic environment. Values are for clearest water.

The depth limits of these zones differ widely according to transparency.

The above discussion has revealed the many important roles

played by light in the world of life. Although there is generally sufficient light on land for plants and animals, the special circumstances of this factor exert a wide control over the activities of many terrestrial organisms. In the aquatic environment light is even more crucial since, in addition to its various periodicities, its rapid changes with depth impose serious limitations on the lives of both animals and plants.

CENTRAL IMPORTANCE OF LIGHT TO THE PHYSIOLOGY OF PLANTS

Although plants do not have a total monopoly on photophysiology, no other group of organisms has the same range of responses to light : photosynthesis and other light-stimulated chemical transformations, photoperiodism, and phototropism, to name the most prominent. Moreover, studies with plants have laid the groundwork for photophysiological research generally and, in some cases, have contributed to or simulated understanding in photochemistry and spectroscopy.

The analysis of a photophysiological phenomenon consists essentially in obtaining answers to two photochemical questions together with some attendant ones : What is the absorbing pigment ? And what are the reactions of the excited pigment ?

The theories of photochemistry and spectroscopy can become quite complicated, especially if we begin on worry about energy levels in excited states and the mechanisms of photochemical reactions. But we shall see that the limited questions of identity and reaction products may be answerable with the evidence of action spectra, absorption or difference spectra, spectra of light emission, and electron spin resonance.

PHOTOCHEMISTRY

The study of known photochemical processes permits us to anticipate the kinds of processes we shall encounter in photosynthesis and other biological responses to light.

Electronic Transitions in Atoms, Molecules, and Crystals

Light can be dought of as traveling in packets called

photons. Photons are often designated as *hv*, which is at once a symbol and a representation of their energy (see below). Every time an ultraviolet or visible photon is absorbed, the absorbing material undergoes an *electronic transition*. The simplest case of an electronic transition occurs in a hydrogen atom (Fig. 3.15). The single hydrogen electron is exicited to a new energy level ; the *entire* energy of the photon is converted into additional kinetic energy of the electron. Just as a violin string may vibrate only in a fundamental tone and higher harmonics, the hydrogen atom is stable only in certain definite energy states, so photons of certain definite energies, and only these energies, are absorbed by hydrogen atoms (Fig. 3.16). It is precisely as if the hydrogen atoms resonated at certain critical frequencies (or wavelengths) of light. If, with a spectograph, we were to observe white light that had passed through hydrogen atoms, we should see absorption at the wavelengths corresponding to transitions of the hydrogen

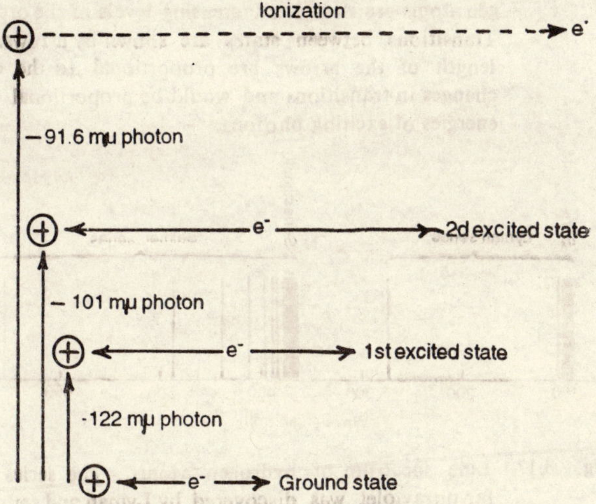

Fig. 3.15. Fxcited states of hydrogen atoms. The single electron of a hydrogen atom oscillates between the nucleus and some outer limit. In excited states the oscillation occurs over a longer radius. If a sufficiently energetic photon is absorbed, the electron escape completely and we say the hydrogen atom is ***ionized.***

atoms. The result is a series of absorption lines (Fraunhofer lines) (Fig. 3.17) ; atoms such as hydrogen are thus said to exhibit "line spectra."

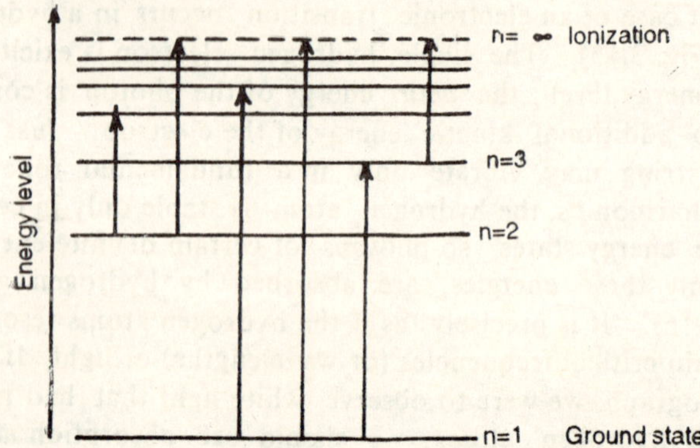

Fig. 3.16. Energy diagram and some possible transitions of hydrogen atoms. The energy levels of excited states in hydrogen atoms are shown as increasing levels of the ordinate. Transitions between states are shown by arrows. The length of the arrows are proportional to the energy changes in transitions and would be proportional to the energies of exciting photons.

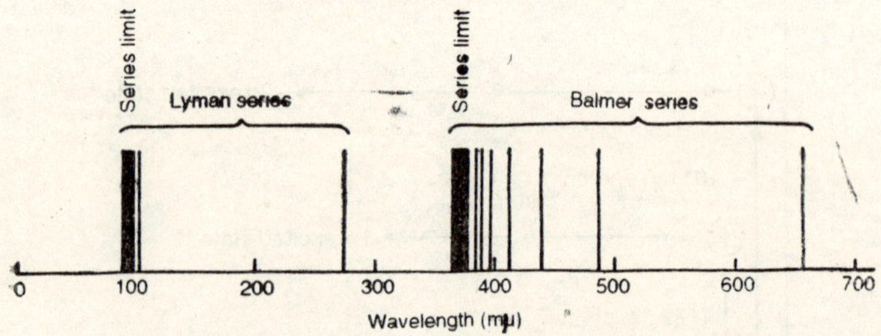

Fig. 3.17. Line spectrum of hydrogen atoms. The series in the far ultraviolet was discovered by Lyman and represents transitions from the ground state. The Balmer series represents transitions from the first excited state.

The general method of wave mechanics, which treats orbital electrons as simple harmonic oscillators, is fairly successful even with complicated atoms such as sodium and iron. In contrast, the transitions of even the simplest molecules are enormously more difficult to dissect, but approximations are possible by starting from the premise that the most stable states for chemical bonds correspond to a minimum energy mode in which an electron wave can oscillate between two nuclei. The same kind of electronic transitions occurs in molecules as in atoms and is responsible for the absorption of visible and ultraviolet light by molecules.

Wave mechanics also provides a rational basis for the phenomena of "allowedness" and "forbiddenness," *i.e.*, the complete or partial exclusion of certain types of transitions. Similar considerations apply to the energy levels of crystals, in which there can exist free electrons, *i.e.*, electron waves oscillating within conduction bands or whole regions of the crystal.

Other Transitions

We have been talking about electronic transitions. There are other kinds of transitions in polyatomic molecules and crystals : (1) vibrational transitions, where the energies are contained in elastic vibration among the nuclei of the molecule or crystal, and (2) rotational transitions, where the energies are associated with the twisting of model along the axis of the bond joining them (Fig. 3.18).

The energies of pure vibrational and rotational spectra are weak, and so the photons that can induce such transitions are restricted to the infrared. (Indeed, infrared absorption spectra are specific indicators of molecular structure for the simple reason that they correspond to the vibrational and rotational modes possible for the molecule.)

The photon energies for these different kinds of transitions are easily calculated. The energy of a photon is equal to the product of Planck's constant and the frequency of light

$$E = h\nu \qquad \text{or} \qquad E = \frac{hc}{\lambda}$$

where c is the velocity of light and λ is wavelength. The energy of a photon thus may be determined from the value of

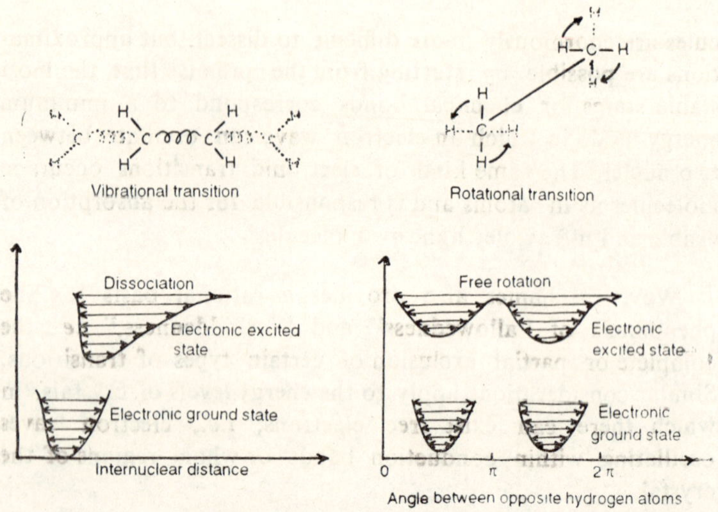

Fig. 3.18. Vibrational and rotational transitions. Some transitions that can occur in the double bond of ethylene are shown as examples of vibrational and rotational transitions.

Planck's constant and the wavelength or frequency of light. When the wavelength is expressed in mμ (or nm), the energy in electron volts is

$$E_{ev} = \frac{1,235}{\lambda_{m\mu}} \quad ...(2)$$

The approximate energy relations among the different kinds of transitions are :

Transition	*Effective Wavelengths, mμ*	*Energy*	
		electron volts	*calories*
Electronic	100—1,000	1—10	300,000—30,000
Vibrational	1,000—10,000	0.1—1	30,000—3,000
Rotational	10,000—100,000	0.01—0.1	3,000—300

In the case of molecules, we usually observe mixed spectra. In these cases, the molecule exists in one of several possible vibrational states associated with the ground electronic state. The molecule may be raised to any of several possible vibrational states associated with an excited electron state (Fig. 3.19). Instead of exhibiting the relatively simple line spectra of atoms, molecules show extremely complex spectra. In the gaseous state, the spectra consist of bands of relatively diffuse lines. In the condensed phase, the multitude of vibrational and rotational states smears into a single blurred peak for each electronic transition.

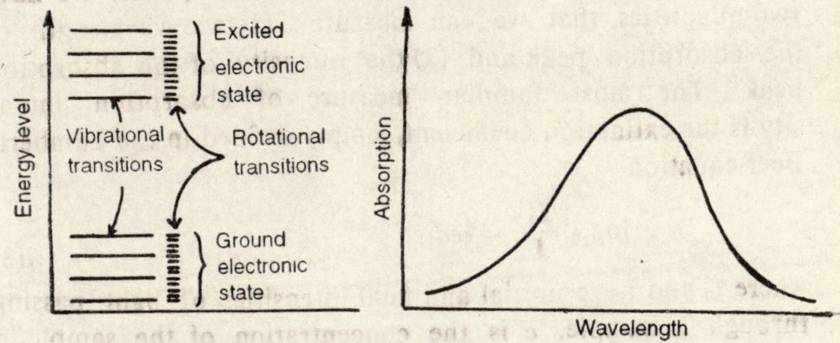

Fig. 3.19. Energy diagram for mixed transitions.

The ultraviolet and visible absorption spectra which have become such a commonplace in chemistry and biochemistry therefore have their physical basis in electronic transitions, greatly modified by vibrational and rotational transitions. Transitions may occur to the first, second, or higher excited states. For example, with chlorophyll *a*, light of 663 $m\mu$ which corresponds to 41 kcal is sufficient to excite the molecule to the first excited state. Light of 410 $m\mu$ (65 kcal) excites chlorophyll to the second excited state.

Once in the excited state, a molecule may be excited still further. Again in the case of chlorophyll *a*, it has been possible with tremendous light intensities to convert most of a given sample to another excited state (in this case the *triplet state*). It remains in this state long enough that one may irradiate the solutions again, this time with monochromatic light, and this time observe transitions to the second triplet state.

Unlike *electronic transitions, vibrational and rotational*

states are functions of the temperature. We may therefore expect that different absorption spectra might be obsërved at different temperatures. In fact, absorption spectra at extremely low temperatures show greatly sharpened peaks.

Extinction Coefficient and Oscillator Strength

If we are to characterize a particular electronic transition for a polyatomic molecule in the condensed phase, we have two quantities that we can measure: (1) the wavelength of the absorption peak and (2) the intensity of the absorption peak. The most familiar measure of absorption intensity is the extinction coefficient, simply defined in the Lambert-Beer equation.

$$\log_{10} \frac{I_0}{I} = \epsilon c d \quad \text{...(3)}$$

where I_0 and I are initial and final intensities of light passing through a sample, c is the concentration of the sample in moles per liter, and d is the light path in centimeters; the extinction coefficient, ϵ, then has the units of liter $mole^{-1}$ $centimeter^{-1}$.

A more fundamental quantity is *oscillator strength*, which is the efficiency with which an oscillator (electron) resonates with the field (light). An ideal oscillator (*e.g.*, a hydrogen atom) absorbs light in an allowed transition with 100 per cent efficiency. We say it has an oscilla or strength $f_A = 1$. Real molecular oscillators absorb light over a band of wavelengths, so that we must integrate (see Equation 4). The integrated oscillator strength f_A is

$$\bar{f}_A = \int_{\lambda_1}^{\lambda_2} f_A(\lambda) d\lambda \quad \text{...(4)}$$

There is a convenient relation between oscillator strength, absorption coefficient (Equation 3) and the index of refraction of the medium, *n*.

$$\bar{f}_A = 4.319 \times 10^{-9} n \int_{\lambda_1}^{\lambda_2} f(\lambda) d\lambda \quad \text{...(5)}$$

Oscillator strength sums the electronic transition over all of the vibrational and rotational substates. This quantity represents, therefore, the total probability or allowedness of the electronic transition occurring. Since the probability of the transition occurring is the same in both directions, the oscillator strength is also related (as is the absorption coefficient) to the reciprocal of the mean lifetime of the excited state. Nearly ideal organic oscillators have strengths of 1 corresponding to absorption coefficients of about 10^6 liters $mole^{-1}$ cm^{-1} and mean lifetimes of about 19^{-9} sec.

Futes of Excited Substances

When a substance absorbs a photon in the visible or ultra-violet, it undergoes an electronic transition and becomes excited.

$$A + h\nu \rightarrow A^* \qquad \text{...(6)}$$

The excited substance may then undergo the following, either singly or in some cases in sequence :

1. Fluorescence.
2. A long-lived or metastable state, such as the triplet state.
3. A chemical reaction.
4. Ionization.
5. Dissociation (if a molecule).
6. Isomerization (if a molecule).
7. Conversion of all or part of its energy of excitation to kinetic energy through what is called internal quenching or a radiationless transition.
8. Transference of the energy of excitation to another atom or molecule at a small distance ; this is called energy transfer ; excition migration, or resonance transfer.

Since all of these processes may possibly occur in intact plants, we shall examine them more closely.

Fluorescence. Atoms in the gas phase may emit the simplest kind of fluorescence, namely reemission of the absorbed photon. This is called resonance fluorescence. In polyatomic molecules and in condensed phases, fluorescence occurs when the species is excited to a higher singlet state. Part of the enemy is lost by internal quenching (alternative 7 above), bringing the species down to the first singlet state. The remaining energy is then reemitted as a photon (Fig. 3.20). Since the probability for a transition is the same in either direction and the average lifetime of a single state (assuming an allowed transition to the ground state) is not much longer than 10^{-9} sec, fluorescence usually occurs within 10^{-9} sec. Longer intervals (up to 10^{-6} sec) have been observed when exciton migration intervenes between excitation and fluorescence.

Long-lived and Metastable states. Although transition from the ground state to forbidden states in highly improbable, it may be quite probable for a molecule to pass from an excited singlet state to a forbidden one (Fig. 3.21). Photons may then leak slowly from the excited species, resulting in *phosphorescence.* A typical forbidden state is one with two unpaired electrons. From the observation that, under the influence of an external magnetic field, atoms with unpaired electrons can exist in three excited states with closely similar energy levels, spectroscopists named such little constellations "triplet states." The term has been used loosely for any excited species which emits light over a long interval, even when the triplet energy splitting has not been demonstrated. This imprecision is unfortunate, especially in biological systems where highly ordered, near-crystalline arrays may occur in which other long-lived or metastable states are possible.

Chemical Reactions. Excited species may enter directly into chemical reactions. The reaction may occur simply because the energy of excitation supplies the energy of activation of the reaction; that is, the energy of excitation "drives" the reaction, producing a net decrease in free energy. A reaction may also occur because the orbitals of the excited state labilize a bond, greatly increasing the possibility of orbital overlap with another reactant. The last-named mechanism is in effect a decreased energy of activation for the reaction. Because of the extremely short lifetimes of singlet states, the probability is

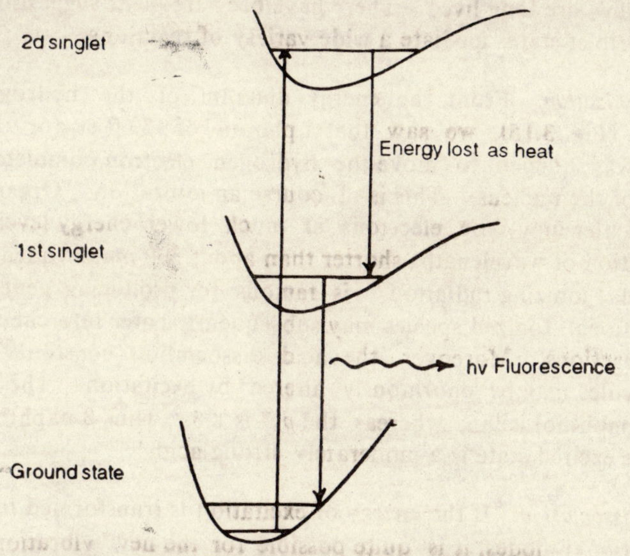

Fig. 3.20. Energy diagram for molecular fluorescence.

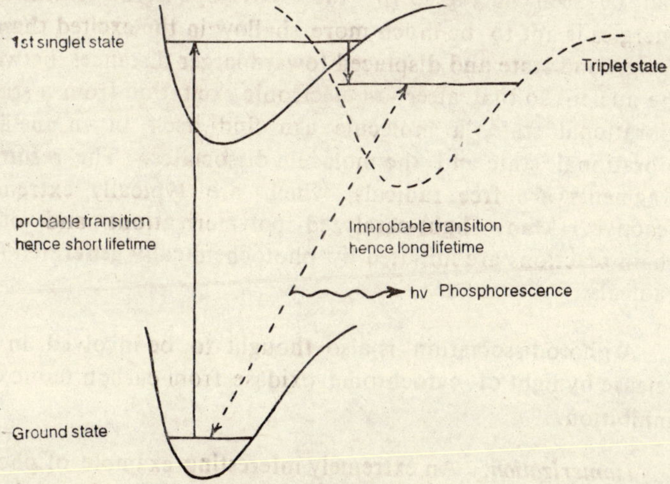

Fig. 3.21. Energy diagram for the metastable state and phosphorescence.

low for their experiencing a favourable collision. Efficient bimolecular reactions involving singlet excitation would require extremely favourable steric factors. Reactions with molecules in forbidden states are quite another matter for the very reason that they are long lived. There have been frequent suggestions that triplet states mediate a wide variety of reactions.

Ionization. From the energy diagram of the hydrogen atom (Fig. 3.15), we saw that a photon of 13.60 ev (or 91.1 $m\mu$) was sufficient to move the hydrogen electron completely free of the nucleus. This is of course an ionization. Organic molecules may lose electrons at much lower energy levels: radiation of wavelengths shorter than about 300 $m\mu$, frequently called "ionizing radiation," is famous for producing genetic mutations. Ionized species may subsequently enter into chemical reactions. Moreover, the acid dissociation constants of molecules may be enormously altered by excitation. The p_k of β-naphthol is 9.5, whereas the p_k^* is 2.5. Thus β-naphthol in the excited state is a moderately strong acid !

Dissociation. If the energy of excitation is transformed into vibrational modes, it is quite possible for the new vibrational level to be sufficiently great to break the chemical bond. As can be seen in Fig. 3.18, the "envelope" for vibrational energies is apt to be much more shallow in the excited than in the ground state and displaced toward larger distances between the nuclei, so that after an electronic excitation from a stable vibrational state, a molecule can find itself in an unstable vibrational state and the molecule dissociates. The resulting fragments are free radicals, which are typically extremely reactive. Many light-catalyzed polymerizations and other chain reactions are initiated by photochemically generated free radicals.

A photodissociation is also thought to be involved in the release by light of cytochrome oxidase from carbon monoxide inhibition.

Isomerization. An extremely interesting example of photochemical isomerization is the interconversion of maleic to fumaric acid (Fig. 3.22). Although these two acids have quite distinct structures, corresponding to the cis and trans configurations around a carbon-carbon double bond, the excited states of the two are nearly identical, so it is possible for an excited fumaric molecule to return to the maleic ground state and *vice versa*. Photoisomerizations may be of great importance in biology.

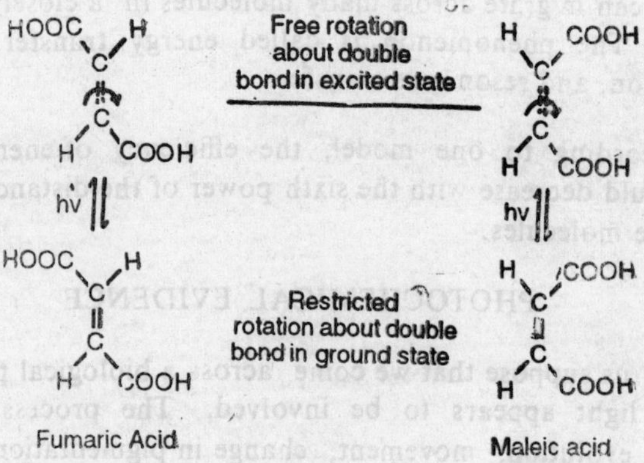

Fig. 3.22. Photoisomerization of maleic and fumaric acids.

Quenching. Even though increased electronic energy resulting from excitation may labilize a bond sufficiently to permit a chemical reaction, the more usual event is that the excitation energy is converted to vibrational energy among all the bonds of the molecule. Through the principle of equal partition of energy, ultimately all of the excitation energy is converted to heat. The initial conversion of electronic to vibrational energy is called internal conversion or quenching. In the liquid phase and at ordinary temperatures, collisions frequently enhance quenching, so that most photochemical reactions, including fluorescence and phosphorescence, are strongly quenched in liquid phases at ordinary temperatures.

Considering the enormous variety of ways in which molecules can lose their excitation energies without entering into a chemical reaction, energy-conserving photochemical reactions such as photosynthesis seem highly improbable !

Energy Transfer. Exactly as a plucked violin string will

induce resonant vibrations in an adjacent string tuned to the same frequency, an excited electron in a molecule will induce an electronic transition in a nearby molecule. More than that, if the resonance energy of the neighbouring molecule is somewhat less than that of the excited molecule but still sufficiently near that their absorption bands overlap, a transference of all of the excitation energy *must* occur. In this way excitation energy can migrate across many molecules in a closely packed array. The phenomenon is called energy transfer, exciton migration, and resonance transfer.

According to one model, the efficiency of energy transfer should decrease with the sixth power of the distance separating the molecules.

PHOTOCHEMICAL EVIDENCE

Let us suppose that we come across a biological process in which light appears to be involved. The process might be oxygen evolution, movement, change in pigmentation, flowering, luminescence, release from CO inhibition, etc. What questions might we appropriately ask about such a system? And by what criteria should we judge answers to these questions? Questions and the kinds of answers to them are the subjects of this section.

The questions we ask are simple enough. What pigment or pigments absorb light? Where are the pigments? What is the excited state. What is the fate of the excited pigment molecules. That is, what transformations or reactions does the pigment experience? How many quanta are required for a unit step in the process? Is there transfer of radiant energy in the system and, if so, how does transfer occur? After the strictly photochemical questions, there are questions that grade rapidly into biochemistry. For example, is the process strictly photochemical, or are there accompanying "dark" reactions, which might be required to regenerate the pigment? How does the product of the light reaction influence the physiological process?

The kinds of evidence required to answer these questions are all based squarely on the principles of photochemistry.

The Bunsen-Roscoe Law

Our first step in appraising a light-influenced process is to determine if the response is proportional to the amount of light absorbed.

$$R = kIt \qquad \text{...(7)}$$

where R may be taken as the extent of a process, I is the light intensity, and t is time. This statement is known as the Bunsen-Roscoe law. It follows directly from the notion of excitation

$$A + h\nu \rightarrow A^* \rightarrow P$$

We can expect that, since every A* has an equal probability of resulting in P, the rate or extent of the reaction should be proportional to A*; the light absorbed is proportional to the concentration of pigment A and the amount of light, which is clearly the product of I and t.

Nonconformity with the Bunsen-Roscoe law means that either A varies over the time intervals chosen, A varies with I, or the extent of the reaction is not proportional to the immediate photochemical product P.

Variation of A with I can occur in the case of optical pumping of lasers, but is rarely encountered in biological systems. In the scheme

$$x \rightarrow A + h\nu \rightarrow A^* \rightarrow P$$

(with an arrow returning from A^* to A)

variation of A with time can mean that the rate of formation and/or regeneration of A is slower than (or of the same order of magnitude as) the rate of excitation (absorption).

$$\frac{d(x \rightarrow A)}{dt} + \frac{d(A^* \rightarrow A)}{dt} \leqslant \frac{d(A \rightarrow A^*)}{dt}$$

We may use flashing light tests to tell us if a "dark" reaction is required for the light-influenced process. By varying the dark interval between flashes, we may learn something about limiting rates of the dark reaction. The effects of variations in the intensity of duration of the flashes give corresponding information about the relative amount of absorbing pigment.

Action Spectra

The formation of the immediate photochemical product P is proportional to A^* and therefore also to the absorption of light by A. From this rather obvious statement, it follows that if a chemical or physiological response to light, R, is proportional to P, then R *is proportional to* A (Equation 8) *for all competent transitions.*

$$R(\lambda)=kA(\lambda) \qquad \ldots(8)$$

A graph of R(λ) is said to be an *action spectrum.* Equation 8 tells us that the *action spectrum should have the same shape as the absorption spectrum of the absorbing pigment.* Equation 8 is subject to certain obvious restrictions : that R is in fact proportional to P, that other pigments do not shade or mask A, that energy transfer does not occur, etc. Stated differently, when the action spectrum does *not* have the same shape as the absorption spectrum of a pigment, determined on other grounds to be *the absorbing pigment*, then one or more of the above restrictive conditions must obtain. For further discussion of some of the pitfalls of action spectra, see Clayton (1965) and Fork and Amesz (1969).

In general, the experimenter must ensure that the Bunsen-Roscoe law is obeyed when measuring action spectra.

Absorption Spectra

The simple requirements that (except in lasers) there must be an absorbing pigment in order for light to produce an effect and that $A \rightarrow A^*$, mean that photochemical processes may result in changes in the pigments of the system; *e.g.*, disappearance of A. This is obviously not a necessity; A^* may be vanishingly small compared with A, and A may be rapidly regenerated. If absorption changes are observed, we may imagine them to be due to either disappearance of A or appearance of either A* or P. The occurrence of absorption changes due to the appearance of products may be further tested by studying the sequence of absorption changes following excitation. The changes observed *in vivo* may correspond to absorption spectra or difference spectra of the participating pigments.

Light Emission

If the reader has gained the impression that measurements of light absorption are powerful and versatile instruments in the physiological armory, he should immediately set down measurements of light emission as an even more sensitive and almost as versatile means of inspecting physiological processes. The reason is purely technical : light absorption requires the comparison of the intensity of one light beam with another and is dependent on the stability of detectors and amplifiers. In contrast, the emission of light may be detected at the highest sensitivities of photocells and photomultipliers where the "control" is simply the dark current of the detector. The sensitivity for measuring fluorescence can be enhanced even further by modulating the exciting radiation and locking the detector to the same frequency.

When an excited molecule emits light, we may measure the rate of decay of light following removal of the exciting radiation. From the lifetime of the excited state, we can determine whether the light is fluorescence or phosphorescence. From measurements of depolarization we can also inter whether exciton migration intervened between excitation and emission.

We may also determine the *emission spectrum*, from which the identity of the excited species may be deduced. And, finally, we may determine the *excitation spectrum* or *action spectrum* of fluorescence, which has the same shape as the absorption spectrum.

A fluorescence emission spectrum resembles a long-wavelength mirror image of the singlet absorption band (Fig. 3.23). This occurs because internuclear distances are greater in the excited state than in the ground state (Fig. 3.18). Consequently an excited molecule is apt to return to a higher vibrational state than the one it occupied in the ground state. The energies of the photons emitted will be less energetic, and hence of longer wavelengths, than those absorbed by the molecule in the ground state.

If the case of exciton migration, the absorbing species will very likely be different from the emitting species. This is typically the case in chlorophyll fluorescence in green plants.

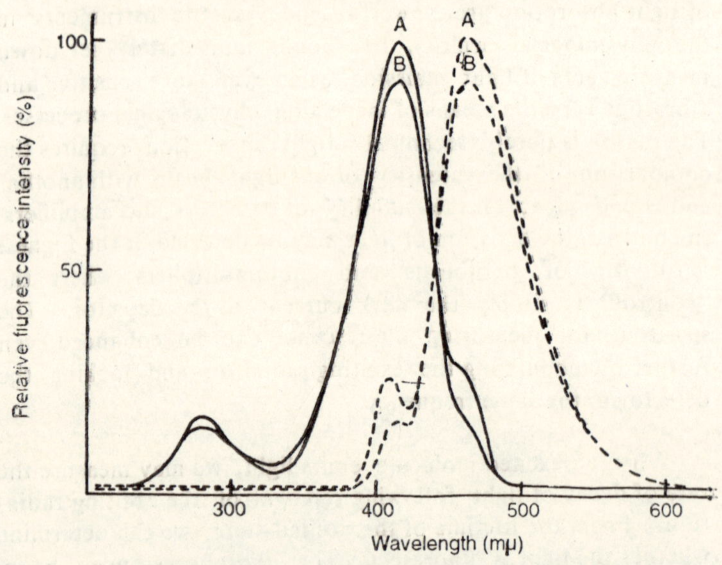

Fig. 3.23. Fluorescence excitation and emission spectra of gibberellins. Because of the Franck-Condon principle—the tendency for the vibrating nuclei of molecules to spend most of their time at the ends of their oscillatory excursions—and because the internuclear distance of excited species is greater than those in the ground state, fluorescence emission spectra tend to form mirror images of absorption (*i.e.*, excitation) spectra and to be shifted to longer wavelengths. Shown above are excitation (solid line) and emission (broken line) spectra of (A) gibberellic acid and (B) barley gibberellin.

A caveat : emitted light energy of fluorescence and phosphorescence is of course not available for photochemistry. The molecule emitting fluorescence or phosphorescence *may* be the same as those involved in a photobiological process, but it may also represent a diversion from the photochemistry of the physiological response. We should not conclude, therefore, that the emitting species are necessarily in the sequence of intermediates leading to the physiological reaction.

Temperature Effects

Since electronic transitions can proceed regardless of the vibrational or rotational state of the absorbing molecules, pure photochemical reactions proceed equally rapidly at all temperatures. Zero temperature dependence becomes then an identifying feature of pure photochemical reactions or the photochemical part of mixed photochemical and dark reactions. Low temperatures can of course modify absorption spectra and therefore action spectra.

Electron Spin Resonance

Metastable excited states result when an excited singlet passes to a state that can be reached from the ground state only by a forbidden transition. The most familiar cause of forbiddenness is the rule that the number of unpaired electrons must be unchanged by a transition. In most molecules (O_2 is an exception), electron spins are scrupulously paired in the ground state. We may write such a spin configuration as $A_{\uparrow\downarrow}$. Allowed transitions preserve spin pairing ; forbidden transitions change it. Thus

$$A_{\uparrow\downarrow} \rightarrow A^*_{\uparrow\downarrow} \quad \text{or} \quad A_{\uparrow\downarrow} \rightarrow A^*_{\downarrow\uparrow}$$

would be allowed, but

$$A_{\uparrow\downarrow} \rightarrow A^*_{\uparrow\uparrow}$$

is forbidden.

Now unpaired spins interact with magnetic fields of

the appropriate strength and may be detected as perturbations of an imposed field as each electron takes a quantum jump to a paired configuration.

This method, known as electron spin resonance (ESR) or electron paramagnetic resonance, can be employed to detect free unpaired electrons. We can infer from an ESR signal that a sample contains free radicals or triplet states and can also infer something about the abundance, the rates of decay, and the environment of the unpaired electrons.

ESR spectrometers display their signals as derivatives with the applied magnetic field strength (H_0) as the abscissa. The *g-value*, the field position at which maximum absorption occurs, is related to the magnetic field strength by the relation

$$g = \frac{h\nu}{\beta H_0}$$

where h is Planck's constant, is the frequency of the microwave radiation (typically 9.5 kMHz), β is the Bohr magneton, and H_0 is the field strength. Most absorption occurs near $g=2$ where H_0 is about 3,400 gauss. Other identifying features of ESR spectra are the width of the signal, ΔH, measured in gauss between points of maximum slope, the amplitude, and the total signal strength, which is proportional to the number of spins.

We know that for many plants the Calvin-Benson hypothesis is approximately correct. In order for carbon dioxide to be reduced to the level of sugar in this scheme, there must be a source of ATP and NADPH. We inferred further that we must look to the light reaction of photosynthesis for a means of generating these two coenzymes. Directly or indirectly there must be reactions

$$ADP + P + h\nu \rightarrow ATP \qquad \text{...(9)}$$

$$NADP^+ + H + h\nu \rightarrow NADPH + H^+ \qquad \text{...(10)}$$

in which photon energy $h\nu$ drives these two reaction. In most green plants the source of hydrogen in Equation 10 is water, and we may rewrite Equation 10 as follows :

$$NADP^+ + H_2O + h\nu \rightarrow NADPH + \tfrac{1}{2}O_2 + H^+ \qquad (\text{...11})$$

In this way we see that oxygen, a familiar product of photosynthesis, arises as a waste product.

From the study of the comparative biochemistry of photosynthesis, van Niel (1949) proposed that bacteria, algae, and higher plants shared a more general process of photosynthesis, which we may write as

$$NADP^+ + H_2A + h\nu \rightarrow NADPH + A + H^+ \quad ...(12)$$

To include the reduction of CO_2 we must write

$$CO_2 + 2H_2A \rightarrow (CH_2O) + H_2O + 2A \quad ...(12a)$$

For bacteria and some algae, the hydrogen conor H_2A need not be water, but can be any of several substances : thiosulphate, H_2S, hydrogen gas, or organic substances such as succinate.

The fundamental questions to be asked are the same as those for any photochemical reaction. What is the absorbing pigment ? What is the fate of the pigment in the excited state ? What are the immediate chemical products of the reaction ? In the present case we must also pursue what may be the essentially biochemical question of the formation of ATP and NADPH from the immediate products of photochemistry.

It has long been recognized that photosynthesis is one of the most remarkable processes in nature. At least 30 percent of the energy of light can be transformed by photosynthesis to chemical energy, an efficiency which is some 10 times greater than any other known photochemical reaction. Because of the enormous theoretical importance of the light reaction for biology, physics, and chemistry, it has attracted some of the most vigorous minds of these three disciplines.

First, a word about experimental systems the two broad categories of plant material that have been commonly used are suspensions of unicellular algae and chloroplasts isolated from higher plants. Students of photosynthesis have employed algae from every algal phylum to advantage, and also photosynthetic bacteria. Unfortunately, the algae have not yielded active chloroplasts so readily ; consequently, *in vitro* studies have rested for the most part on chloroplasts from a few especially cooperative plants, such as spinach, swiss chard, and poke weed (*Phytolacca americana*). The chromatophores of bacteria, such as *Rhodospirillum rubrum*, have also been helpful, although these may be strongly modified during cell disruption.

Because gas exchanges can be measured sensitively and conveniently, the rates of the light reaction are usually measured as CO_2 absorption and oxygen consumption. They are not necessarily equal. In some cases ATP and NADPH formation have been measured directly.

Hypothesis

The simplet photochemical hypothesis of photosynthesis was that chlorophyll is the absorbing pigment and that excited chlorophyll causes the dissociation of water to yield reducing equivalents and oxygen.

Chl*
$h\nu$: Chl → Chl*; Chl* → Chl
HOH
{H} → NADPH
{OH} → O_2

Braces were placed discreetly about the H and OH to indicate our innocence of their state. The scheme was elaborated further by the addition of X and Y to serve as intermediate acceptors for H and OH. For example,

$$HOH + XY \rightarrow XH + YOH$$

The supposition of X and Y has had some heuristic value, but has not of itself advanced our understanding of photosynthesis.

The fundamental notion here is that of charge separation. The first photochemical act can be thought of as the separation of a negative and positive charge, as could occur in a photoionization. If the photosynthetic machinery can be thought of as a crystalline array. we can substitute the idea of electrons and holes of a semiconductor. The proposed action of the excited pigment is analogous to that of a compressed spring (Fig. 3.24). Excitation energy compresses the spring; the spring then pulls the charges apart. The negative and positive charges or electrons and holes are then stabilized as reductant (NADPH) and oxidant (O_2). or alternatively they recombine along an electron transport pathway, which is coupled to the phosphorylation of ADP.

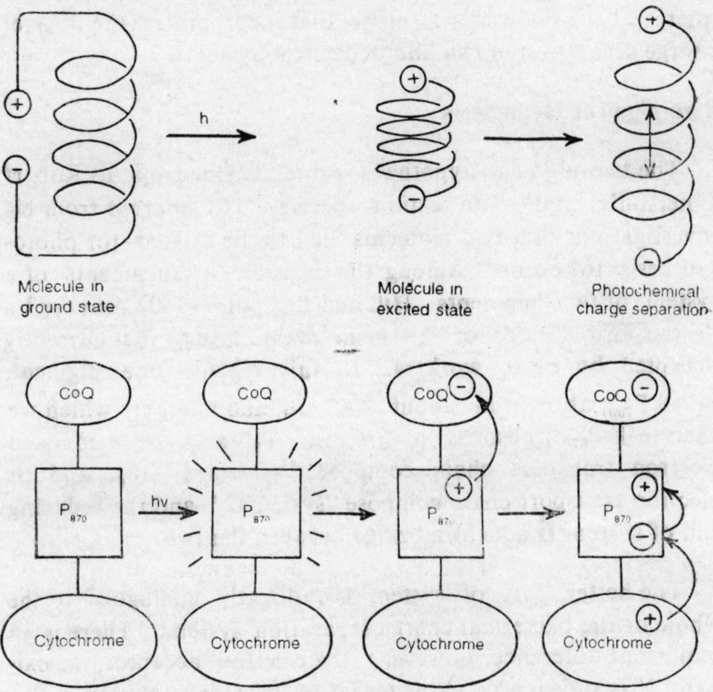

Fig. 3.24. Simple model of photochemical charge separation.

This simple picture serves as a fairly consistent model for bacterial photosynthesis. According to one hypothesis, a long wavelength form of bacteriochlorophyll termed P_{870} (short for "pigment absorbing at 870 mμ") is the photoreactive pigment. It forms a complex with an electron acceptor or oxidant, possibly coenzyme Q, and a cytochrome (Fig. 3.24). Upon excitation, P_{870} photoionizes and transfer an electron to COQ. An electron is then donated to the now-oxidized P^{+}_{870} by the accompanying cytochrome. The separated charges then recombine along the postulated electron transport chain with the coupled generation of ATP.

The idea of photochemical charge separation has been found as fruitful with green plants as with bacteria, but here the hypothesis of a single photoreactive pigment supplying energy to a single electron transport system is clearly inadequate Let us consider a model that incorporates the idea of charge separation in *two* photochemical systems.

Two-pigment Hypothesis

The two-pigment hypothesis is directly traceable to Robert Emerson's studies on action spectra. He inferred from his investigations that two pigments had to be excited for photosynthesis to occur. Among the possible arrangements of a system of two pigments, Hill and Bendall (1960) proposed a "series formulation" or "z-scheme" very close to that currently accepted by most workers. In this scheme one pigment, called P_{700}, absorbs at about 700 mμ, and another, which we shall call P_{670}, absorbs at 670 mμ. P_{700} and an associated electron transport chain compose "system I," P_{670} and its electron transport chain compose "system II," and the reducing end of system II acts as a bridge between the two.

The better part of system I is directly analogous to the whole of the bacterical charge separation system. There is an important difference, however : the electron acceptor, or oxidant, X is sufficiently electronegative to reduce NADP. The movement of electrons and "holes" or positive charges proposed for system I is

$$\text{cyto } f \overset{\oplus}{\leftarrow} \overset{*}{P}_{700} \overset{e^-}{--\rightarrow} X \overset{e^-}{--\rightarrow} \text{ferredoxin} \overset{e^-}{--\rightarrow} \text{NADP}$$

Ferredoxin is an iron-containing protein with an unusual structure.

System II is thought to consist of components with more positive (that is, more oxidizing) redox potentials, whose func-

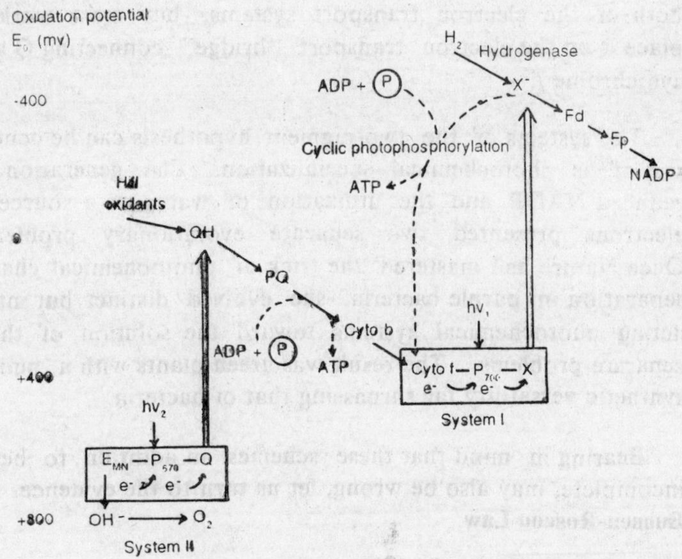

Fig. 3.25. **A version of the two-pigment hypothesis. This scheme is one of the possible representations of photosynthesis in green plants. System I is analogous to the scheme for bacterial photosynthesis (Fig. 3.24). P_{700}, a long wavelength form of chlorophyll, is complexed with cytochrome *f* and electron acceptor X. When P_{700} is excited, the electron path is $P_{700} \rightarrow X \rightarrow$ ferredoxin $\rightarrow$ NADP. P_{700} subsequently receives an electron from cytochrome *f*. In system II excitation leads to charge separation with the electron ultimately given to cytochrome *f* and the positive charge reacting with OH^- to yield O_2.**

tion is to oxidize OH^- to O_2. This strongly oxidizing system contains a manganese component, labeled here as E_{M_n}. The electron acceptor is called "Q" for "quencher." The electron transport system following Q probably contains plastoquinone (PQ), which then reduces cytochrome. PQ is one of a number of CoQ-like quinones found in green plants. Note that the role proposed for quinones is substantially different from that in bacterial photosynthesis. The electron and hole transport system of II is

$$OH^- \leftarrow \overset{\oplus}{E_{M_n}} \leftarrow \overset{\oplus}{P_{670}^{*}} \overset{e^-\,e^-}{\longrightarrow} Q \overset{e^-}{\rightarrow} PQ \overset{e^-}{\rightarrow} \text{cyto } b \rightarrow \text{cyto } f$$

Phosphorylation may in principle be coupled to one or both of the electron transport systems, but most workers place it on the electron transport "bridge" connecting Q and cytochrome *f*.

The systems of the two-pigment hypothesis can be conceived of as photochemical specialization. The generation of reduced NADP and the utilization of water as a source of electrons presented two separate evolutionary problems. Once Nature had mastered the trick of photochemical charge separation in purple bacteria, she evolved distinct but interacting photochemical systems toward the solution of these separate problems. The result was green plants with a photosynthetic versatility far surpassing that of bacteria.

Bearing in mind that these schemes, in addition to being incomplete, may also be wrong, let us turn to the evidence.

Bunsen-Roscoe Law

Obedience to the Bunsen-Roscoe law is an essential characteristic of true (or at least uncomplicated) photochemical reactions. If we are to study the photochemistry of processes in which a photoreaction is accompanied by "dark" chemistry, it is essential to limit ourselves to experimental conditions under which the dark reactions are not limiting to the overall process.

Plants typically show a fairly wide range of light intensities over which the rate of photosynthesis (measured as CO_2 reduction) is linear (Figs. 3.26 and 3.27). Beyond some critical level of light which depends on the concentration of carbon dioxide, temperature, nutrition, etc., photosynthesis is no longer increased. In 1905 Blackman proposed that the fact of interaction between light intensity and CO_2 levels was evidence for dark or enzymatic reactions intervening in photosynthesis.

More conclusive evidence for dark reactions was provided by Otto Warburg, who found that the yield from high intensities of light was considerably enhanced by giving the light in brief flashes interspersed with relatively long dark intervals.

Robert Emerson and William Arnold (1932) pursued this and found that the photochemical yield was constant when dark intervals of 20 msec or longer intervened between light

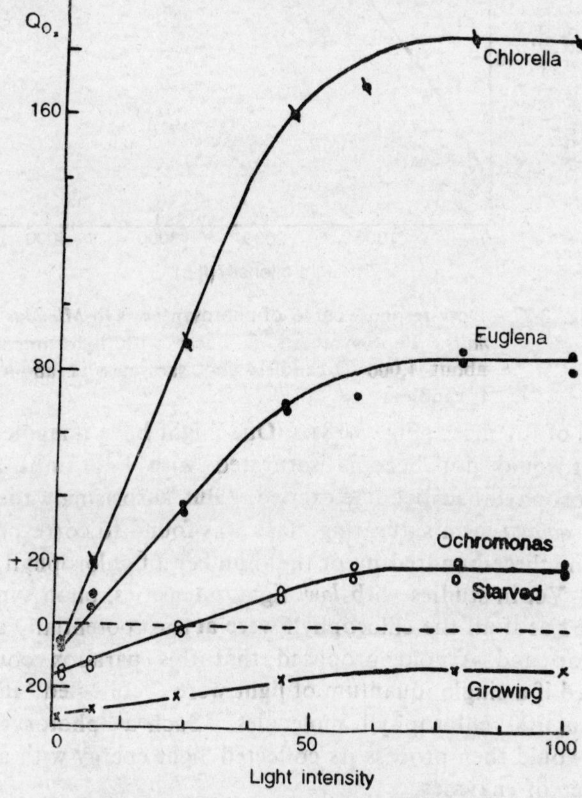

Fig. 3.26. Photosynthesis in several algae as functions of light intensity. Photosynthesis in *Chlorella* and *Euglena* is saturated at less than 100 ft-candles and the maximum rate is many times greater than the rate of respiration (oxygen absorption at zero light intensity). *Ochromonas malbamensis* by comparison photosynthesizes at rates less than or only slightly greater than its respires. *Ochromonas* may employ its photosynthetic apparatus only to generate O_2 in stagnant environments.

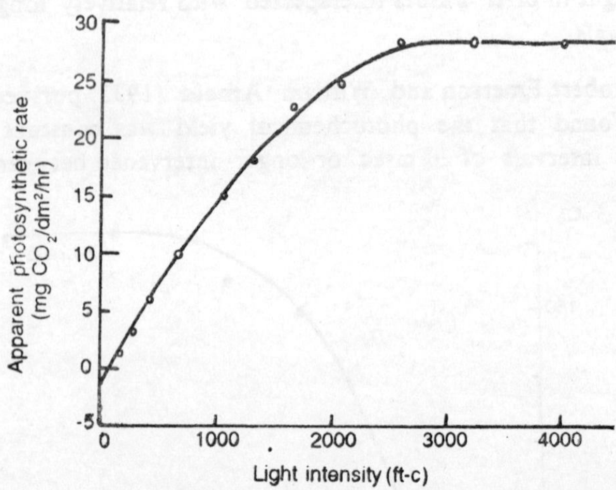

Fig. 3.27 Dose-response curve of photosynthesis in *Mimulus cardinalis*. Photosynthesis is linear with light intensity to about 1,000 ft-candles and saturates at about 2,500 ft-candles.

flashes of 0.1 msec (Fig. 3.28). One might have imagined that a plant would not become saturated with light until all of its chlorophyll had become excited. But surprisingly the *number of quanta* in a saturating flash was found to correspond to only one three-hundredth of the number of chlorophyll molecules. Yet in studies with *low* light intensities, photosynthesis behaved as if all the chlorophyll were at least potentially active. Emerson and Arnold proposed that this paradox could be resolved if a single quantum of light were "collected" in units of about 300 chlorophyll molecules. Such a "photosynthetic unit" would then process its collected light energy with a common set of enzymes.

For our present purposes we may conclude that dark reactions do intervene and that the Bunsen-Roscoe law is obeyed under conditions of brief light flashes followed by relatively long dark periods.

Action Spectra

An example of an action spectrum in a green alga is shown

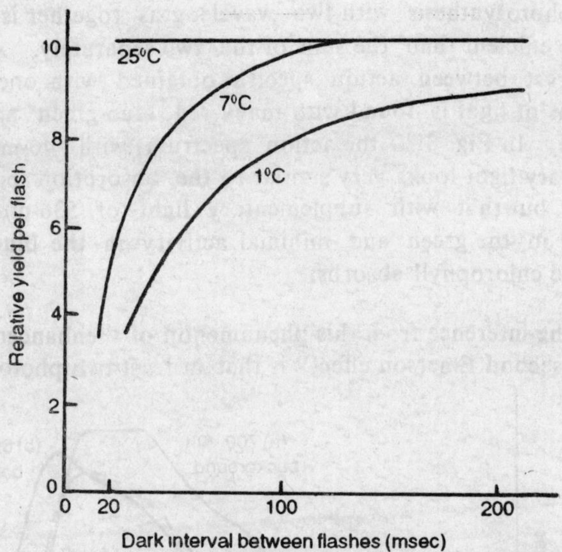

Fig. 3.28. Photochemical yield in flashing light as a function of the dark time. At 25°, the yield is constant even when the dark interval is decreased to 20 msec. At 7°, the time required for completion of the dark reaction is about 100 msec. At 1°, the time for the dark reaction is over 200 msec.

in Fig. 3.29. With a *single* beam of light, the effectiveness of light (dotted line) shows two principal peaks, one near 435 mμ and one near 675 mμ ; the action then falls sharply to zero near 700 mμ. The action spectra obtained with a single beam correspond only in a general way to the absorption spectrum of chlorophyll. Action is higher in green light (500 to 600 mμ) than we should have expected from chlorophyll absorption, and light beyond 700 mμ is much less effective than the corresponding absorption of chlorophyll *in vivo*. The data are clearly at variance with the single chlorophyll pigment hypothesis first considered.

The two-pigment hypothesis has its origin in Emerson's, studies of action spectra obtained with *two* wavelengths (Fig. 3.30). Two features are observed : (*a*) light in the far red which cannot support photosynthesis alone *enhances* the photosynthesis when a shorter wavelength is added and (*b*) photosynthesis with two wavelengths together is typically more efficient than the sum of the two separately. A sharper contrast between action spectra obtained with one and two beams of light is found with many red, blue-green, and brown algae. In Fig. 3.30 the action spectrum with 546-mμ supplementary light looks very similar to the absorption spectrum of chl *a*, but that with supplementary light of 536 mμ shows a peak in the green and minimal activity in the blue and red where chlorophyll absorbs.

The inference from this phenomenon of "enhancement" or the "second Emerson effect" is that at least two photochemical

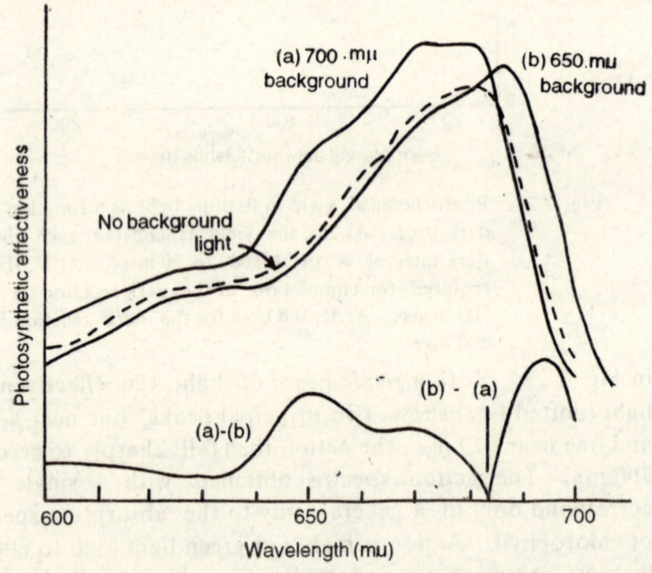

Fig. 3.29. Action spectrum for photosynthesis and enhancement of photosynthesis in the green alga *Chlorella*. Photosynthetic O_2 evolution was measured with and without background illumination of 700 or 650 mμ. Without any background illumination, that is, with a single beam of light, photosynthetic efficiency is very low near 700 mμ. The addition of a background beam at 650 mμ increases the effectiveness of far-red light. This increased efficiency caused by a second beam of light is called **"enhancement."**

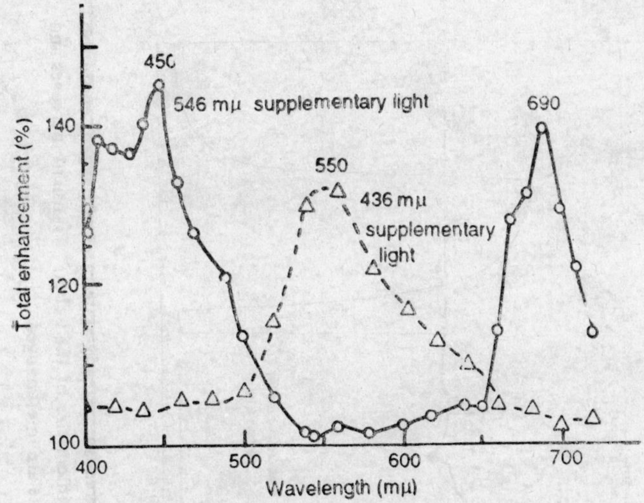

Fig. 3.30. Action spectrum for enhancement in the red alga *Porphyra perforata*. Enhancement was measured as additional oxygen production obtained by shining a beam of light with a background light of 546 or 436 mμ. Enhancement occurs with 546-mμ background in the red and blue with maxima characteristic of chlorophyll *a*. Enhancement spectrum with 436-mμ light shows maximum at 550 mμ, which is near the absorption maximum of the phycobilins.

Note that the two enhancement spectra are more distinct from one another and the percentage increases are larger than with green plants, in which chlorophylls *a* and *b* have similar absorption spectra.

reactions must occur in photosynthesis ; that is, P_{700} *and* P_{670} must be excited in order for photosynthesis to be completed. An alternative view, that the second beam of light is required for a physical process such as excitation to a higher state, is vitiated by the finding of Blinks (1960) that the two beams of light may be separated in time. Let us look at this crucial piece of evidence more closely.

Blinks discovered that, if two beams or light at two wavelengths are given successively rather than simultaneously, transient changes in the rate of O_2 production, called *chromatic transients*, are observed. If darkness intervenes between the

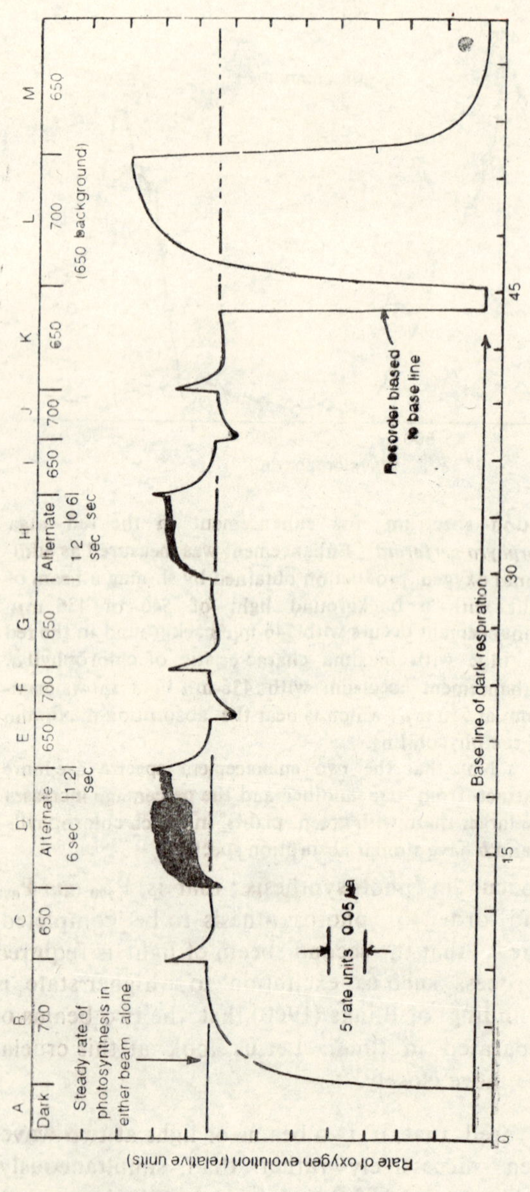

Fig. 3.31. "Chromatic transients" of photosynthesis in *Chlorella pyrenoidosa*. Steady-state rates of oxygen evolution are made equal in different colours of light by adjusting the intensities of the lights. Transient increases and decreases in the rates of photosynthesis occur when the lights are interchanged.

two beams, the chromatic transients still occur, but the magnitudes decrease with the length of the dark interval. The changes observed upon illumination at 680 mμ show a halflife of 5 sec in the dark (Figs. 3.31 to 3.33). The lifetimes of these transients can be interpreted to mean that steady states of the pigment systems are reached within a relatively few seconds and that the intermediates which link the two systems have a lifetime of about the same duration.

The action spectra for enhancement of photosynthesis obtained with two simultaneous light beams agree nicely with the action spectra for exciting chromatic transients (Figs. 3.32 and 3.33). One can say, then, that enhancement occurs when

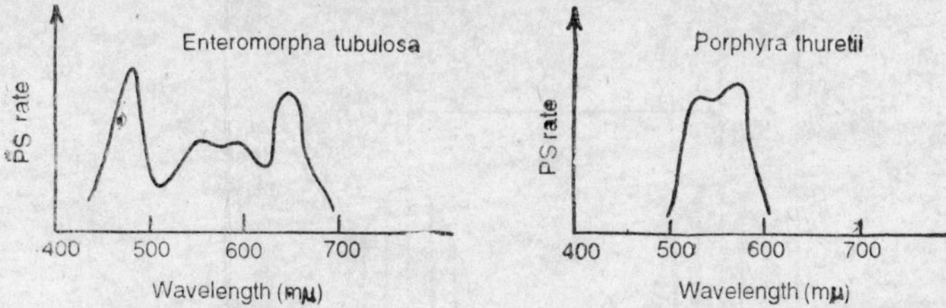

Fig. 3.32. Action spectra for chromatic transients. The action in these experiments was the transient increase in the rates of oxygen evolution that occurs upon shifting from 700-mμ light to light of a lower wavelength. In each case the intensity of the second beam was adjusted so that the *steady state* of photosynthesis was equal to that of the 700-mμ light alone.

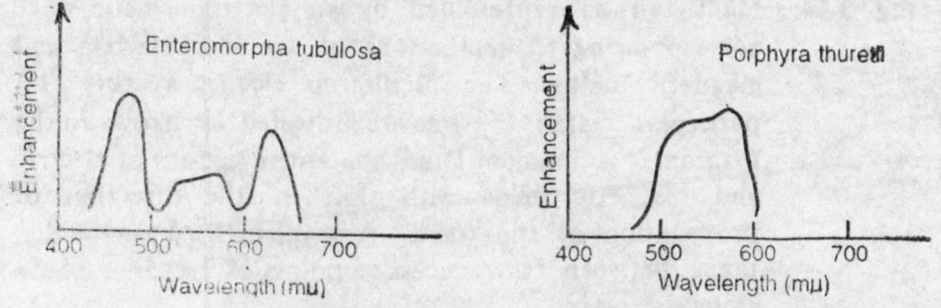

Fig. 3.33. Action spectra for enhancement. Although these action spectra look the same as those in Fig. 3.32, the action in these experiments was the *additional* oxygen evolution that occurs when a second beam of light is imposed simultaneously with a 700 mμ beam.

the two beams of light are given either simultaneously or within a few seconds of each other.

There are two additional bits of intersecting evidence. A number of algae show "photoreduction," in which H_2 serves as $[H_2A]$ in Equation 12, and in which no O_2 is evolved. The action spectrum for photoreduction corresponds to the P_{700} system and does *not* show enhancement. From this it was inferred that P_{670} is concerned principally with driving O_2 evolution. Enhancement is also absent from bacterial photosynthesis, where again O_2 evolution does not occur.

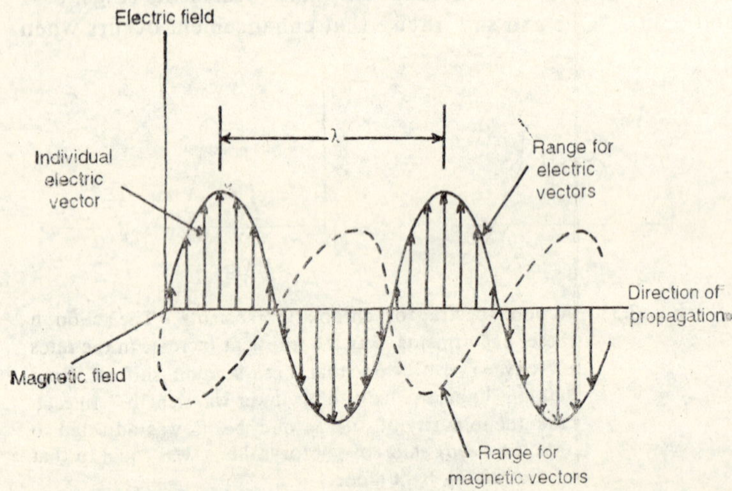

Fig. 3.34. Light can be represented by an electromagnetic wave corresponding to oscillations of the local electric and magnetic fields. The oscillating electric vectors at a particular instant in time are indicated by arrows in the diagram. A moment later, the entire pattern of electric and magnetic fields with shift in the direction of propagation of the wave. A wavelength (λ) is the distance between two successive points of the same phase along the wave.

Absorption of light by molecules

In our discussion of light utilization we will first consider the absorption process itself. Only light that is actually absorbed can be effective in producing a chemical change, a principle embodied in the Grotthus-Draper law of photochemistry. This is true whether radiant energy is to be converted to some other form and then stored or simply used as a trigger. Another important principle of photochemistry is the Stark-Einstein law, which specifies that each absorbed photon activates but one molecule. Einstein further postulated that all the energy of the light quantum is transferred to a single electron during the absorption event, resulting in the movement of this electron to a higher energy state. To understand light absorption, we must therefore first consider some of the properties of electrons.

Role of electrons in absorption event. From a classical viewpoint, an electron is a charged particle that moves in some orbit around an atomic nucleus. We can readily appreciate that the energy of the electron would depend both on the location of the orbit in space and on how fast the electron moves in its orbit. Thus, the increase in energy of an electron upon absorbing a photon could be used either to transfer that electron into an orbit which lies at a higher energy than the original orbit or to cause the electron to move more rapidly about the nucleus than it did before the light arrived. The locations of various possible electronic orbits and the speeds for electrons moving in them are both limited to certain discrete or "allowed" values, a phenomenon which has been interpreted by quantum mechanics. This means that the energy of an electron in an atom or molecule can change only by certain specific values. Consequently, the total electronic energy of an atom or molecule occurs in discrete amounts. Only light of certain wavelengths will have the proper quantum energy to cause the electron to move from one possible energetic state to another.

During light absorption an interaction occurs between the electromagnetic field of the light and some electron. Because electrons are charged particles, they will experience a force when they are in an electric field. But light consists of periodic variations in both the local electric and magnetic fields in the medium through which it is passing (Fig. 3.34). The oscillating electric field or light thus represents a periodic driving force acting on the electrons in matter. This electric field—a vector quantity which has a certain direction in space, such as along the *y*-axis in Fig. 3.34—causes or induces electrons to move. If the frequency of the electromagnetic radiation in such as to cause a large sympathetic oscillation or beating of some electron, that electron is said to be in *resonance* with the light wave. Such a resonating electron corresponds to a so-called *electric dipole* in the molecule, as the electron is forced to move first in a certain direction and then in the opposite one in response to the oscillating electric field of the light. (An electric dipole is simply the local separation of positive and negative charge.) The displacement of the electron back and forth requires energy—in fact, it may take the entire energy of the photon, in which case the quantum is captured or absorbed. The direction and magnitude of the induced electric dipole will depend on the resisting, or restoring forces on the electron provided by the rest of the molecule. These restoring forces depend on the other electrons and the atomic nuclei in the molecule ; consequently, they are not the same in different types of molecules. The actual electric dipoles which can be induced in a particular molecule are characteristic of that molecule, which helps explain why each molecular species has its own unique absorption spectrum.

The probability for the absorption of light depends on both its wavelength and the relative orientation of its electromagnetic field with respect to the possible induced oscillations of the electrons in the molecules. Absorption of photons without ejection of electrons from the absorbing species can take place only if the following two conditions exist : (1) the light has the proper energy to get to a discrete excited state of the molecule, *i.e.*, a specific wavelength, and (2) the electric field vector associated with the photon (Fig. 3.34) has a component parallel to the direction of some potential electric dipole in the molecule so that an electron can be induced to oscillate. In other words, the electric field of light must be in the correct direction so that it is able to exert a force on some electron. It must exert this force in the direction of a potential electric dipole so that the electron can be induced to move and thus be able to accept the energy of the photon. The probability for absorption is actually proportional to the square of the cosine of the angle between the electric field vector of light and the direction of the induced electric dipole in the molecule. Within these limits set by the wavelength and the orientation, light energy can be captured by the molecule, placing it in an excited state.

Electron spin and state multiplicity. Light absorption is affected by the arrangement of electrons in an atom or molecule, which depends among other things on a property of the individual electrons known as their "spin." We can consider that each electron is a charged particle spinning about an axis in much the same way as the earth spins about its axis. Such rotation has an angular momentum or *spin* associated with it. The magnitude of the spin of all electrons is the same; but since spin is a vector quantity, it can have different directions in space. For an electron, only two orientations are actually found, *viz.*, the spin of the electron can be aligned either parallel or antiparallel to the local magnetic field. Even in the absence of an externally applied magnetic field, such as that of the earth or some electromagnet, there is a local internal magnetic field provided by both the moving charges in the nucleus and the motion of the electrons. There is therefore always a magnetic field with which the electron spin can be aligned.

It has proved convenient to express the spin of electrons in units of $h/2\pi$, where h is Planck's constant. In these units, the projection along the magnetic field of the spin for a single electron is either $+\frac{1}{2}$ (*e.g.*, when it is parallel to the local magnetic field), or $-\frac{1}{2}$ (when the spin is in the opposite direction). The total spin of an atom or molecule is the vector sum of the spins of all the electrons, each individual electron having a spin of either $+\frac{1}{2}$ or $-\frac{1}{2}$. The magnitude of this total spin is given the symbol S, an extremely important quantity in spectroscopy.

In order to talk about the spectroscopic properties of various molecules, we will find it convenient to introduce a new term known as the *spin multiplicity*. The spin multiplicity of an electronic state is defined as 2S+1, where S is the magnitude of the net spin for the whole atom or molecule. For example, if S equals zero—which means that the spin projections of all the electrons taken along the magnetic field cancel each other—then 2S+1 is one, and the state is called a *singlet*. On the other hand, if S equals one, the state is a *triplet* (2S+1 then equals 3). Singlets and triplets are the two most important spin multiplicities encountered in biological systems, and we will consider both in further detail below. When referring to an absorbing species, the spin multiplicity is generally indicated by S for singlet and T for triplet, as we will do here. (When S equals $\frac{1}{2}$ as would occur if there were one unpaired electron in a molecule, 2S+1 is 2 ; such "doublets" occur for free radicals.)

Electrons are found of course only in certain "allowed" regions of space ; the particular locus in which some electron can move is referred to as its *orbital*. In the 1920s W. Pauli observed that when an electron is in a given atomic orbital, a second electron having its spin in the same direction is excluded from that orbital. This led to the enunciation of the *Pauli exclusion principle* of quantum mechanics : when two electrons are in the same orbital, their spins must be in opposite directions. When a molecule has all of its electrons paired in orbitals with their spins in opposite directions, the total spin of the molecule is zero (S=0), and the molecule is in a singlet state (2S+1=1), cf. Fig. 3.35*a*. The ground, or unexcited state of most molecules is such a singlet, *i.e.*, all the electrons are in pairs in the lower energy orbitals. When some electron is excited to an unoccupied orbital, two spin possibilities exist. The spins of the two electrons (which are now in different orbitals) may be in opposite directions, as they were when paired in the ground state (Fig. 3.35*b*). This electronic configuration is still a singlet state. On the other hand, the two electrons may have their spins in the same direction, a triplet state (Fig. 3.35*c*). (Since the electrons are in different orbitals, their spins car be in the same direction without violating the Pauli principle). An important rule—first enunciated by Hund, based on empirical observations, and later explained using quantum mechanics—is that the level with the greatest spin multiplicity has the lowest energy. Thus, an excited triplet state is lower in energy than the corresponding excited singlet state, as illustrated in Fig. 3.35*c*; we will consider this phenomenon for chlorophyll.

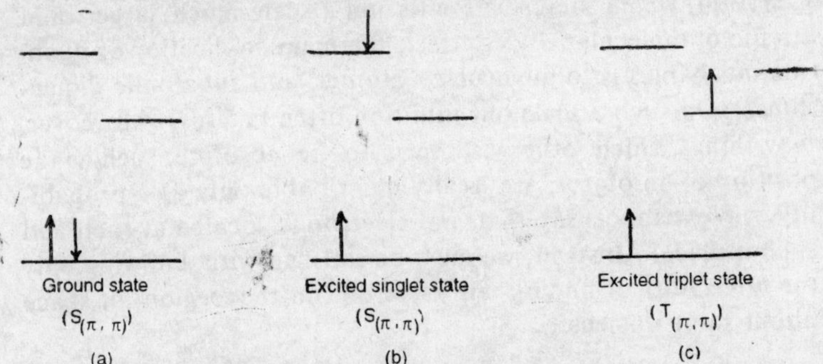

Fig. 3.35. Effect of light on a pair of electrons in a molecular orbital. The arrows indicate the directions of the electron spins with respect to the local magnetic field. (*a*) In the ground state, the two electrons in a filled orbital have their spins in opposite directions. (*b*) The absorption of a photon can cause the molecule to go to an excited *singlet* state where the spins of the electrons are still in opposite directions. (*c*) In an excited *triplet* state, the spins of the two electrons are in the same direction.

Molecular orbitals. For a discussion of the light absorption event involving the interaction of an electromagnetic wave with some electron, it is easiest to visualize the electron as a small point located at some specific position in the atom or molecule. For a classical description of the energy of an electron, we can imagine the electron moving in some fixed trajectory or orbit about the nucleus. However, to describe the role of electrons in binding together atoms to form a molecule, it is most convenient to imagine the electron spatially distributed like a cloud of negative charge surrounding the nuclei of adjacent atoms in the molecule. In this last description, involving probability considerations introduced by quantum mechanics, we say that the electrons are located in *molecular orbitals*.

In describing the electronic states of molecules we need to use the language of molecular orbitals in order to discuss adequately the absorption of light by chlorophylls, carotenoids, and other pigments in plants. We must therefore introduce many new terms and unfortunately, many new concepts which can truly be mastered only by a study of quantum mechanics. Our common experience yields relations such as Newton's laws of motion, which describe events on a scale much larger than atomic or molecular. By contrast, the main application of quantum mechanics is to molecular, atomic, and subatomic dimensions. On such a scale our intuition often fails us. Moreover, may things which otherwise seem to be absolute, such as the position of an object, are really describable only as a probability. We cannot say that an electron is located at such and such a place. Instead, we must be satisfied with knowing only the *probability* of finding an electron in the region of space about some nucleus.

A logical way to begin our discussion of molecular orbitals is to consider the probabilities of finding electrons in given regions about a single atomic nucleus. The region of space in which an electron moves is of course its orbital. The simplest case is the hydrogen atom which contains a single electron moving about a nucleus. If we were to determine the probability of finding that electron in the various regions of space about its nucleus, we would find that it spends most of its time fairly close (within a few Angstroms) of the nucleus, and that the probability distribution is *spherically symmetric* in space. In other words, the chance of finding the electron is the same in all directions about the nucleus. In atomic theory, this spherically symmetric distribution of the electron about the hydrogen nucleus is called an *s* orbital, and the electron is referred to as an *s* electron (see Pauling, or other elementary chemistry texts).

The next simplet case is the helium atom. Helium has two *s* electrons moving in the *s* orbital about its nucleus. By the Pauli exclusion principle these two electrons must have their spins in opposite directions, as is indeed the case. What happens for lithium, which contains three electrons ? Two of its electrons are in the same type orbital as that of helium. This orbital is known as the K shell, and represents the spherically symmetrical orbital closest to the nucleus. The third electron is excluded from the K shell but occurs in another orbital with spherical symmetry whose probability description indicates that on the average its electrons are further away from the nucleus than is the case for the K shell. Electrons in this new orbital are still *s* electrons, but they are now referred to as being in the L shell. Beryllium, which has an atomic number of four, can have two *s* electrons in its K shell and two more in its L shell. What happens when we have 5 electrons moving about a nucleus, as in an uncharged boron atom? It turns out that the fifth electron is in an orbital that is not spherically symmetric about the nucleus. Instead, the probability distribution is in the shape of a dumbbell, although it is still centered about the nucleus (similar in *appearance* to the distribution depicted in Fig. 3.36*b*). This new orbital is referred to as a *p* orbital and electrons in it are called *p* electrons. The probability of finding a *p* electron in various regions of space is greatest along some axis through the nucleus and vanishingly small at the nucleus itself.

How do all these various atomic orbitals relate to the spatial distribution of electrons in molecules? A molecule contains more than one atom (except for "molecules" like helium or neon), and certain of the electrons can move between the atoms. Such interatomic motion is of course crucial for holding the molecule together. Fortunately, it still proves possible to describe the spatial localization of electrons in molecules by using suitable linear combinations of the spatial distributions represented by various atomic orbitals centered about the nuclei involved. In fact, molecular orbital theory is concerned with giving the correct quantum-, or wave-mechanical description of the probability of finding electrons in various regions of space in *molecules* by using the already known descriptions for the probability of finding electrons in various locations making up their *atomic* orbitals.

Some of the electrons in molecules are localized about a single nucleus, while others are *delocalized*, or shared between nuclei. For the delocalized electrons, the combination of the various atomic orbitals used to describe the spatial positions of the electrons in three dimensions is consistent with a sharing of the electrons between adjacent nuclei. In other words, the molecular orbitals of the delocalized electrons spatially overlap more than one nucleus. This sharing of electrons is responsible for the chemical bonds that prevent the molecule from separating into its constituents atoms, *i.e.*, the negative electrons moving between the positive nuclei hold the molecule together by attracting the nuclei of different atoms. Moreover, these delocalized electrons are the ones usually involved in light absorption by molecules.

The lowest energy molecular orbital is a σ orbital, which is analogous in symmetry properties to the *s* atomic orbital. Nonbonding, or lone-pair electrons contributed by atoms like oxygen or nitrogen occur in *n* orbitals and retain their atomic character in the molecule. These ***n*** electrons are essentially physically separate from the other electrons in the molecule and do not take part in the bonding between nuclei. We shall devote most of our attention to electrons in π molecular orbitals, which are the molecular equivalent of *p* electrons in atoms. Such π electrons are delocalized in a bond joining two or more atoms, and they are of prime importance in light absorption. In fact, most photochemical reactions and spectroscopic properties of biological importance result from the absorption of photons by the π electrons.

The excitation of a π electron by light absorption can lead to an excited state of the molecule in which the electron moves into a π^* orbital, the asterisk referring to an excited, or high-energy molecular orbital. The probability distributions for electrons in both π and π^* orbitals are illustrated in Fig. 3.36 (the circumscribed regions indicate where the electrons are most likely to be found). Fig. 3.36, shows a π orbital delocalized between two nuclei ; the same clouds of negative charge electrostatically attract both nuclei. Such sharing of electrons in the π orbital helps join the atoms together, hence we refer to this type of molecular orbital as *bonding*. As shown in Fig. 3.36*b*, electrons in a π^* orbital do not help join atoms together, rather, they tend to decrease bonding between atoms in the molecule, since the clouds of negative charge in adjacent atoms are located in a position to repeal each other. A π^* orbital is therefore referred to as *antibonding*. The decrease in bonding in going from a π to a π^* orbital results in a less stable (higher energy) electronic state for the molecule, so that a π^* orbital is at a higher energy than a π orbital.

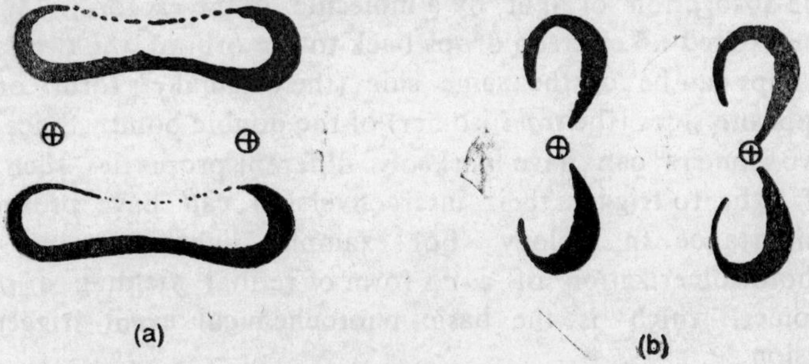

Fig. 3.36. Typical π and π^* orbitals, indicating the spatial distribution about the nuclei ⊕ where the greatest probability of finding the electrons occurs : (*a*) π orbital (bonding) (*b*) π^* orbital (antibonding).

In photochemistry, the energy required to move an electron from the attractive (bonding) π orbital to the antibonding π^* orbital is obtained by the absorption of a photon. For molecules such as chlorophylls and carotenoids, which we will consider later, the π^* orbitals are often only a few electron volts higher in energy than the corresponding π orbitals.

Light energy can cause molecular changes known as *photoisomerizations*. The three main types of photoisomerization are (1) *cis-trans* isomerization about a double bond, (2) a double bond shift, like that implicated in the interconversion between the two forms of phytochrome that we will present below, and (3) a molecular rearrangement involving changes in carbon-carbon bonds, *e.g.*, ring cleavage or formation. *Cis-trans* photoisomerization—which illustrates the consequences of the different spatial distribution of electrons in π and π^* orbitals—makes possible the generally restricted rotation about a double bond.* A double bond has two π electrons ; light absorption leads to the excitation of one of them to a π^* orbital. The attraction between the two carbon atoms caused by the remaining π electron is mostly canceled by the antibonding, or repulsive contribution from this π^* electron (cf. Fig. 3.36). Hence, the molecular orbitals of the original two π electrons which prevented rotation about the double bond have been replaced by an electronic configuration permitting relatively easy rotation about the carbon-carbon axis. Let us consider the absorption of light by a molecule in the *cis* form. When the excited π^* electron drops back to a π orbital, the two large groups can be on the same side (the original *cis* form) or on opposite sides (the *trans* isomer) of the double bond. Since the two isomers can have markedly different properties, such use of light to trigger their interconversion can have profound importance in biology. For example, light can cause the photoisomerization of a *cis* form of retinal yielding a *trans* isomer, which is the basic photochemical event triggering vision.

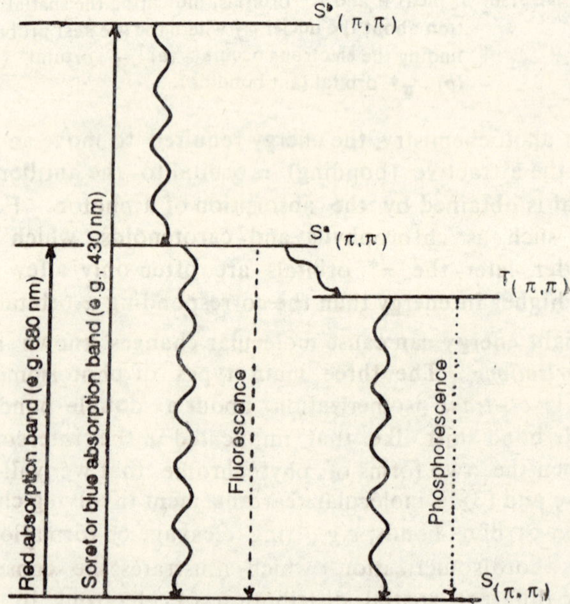

Fig. 3.37. Energy level diagram indicating the principle electronic states and some of the transitions of chlorophyll. Straight vertical lines represent the absorption of light; wavy lines indicate radiationless transitions; broken lines indicate those de-excitations accompanied by radiation.

Light absorption by chlorophyll. To illustrate certain of the terms introduced above to describe the absorption of light by molecules we will use chlorophyll. The principal energy levels and electronic transitions of chlorophyll are presented in Fig. 3.37. Chlorophyll is a singlet in the ground state, as are practically all pigments of importance in biology. Upon the absorption of a photon, some π electron is excited to a π^* orbital. When this excited state is a singlet, it is represented by $S_{(\pi.\pi^*)}$ (Figs. 3.35 and 3.37. The first symbol in the subscript is the type of electron (here a π electron) which has been excited to the antibonding orbital indicated by the second symbol (here π^*). We will represent the ground state by $S_{(\pi,\pi)}$, indicating that no electrons are in excited, or antibonding orbitals. If the orientation of the spin of the excited π electron has become reversed during excitation, it would be in the same direction as the spin of the electron which remained in the π orbital (cf. Fig. 3.35). (Each filled orbital contains two electrons whose spins are in opposite directions.) In this case, the net spin of the molecule in the excited state would be 1, known as a triplet. The excited triplet state of chlorophyll is represented by $T_{(\pi,\pi_*)}$ in Fig. 3.37.

Chlorophyll has two principal excited singlet states which differ considerably in energy. One of these states, designated $S^a{}_{(\pi,\pi^*)}$ in Fig. 3.37, can be excited by red light, *e.g.*, a wavelength of 680 nm. The other state, $S^b{}_{(\pi,\pi^*)}$, lies higher in energy and is excited by blue light (*e.g.*, 430 nm). The electronic transitions caused by the absorption of photons are indicated by solid vertical arrows in Fig. 3.37, while the vertical distances correspond approximately to the differences in energy involved. Excitation of a singlet ground state to an excited triplet state is usually only about 10^{-5} times as probable as going to an excited singlet state, so that the transition from $S_{(\pi,\pi)}$ to $T_{(\pi,\pi^*)}$ has not been indicated for chlorophyll in Fig. 3.37. In order to go from $S_{(\pi,\pi)}$ to $T_{(\pi,\pi^*)}$, the energy of the electron must be substantially increased and the orientation of the electron spin must be simultaneously reversed. Since the coincidence of these two events is rather improbable, very few chlorophyll molecules are directly excited to $T_{(\pi,\pi^*)}$ from the ground state by the absorption of light.

We will now briefly compare the two most important excitations in photobiology, *viz.*, the transitions of *n* and π electrons to π* orbitals. The *n* electrons have very little spatial overlap with other electrons, and they tend to be higher in energy than the π electrons, which are reduced in energy (stabilized) by being delocalized over a number of nuclei. Therefore, the excitation of an *n* electron to a π* orbital generally takes less energy than the transition of a π electron to the same antibonding orbital. Hence, a $S_{(n,\pi^*)}$ state usually occurs at a lower energy—a longer wavelength is required for its excitation—than does the analogous $S_{(\pi,\pi^*)}$. Another difference is that π and π* orbitals may overlap spatially (cf. Fig. 3.36), while *n* and π* orbitals generally do not. Consequently, the *n* to π* transition is not as favoured or probable as the π to π* one. Thus, transitions to $S_{(\pi,\pi^*)}$ states tend to dominate the absorption properties of a molecular species (as evidenced, for example, by an absorption spectrum) compared with excitations yielding $S_{(n,\pi^*)}$ states.

As our final topic in this section, we will consider the effect of molecular environment on *n*, π, and π* orbitals. The *n* electrons generally interact strongly with water, *e.g.*, by participating in hydrogen bonding, and therefore the enery level of *n* electrons is considerably lower in an aqueous environment than in an organic solvent. The energy of π* electrons is also lowered by water, but to a lesser extent than for *n* electrons ; the π orbitals are the least affected by the solvent. Thus, a transition from an *n* to a π* orbital takes more energy in water than in an organic solvent. On the other hand, the π to π* transition takes less energy in an aqueous environment than in an organic one. As an example, let us consider the transition of chlorophyll to an excited singlet state in different solvents. The excitation to either excited singlet state takes less energy in an aqueous environment than it does in an organic solvent, *i.e.*, when chlorophyll is in water the required photons have a longer wavelength than when the chlorophyll is dissolved is an organic solvent like acetone. This occurrence of a lower energy requirement in water is consistent with our statement that the transition for chlorophyll corresponds to the promotion of a π electron to a π^* orbital.

Light Emission

If the photosynthetic pigments can be persuaded to emit some of their excitation energy, we can hope to identify them both from their fluorescence emission spectra and action (that is, fluorescence excitation) spectra. In fact plants can be seen to fluoresce, especially when photosynthesis is inhibited or prevented, just as if the excitation energy were switched from photochemistry to light emission.

The emission spectrum of green leaves and algae at room temperature normally shows a single maximum at 685 mμ. At liquid nitrogen temperature the emission band splits into two or three bands at 685, 695, and 720 mμ. The location and widths of the absorption bands at 685 and 695 mμ are characteristic of chlorophyll *a* in organic solvents. In fact fluorescence of chl *b*, carotenoids, or phycobilins is rarely observed *in vivo*. There are, lowever, long-wavelength fluorescence bands which might be due to long-wavelength forms of chl *a* in P_{700}.

In sharp contrast to the unanimity of the emission spectra of plants, the excitation spectra are far more diverse. Chl *a* fluorescence can be induced by exciting chl *b*, carotenoids, phycobilins, and chl *a* itself. If these other pigments are excited and chl *a* emits the fluorescence, we can infer only that there must be energy transfer (exciton migration) among them and that chl *a*, or at least some form of it, is the energy sink.

Quantitatively, the excitation spectra show that most fluorescence comes from system II. Indeed, the oxidant of system II was named "quencher" (Q) because in its normal, oxidized state it competes with or quenches fluorescence for the energy of P_{670}. Consistent with this view, one finds that fluorescence is decreased by absorption of light by system I, as this causes QH to revert rapidly to Q, and is promoted by DCMU which blocks the oxidation of QH.

As suggested by Duysens and Amesz (1962) the model can be refined further : the two pigment systems are each arrays of pigments with overlapping dipoles. System I is composed of about 300 (chl *a* 683) : 1 (cyt *f*) : 1 (chl *a* 700) at the energy sink, or collection center. System II is composed of chl *b* or, in the case of blue-greens, red, and brown algae, phycobilins. The energy sink in this case is another form of chlorophyll, chl *a* 673, at the collection center. Hundreds of pigment molecules might thus be combined into two separate kinds of units with overlapping dipoles [as in the photosynthetic units of Emerson and Arnold (1932)], so that a photon absorbed by any one of the molecules in the unit is passed on to a few molecules of chlorophyll *a* at the collection center.

The average distance between chlorophyll molecules in chloroplasts is computed to be 15 to 20 Å. At this distance the model for energy transfer predicts nearly 100 per cent efficiency. Additional light-gathering or "antenna" pigments include β-carotene, which probably transfers only to system I. Xanthophylls appear not to transfer at all.

What if one collection center becomes inoperative, due, for example, to a bleached reaction center ? Can photons be passed from one unit to another of the same kind ? The answer appears to be yes. On the basis of fluorescence measurements, units of system I or system II can pass photons to other units with an efficiency of about 0.5.

A more difficult problem, both conceptually and experimentally, is that called "spill-over," the transfer of energy between different kinds of units. If system II were to receive more light energy than it could efficiently use, could it relay the excess energy to system I ? The seeming necessity for some regulated spill-over comes from the nearly constant quantum yield of photosynthesis over a range of wavelengths where the two systems would be expected to absorb very unevenly. The experimental evidence points both ways.

Experiments with chromatophores of *Chromatium*, one of the photosynthetic bacteria, support this general scheme. These bacteria have at least two kinds of bacteriochlorophyll, B_{890} and B_{850}. When the chromatophores have been extracted with detergents, fluorescence of B_{850} is enhanced whereas that of B_{890} is decreased, as if the detergent had interrupted exciton migration to the long-wavelength energy sink.

Blue fluorescence characteristic of NADPH or NADH emission also occurs. The action spectra for NADPH fluorescence corresponds to that of P_{700}.

Electron Spin Resonance

Electron spin resonance can be used to detect unpaired electrons, such as those that might exist in free radicals or triplet states. Signals from photosynthesizing cells were first detected by Commoner's group.

Two finds of ESR signals (originally denoted I and II) are seen in green plants (Fig. 3.38). Signal I has a *g*-value of 2.0025 and a $\triangle$H of 7 to 9 gauss. A closely similar signal occurs in photosynthetic bacteria. The occurrence of signal I correlates exactly with P_{700} (or the corresponding bacterial reaction centers) with respect to the kineties of their response to light, their response to redox potential and their deletion from *Scenedesmus* mutants and at least approximately with the estimated numbers of reaction centers. Although chlorophyll alone will give a similar signal *in vitro*, there is no correlation *in vivo* with bulk chlorophyll

Signal II just as fortuitously correlates with system II. It has a *g*-value of 2.0046, and its $\triangle$H value is about 20 gauss. It is induced by light, as is signal I, but decays over several hours in the dark. Signal II is absent from bacteria. From the absence of signal II from *Screnedesmus* mutants lacking plastoquinone

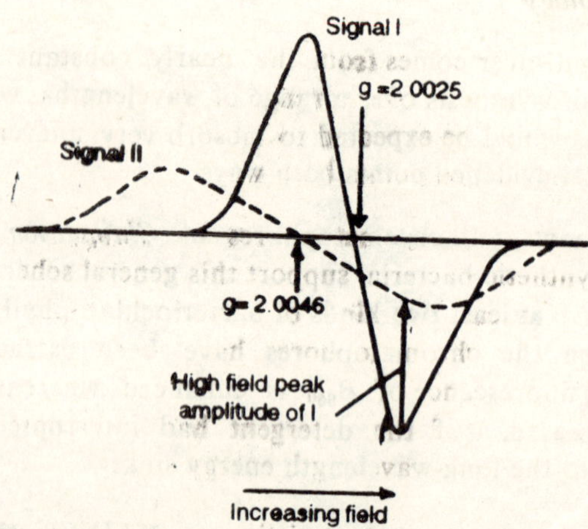

Fig. 3.38. EPR signals from *Chlamydomonas reinhardi*. The two kinds of electron paramagnetic resonance signals detected in photosynthetic organisms are distinguished by their decay curves. Signal I decays within seconds whereas signal II decays slowly, over the course of hours. Signal II is associated with oxygen evolution and signal I is associated with the P_{700} form of chlorophyll *a*.

and the similarity of the ESR signal of plastoquinone *in vitro*, it is not unreasonable to think that a semiquinone of plastoquinone may be the source of signal II.

Conclusions on Photochemistry

The accumulated evidence is consistent with the overall view of the two-pigment hypothesis. The participation of a variety of chloroplast pigments as collectors of photon energy is clearly indicated, but special forms of chl *a*, probably bound to specific proteins, appear to be the photoreactive components. Thus the identities of the absorbing pigments are tentatively and approximately established.

Photoionization of the excited chl *a* centers, followed by stabilization of the charge separation, is established with about the same level of uncertainty. It seems clear from the lack of temperature dependence of the absorption changes corresponding to cytochrome oxidation that electron transport is set in motion as an early event after excitation. However, we must reserve judgment on the precise nature of the photoreaction.

The behaviour of system I, including P_{700}, is reasonably clear for the simple reason that it can be set in motion by light which leaves system II cold, so to speak.

The most promising approaches to the isolation of system II are through excitation of the phycobilins of red algae where the overlap between systems I and II is minimal, and through the use of algal mutants, in which the interaction has been rendered inoperative.

Until complete stories are filed on several organisms, we must view our formulations with cheerful skepticism.

Biochemical Evidence

We should now assemble the biochemical evidence for the pathways of ATP and NADPH formation. Such evidence has been obtained largely with isolated chloroplasts and fragments of chloroplast lamellae, loosely called grana, from higher plants. Additional evidence has come from the method of mutations. Robert Hill (1939) first demonstrated that oxygen could be liberated from water by illuminated chloroplast provided with a variety of hydrogen acceptors.

$$B + H_2 \xrightarrow[\text{chloroplasts}]{hv} BH_2 + \tfrac{1}{2}O_2$$

where B could be ferricyanide, benzoquinone, or any of a variety of dyes. Judging by the action spectra of the Hill reaction, ferricyanide must normally accept electrons from system II. It thus provides a lever for separating the O_2-liberating end of photosynthesis from the photochemistry of system I. But the Hill reaction is at once more complicated and more general than system II.

It now seems that Hill oxidants, depending on their redox potentials, may act as electron acceptors either to X^- or to the P_{670} electron acceptor or to other intermediates in electron transport. Hill's general strategy of feeding oxidants to photoreducing systems helped others to identify the electron trans-

port path from X to NADP through the isolation and characterization of ferredoxin (Fd) and its flavoprotein reductase (fp). Ferredoxin has a most unusual struture for a protein (Fig. 3.39).

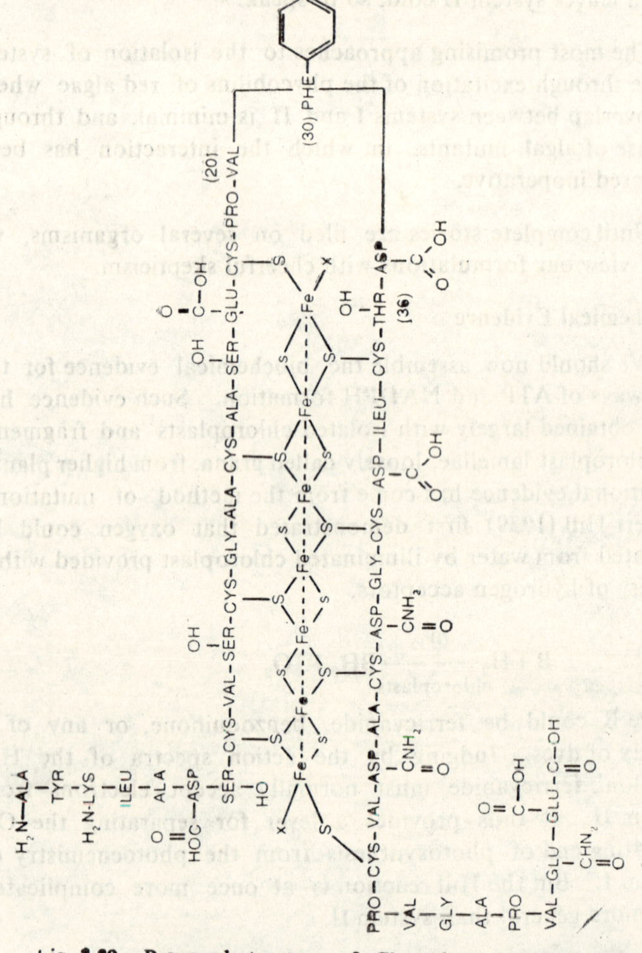

Fig. 3.39. Proposed structure of *Clostridium ferredoxin*, Tanaka *et. al*. (1965) proposed a structure for the ferredoxin of *Clostridium pasteurianum*, a carrier which is involved in nitrogen fixation and is similar to ferredoxin from higher plants. Plant ferredoxin differs in having much more tyrosine, less iron, and less sulphydryl.

These steps have been worked out by San Pietro, Jagendorf, Arnon, and others.

Formation of the other required photoreaction product, ATP can also be demonstrated in isolated chloroplasts and bacterial chromatophores. The process, called *photophosphorylation*, was first described by Arnon *et. al.* (1954) with chloroplasts and by Frenkel (1954) with fragments ("chromatophores") of photosynthetic bacteria. Arnon identified both "cyclic" and "noncyclic" photophosphorylation. In the cyclic process, phosphorylation is somewhat coupled to an electron transport chain that starts with one of the photoacts (probably P_{700}); the electrons proceed around a closed path, returning to chlorophyll. In the noncyclic process, the electron chain terminates in a pool of oxidant, *e.g.*, NADP.

Which of these two processes fuels the Calvin-Benson cycle of CO_2 fixation ? Tanner *et. al.* (1969) have approached this question by showing that noncyclic (but not cyclic) photophosphorylation proceeds *pari passu* with CO_2 fixation with respect to light intensity and the response to certain kinds of inhibitors. While cyclic phosphorylation surely exists *in vivo* and indeed seems to be the sole contributor of ATP to several transport processes, its magnitude appears to be too small to contribute significantly to normal rates of photosynthesis.

The difficulty, as in the case of absorption changes, is that different workers obtain substantially different results depending on the conditions and the choice of experimental material. Action spectra for photophosphorylation obtained in different laboratories show peaks at the following wavelengths : 650, 670, 677, 680, and 710 mμ. In other words, the action spectra correspond to system I, system II, and both !

The simplest conclusion for the present is that photophosphorylation may be coupled to any of the electron transport chains in the chloroplast.

4

Solutes in Plants

From the time that a young plant starts to grow until its death, a more or less continuous movement of solutes is in progress through the conducting elements of every organ of the plant. In a very young seedling, foods are usually translocated upward in the growing stems and downward in the developing roots from the storage tissues of the seed. As soon as the rate of photosynthesis in the developing seedling becomes sufficiently high, at least part of the photosynthate moves in a downward direction from the leaves toward the roots while some of it is often moving upward toward apical meristems. Furthermore, as soon as the developing roots make effective contact with the substrate, absorption of mineral salts begins, followed by translocation of a large part of them in a generally upward direction through the plant. At least some of the solutes absorbed from the soil, especially those containing nitrogen, phosphorus, or sulphur, often react within the root cells with organic compounds which have descended into the roots from the leaves. The resulting chemically more complex compounds, such as amino acids and acid amides, may then be translocated in the reverse direction from the roots into the aerial parts of the plant.

Mineral salts absorbed by the roots are mostly translocated to young leaves and other growing organs of the plant. All of them do not remain, however, in the organ into which they are first translocated. A considerable proportion of the mineral salts which move into a leaf or a flower petal, for example, may sooner or later move out of such a lateral organ back into the stem, and become redistributed to other, usually younger parts of the plant.

The translocation patterns within a plant often exert significant effects upon its behaviour. At the time that young fruits and seeds are developing, for example, there appears to be a general migration of organic materials from all parts of the plant toward the enlarging fruits and seeds. This movement may so nearly monopolize the food resources of the plant as to check severely the growth of vegetative organs. In cotton plants, for example, there is a conspicuous decrease in vegetative growth when the plant is fruiting heavily. A large part of the carbohydrate synthesized in photosynthesis and of the nitrogenous compounds in the above ground organs of the plant moves into the enlarging fruits. The supply of carbohydrates reaching the root system becomes insufficient to maintain rapid respiration and root elongation. As a result here is a decrease in the rate of absorption of mineral salts, which in turn results in checking or even stopping the vegetative growth of the plant.

The patterns of translocation in plants are complex and may be different at different stages in the life history of the plant. Many kinds of solutes are being translocated in various directions within the plant. Nevertheless certain predominant translocation routes can be recognized as follows : (1) downward translocation of organic solutes from leaves to other parts of the plant, (2) upward translocation of organic solutes to growing or storage regions, (3) upward translocation of mineral salts from roots to aerial organs, (4) outward translocation of mineral salts from leaves and other lateral organs into stems, (5) lateral or cross transfer of solutes within stems.

Downward Translocation of Organic Solutes. Downward translocation of organic compounds unquestionably occurs for the most part through the phloem tissues. Much of the evidence indicating that organic solutes move toward the basal portions of the plant in the phloem has been obtained by ringing experiments. "Ringing," when the term is employed without qualification, refers to the removal of a narrow continuous band of tissues external to the xylem. Since ringing entirely encircles the stem, all tissues external to the xylem are completely intercepted. This operation is also called "girdling."

In a girdled tree carbohydrates and other organic compounds slowly accumulate in the tissues above the ring and slowly decrease in quantity in the tissues below the ring as they are utilized in respiration and assimilation. Unless special conditions intervene, such, for example, as development of sprouts on the tree trunk below the ring, a girdled tree ultimately dies because of starvation of the roots, showing clearly that no appreciable amounts of foods are conducted downward through the xylem. Accumulation of oranic compounds above a ring has also been demonstrated in certain herbaceous plants such as cotton. Experimental results of this kind indicate that the downward translocation of carbohydrates and other organic compounds occurs through the phloem tissues.

Chemical analyses show that cells of the phloem are relatively rich in carbohydrates and organic nitrogenous compounds. This finding is consistent with the concept that translocation of organic compounds occurs through the phloem but is not proof, since storage tissues also contain relatively high concentrations of foods.

We know that the upward movement of water takes place in the xylem. For this reason it is difficult to conceive of the xylem as an important avenue of downward translocation of organic substances. Dyes injected into stems at times of rapid water movement usually move both upward and downward—sometimes to approximately equal distances. Such results have sometimes been cited as evidence of downward currents in the xylem. Downward movements of injected dyes probably result, however, from the effects of tension or from a subatmospheric pressure in vessels occupied only by gases. They cannot be accepted as valid evidence that downward movements of solutes ordinarily occur in the xylem of intact stems.

Further evidence that the phloem is the tissue in which organic compounds are translocated has come from experiments in which short segments of the stem have been killed. Since the conducting elements of the xylem are nonliving, the passage of solutes through the xylem is not prevented by the death of the cells in the stem. The conducting elements of the phloem, on the other hand, are living, hence any phase of translocation which is interrupted be killing the cells in a stem segment can be assumed to be occurring in the phloem. Rabideau and Burr (1945) labeled the carbohydrate formed in bean plants by allowing one leaf on the plant to photosynthesize in an atmosphere containing carbon dioxide made with the identifiable C^{13} isotope. They then showed that movement of carbohydrates occurred readily in the stem, both upward and downward, from the node at which the leaf petiole was attached. If short segments of the stem above and below the node at which the petiole was attached were killed by treatment with hot wax, however, no upward or downward movement of carbohydrates occurred. Phosphates, containing radioactive phosphorus (P^{32}), however, were readily conducted through such killed stem segments, indicating that their translocation occurred through the xylem (*cf.* later discussion of upward translocation of mineral salts.) Bonner (1944) and Went (1944) obtained similar results in investigations of the translocation of organic compounds through tomato stems, short segments of which had been killed with superheated steam.

Upward Translocation of Organic Solutes. Under many conditions an upward translocation of organic solutes takes place in plants. This occurs, for example, in the stems of woody species when the buds resume growth in the spring. The tissues of the new shoots are constructed largely out of foods which move in an upward direction from the storage tissues of the stems, as during the early stages in their expansion the leaves do not photosynthesize at a rate sufficient to supply all the carbohydrates used in the growth of the shoot which bears them. Upward translocation of foods from the older leaves on a given shoot to developing leaves situated closer to its apex also probably occurs. As the leaves mature there is a reversal in the direction of translocation of carbohydrates ; they are then translocated from the leaves into the stems in which they move to other organs of the plant.

A number of other examples of the upward transport of foods in plants can be cited. Opening flowers and developing fruits are often attached to stems in such a position that some or all of the organic compounds translocated into them move through the stems in an upward direction. In the early stages of the development of seedlings, upward translocation occurs from the endosperm or cotyledons toward the apical portions of the plant in which rapid growth is taking place. Likewise upward transport of foods invariably occurs during the earlier stages of shoot growth from bulbs, tubers, rhizomes, and other types of underground organs.

The "classical" view that upward translocation in plants takes place in the xylem was long accepted as referring to mineral salts and organic compounds as well as to water. The concept that upward transport of carbohydrates occurs principally through the xylem is based largely on phenomena which have been observed in certain woody species. Large quantities of soluble and insoluble carbohydrates are present in the wood parenchyma, wood rays, and (in the younger stems) pith cells of many varieties of trees and shrubs. At certain seasons soluble carbohydrates are also found in the xylem conduits, as illustrated by data of Anderssen (1929, Fig. 4.1). As found by this investigator the concentrations of both sucrose and free reducing substances (probably largely hexoses) in the xylem sap of pear trees were highest in the winter and early spring. Both fell to a zero value during the summer months, and increased slowly during the autumn. The sugars found in the conducting elements undoubtedly come from the storage tissues of the pith or xylem. The relatively high soluble carbohydrate content of the xylem tissues in the winter and spring probably results largely from shifts in the starch$\rightleftarrows$sugar equilibrium toward the sugar side, as a result of the relatively low temperatures prevailing during these seasons.

The fact that, in the early spring, the soluble carbohydrate content of the xylem sap is relatively high and by the time the shoots are well developed has dropped to a low or zero value seems to indicate that soluble carbohydrates have been con-

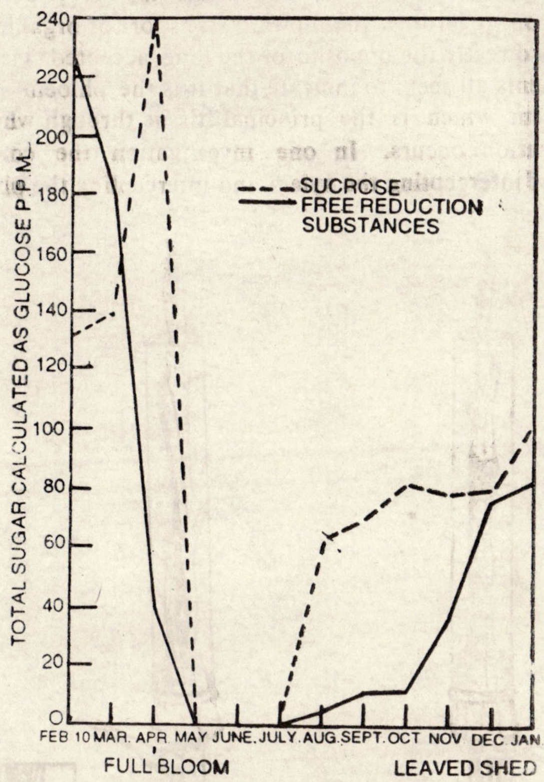

Fig. 4.1. Seasonal variations in the carbohydrates content of the xylem sap of pear trees. Data of Anderssen (1929).

ducted upward through the xylem to the developing buds. Although this assumption appears to be superficially plausible, there are good reasons for doubting if this phenomenon actually is very strong evidence that upward translocation of organic solutes occurs in the xylem. In the first place the concentration of organic solutes in the xylem sap is always very low, seldom exceeding 2 or 3 per cent. Furthermore, the highest concentrations of sugar occur in the sap during the winter when there is little or no upward translocation of water, and it is not at all certain that these solutes do not largely disappear from the xylem sap in the spring before its upward flow begins to take place at an appreciable rate.

Investigations by Curtis, summarized in 1935, point to a conclusion regarding the upward transport of organic solutes which is directly the opposite of the long accepted view. His experiments all seem to indicate that it is the phloem and not the xylem which is the principal tissue through which such translocation occurs. In one investigation the contrasting effects of intercepting the xylem and intercepting the phloem of

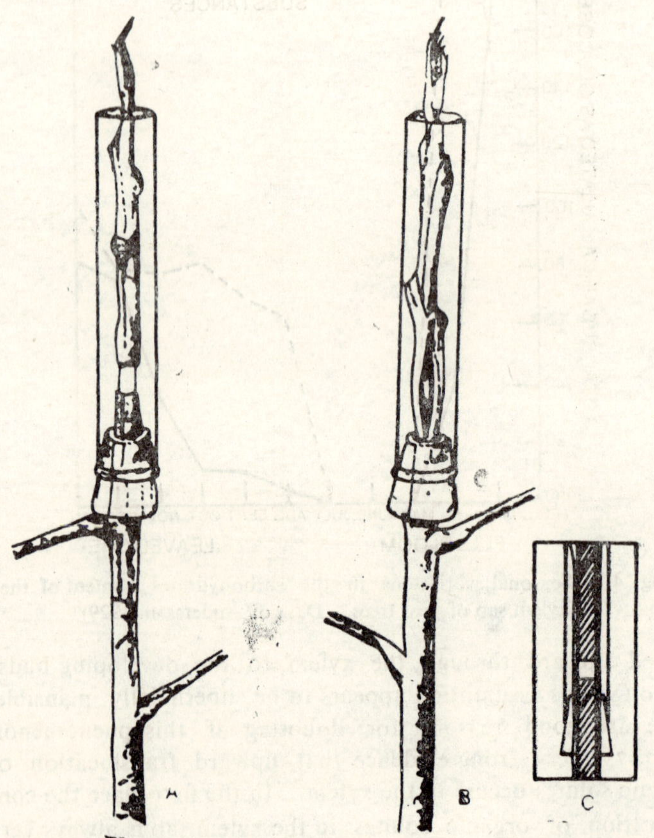

Fig. 4.2. Diagrams to show (A) stem with phloem removed. (B) with xylem removed, the cut portion of each stem being enclosed in a water jacket, (C) sectional view of (B).

woody stems upon upward transport of organic solutes were studied (Curtis, 1925). In these experiments a number of growing shoots were first defoliated. Some received no further treatment, thus serving as checks. In others (Fig. 4.2) a ring of the tissues external to the xylem was removed, and in still others a segment of the xylem was excised, leaving the phloem and cortical tissues intact. Every stem which was cut into was enclosed in a glass cylinder as shown in the figures. This cylinder was filled with water in order to keep the exposed tissue surfaces moist and in order to supply water to the top of the stems in which the xylem was cut. The outcome of experiments in which the water jacket was rinsed out once each day with distilled water was essentially the same as in those experiments in which this was not done, indicating that translocation of solutes did not occur through the water.

The results of some of these experiments are presented in Table 4.1. In variably the stems in which the xylem was cut showed greater elongation than those in which the phloem was cut, indicating greater upward translocation of foods in the former than in the latter.

Shoot elongation is a somewhat indirect measure of translocation, but the conclusions drawn from such observations were supported in a number of experiments by dry weight determinations and analyses for sugar. As shown in Table 4.1, the dry weight and total sugar content per stem were invariably least in the ringed stems. This was also true for the percentage of sugar in terms of fresh weight or dry weight.

In general the evidence seems to warrant the conclusion that the phloem is the principal tissue in which upward translocation of organic solutes occurs, although it seems probable that small quantities of such compounds are, on occasion, translocated in an upward direction through the xylem.

Upward Translocation of Mineral Salts. The term "mineral salts" is necessarily employed in this discussion somewhat loosely. For present purposes, nitrogen, which may be translocated upward in either inorganic or organic combination is

Table 4.1. Comparative effects of cutting the xylem or phloem on growth, dry weight, and sugar content of defoliated shoots.

Species	*Treatment*	*Ave. total growth, mm.*	*Dry weight, per cent of fresh growth*	*Total sugar per stem, mg.*	*Sugar, per cent fresh weight*	*Sugar, per cent dry weight*
Mock Orange (*Philadelphus pubescens*) June 13—June 19	Check	63.6	10.8	3.08	0.12	1.12
	Phloem cut	7.8	9.0	0.08	0.03	0.35
	Xylem cut	49.2	10.8	5.32	0.22	2.03
Mock Orange (*Philadelphus pubescens*) June 25—July 1	Check	105.3	13.0	2.10	0.094	0.72
	Phloem cut	19.7	9.4	1.63	0.087	0.93
	Xylem cut	47.4	11.8	4 83	0.231	2.08
Sumac (*Rhus typhina*) June 26—July 1	Check	63.0	22.3	4.17	0.33	1.48
	Phloem cut	15.8	17.2	3.05	0.67	3.89
	Xylem cut	49.5	20.5	3.90	0.42	2.05

included among mineral salts. Upward translocation of sulphur and phosphorus also probably occurs in both organic and inorganic combination.

For many years it was universally agreed that upward translocation of mineral salts occurred through the xylem although it now appears that the situation is not quite this simple.

Studies of the sap from xylem vessels show that it usually contains at least traces of both organic and inorganic solutes. In proportion to the total quantities utilized, however, the concentration of inorganic constituents in the xylem sap is usually relatively higher than that of organic solutes. Furthermore, appreciable concentrations of mineral salts are commonly present in the sap of vessels at seasons when upward flow of water is occurring at its most rapid rates. At such times the xylem sap contains little or no organic material in solution. Presence of dissolved mineral salts at such seasons is presumptive evidence that at least some of them are translocated upward in the plant through the xylem.

Clements and Engard (1938), Phillis and Mason (1940), and others have shown that ringing stems of various species does not prevent upward movement of mineral salts through the plant. Results of such experiments show that upward translocation of mineral salts can occur in the xylem.

Curtis employed the technique of intercepting the xylem versus intercepting the phloem (Fig. 4.2) in attempting to ascertain whether mineral salts moved into growing, defoliated stems of sumac through the xylem or through the phloem. Relative to the quantity which moved through check stems, a much larger proportion of mineral salts or nitrogenous compounds was translocated through the stems in which the xylem was intercepted than through those in which the phloem was intercepted. Movement of mineral salts into such woody shoots can therefore occur in the phloem. Somewhat similar experiments performed by Mason and Phillis (1940) on cotton plants also led to the conclusion that upward translocation of nitrogenous compounds can occur in the phloem.

Experiments on ringing often show that mineral salts can be conducted in the xylem, and experiments on intercepting wood often show that they can be conducted in the phloem, but no such experiments show conclusively which is the predominant pathway of such conduction in intact plants.

Stout and Hoagland (1939) were among the first to employ the radioactive tracer technique in experiments designed to ascertain the path of upward movement of mineral salts in plants. Small plants of cotton, geranium, and willow, rooted in sand or solution cultures, were used. Certain branches of each plant were "stripped" by cutting longitudinal slits 9 in. long on opposite sides of the stem, and then carefully pulling the bark away from the wood, but leaving it attached at the ends. A sheet of paraf fined paper was then

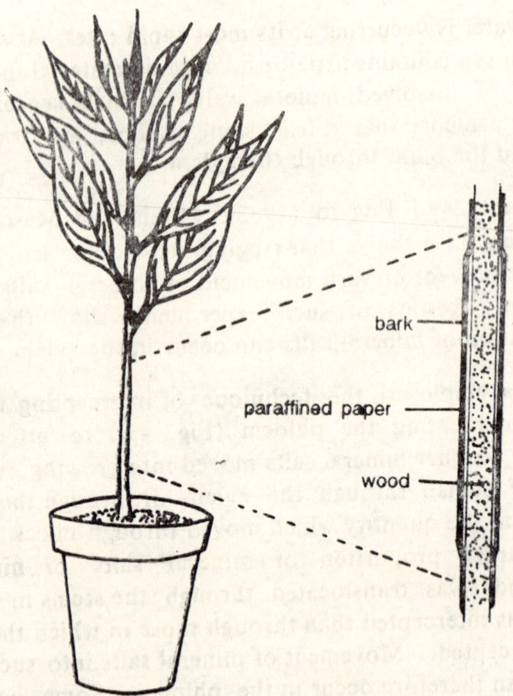

Fig. 4.3. Diagram to illustrate technique employed in the experiment of Stout and Hoagland.

introduced between the phloem and the xylem (Fig. 4.3). This treatment resulted in no visible signs of injury to the plants during the course of an experiment. Radioactive ions of potassium, sodium, phosphate, or bromide were introduced into the rooting medium. After a period of not more than a few hours, under conditions favourable to transpiration, distribution of the tracer ions in the stem was ascertained by measuring the quantity of radioactive elements in ashed segments of the xylem and phloem from the stem above, below, and in the region where the xylem and phloem were separated with paraffined paper.

As shown by the results from a representative one of their experiments (Table 4.2), radioactive potassium (K^{42}) was found to be relatively abundant in both the bark and the wood above

Table 4.2. Distribution of radioactive potassium in the stem of willow after as absorption period of 5 hours.

		Branch stripped $1\frac{1}{2}$ hr. before absorption period		*Intact branch*	
		ppm in bark	*ppm in wood*	*ppm in bark*	*ppm in wood*
Above strip	SA	53	47	64	56
	S6	11.6	119		
	S5	0.9	122		
Stripped section	S4	0.7	142	87	69
(see Fig. 4.3)	S3	<0.3	68		
	S2	<0.3	108		
	S1	20	103		
Below strip	SB	84	58	74	67

and below the section of the willow stem in which the xylem and phloem had been separated by paraffined paper. Within this latter segment of the stem, however, almost all of the tracer element was located in the xylem. In the intact branch there was no such marked difference in the distribution of the radioactive potassium.

Similar results were obtained with the other plants used, and with each of the radioactive elements employed as a tracer.

In these experiments it is obvious that mineral salts absorbed by the roots were transported upward at a relatively rapid rate through the xylem. During their upward passage some of the mineral salts moved laterally from the xylem into the phloem, thus accounting for the relatively large amounts of tracer element in the phloem. When, however, the lateral movement of salts from xylem to phloem was intercepted by interposing an impermeable barrier between these two tissues, practically all of the tracer element was found in the xylem.

These results demonstrate that upward movement of mineral salts occurs in the xylem, but the possibility of some such movement in the phloem also is not excluded even though it is clear that lateral movement from xylem to phloem takes place readily. The data in Table 4.2 furnish some evidence of upward and downward translocation in the phloem. Bark sections S1 and S6, adjacent to the unstripped portions of the stem, are both higher in radioactive potassium that sections S2 to S5. This can be accounted for only by assuming that a slow movement of radioactive potassium has occurred into these sections from adjacent unstripped portions of the bark. Somewhat similar experiments of Gustafson (1939) also indicate that a limited amount of upward translocation of radioactive phosphate may occur through the phloem.

There is no doubt that upward translocation of mineral salts occurs in the xylem, and it seems virtually certain that this is the main pathway along which their general upward movement from roots to leaves occurs. Conditions which are especially favourable to mineral salt translocation in the xylem include high rates of transpiration, high concentrations of mineral salts in the substrate, and the prevalence in the root cells of metabolic conditions which favour rapid movement of mineral salts from the absorbing cells to the xylem.

Some upward translocation of minineral salts also occurs in the phloem under certain circumstances. Such movement appears to occur at rates that are usually slower than in the xylem. As described in the next section, upward movement of mineral salts in the phloem of younger stems appears to be a usual occurrence when they are "exported" from leaves into the stem. There are some indications that mineral elements are more likely to move in the phloem when in organic than when in inorganic combination. Entrance of such elements as nitrogen, phosphorus, and sulphur into ornanic combinations in the root cells may favour their upward translocation in the phloem.

Outward movement of solutes other than carbohydrates from leaves and other lateral organs. As the previous discussion has indicated, movement of mineral salts to the leaves appears to occur principally, if not wholly, in the xylem. Not all of the mineral salts which enter a given leaf remain in that leaf.

however, some of them being "exported" back into the stem, whence they are translocated to other parts of the plant. This is shown by the results of chemical analyses of leaves at different times of the day and also by results of periodic chemical analyses of leaf and stem tissues during the period just prior to leaf abscission. Nitrogen, phosphorus, potassium, sulphur, magnesium, and chlorine may all be exported from leaves in one form or another, but calcium, boron, iron, and manganese appear to be virtually immobile. Nitrogen, phosphorus, and sulphur probably move out of leaves at least partly in organic combination, while potassium, magnesium, and chlorine probably move out mostly in the form of inorganic ions. Translocation of mineral elements out of flower petals just prior to their abscission also occurs. In general, those mineral elements which are readily redistributed in plants are the ones which are the most readily, exported from leaves.

The earlier evidence of the export of substances other than carbohydrates from leaves and other lateral organs has been augmented and strengthened by more recent studies in which radioactive elements have been used. To study this phenomenon by means of tracer elements it is necessary to apply the tagged ions directly to the leaf tissues rather than to let them enter the plant in the usual way through the root system. The manner in which the radioactive elements are introduced into the leaf tissues may determine the channel through which they pass in moving from the leaf to the stem. If the solution containing the tagged compound is brought into contact with the open ends of several xylem vessels it may be drawn rapidly into the stem by way of the xylem. This is particularly likely to happen if an internal water deficit exists in the leaf at the time the solution is applied. On the other hand, when the marked ions are allowed to reach the vascular system by cell-to-cell transfer in the mesophyll, the result obtained is very different.

Colwell (1942) found that radioactive phosphorus (P^{32}), introduced as phosphate into intact leaves of squash plants under suitable conditions, moved out of such leaves into the stem readily but did not move out of similar leaves if the petioles had first been scalded. This is evidence that outward movement of the phosphorus-containing compounds occurred in the phloem.

Biddulph and Markle (1944) introduced radioactive phosphorus as phosphate into the leaves of cotton plants in such a way as to insure that the tagged ions would move into the phloem tissue of the leaf. The xylem and phloem of the stem were separated by a membrane of waxed paper in a segment extending for 4 in. immediately below the point of leaf attachment. The phosphate ions were allowed to migrate through the tissues of the leaf and stem for from 1 to 3 hr. before their distribution in the stem was determined. Examination of the treated portion of the stem revealed that almost all of the radioactive phosphate was located in the bark. The phosphate had moved both up and down after reaching the stem, but only traces of the tagged element were found in the xylem where it was separated from the phloem by the waxed paper. It is evident from these results that the phosphate was translocated out of the leaf and longitudinally within the stem almost entirely in the tissues of the phloem.

When the radioactive phosphate was applied to leaves of plants with intact stems, the marked ions were found in both the xylem and the phloem of the stem and both above and below the point of leaf attachment. Again, much more of the radioactive element was present in the bark than in the wood which suggests that the tracer ions reached the latter by lateral diffusion from the phloem.

Studies of the outward movement of radioactive phosphate from the leaves of bean plants indicated that more such movement occurred during the daytime than at night. The direction of the movement of phosphate was principally downward, under the conditions of these experiments, but some lateral diffusion of the radioactive ions into the xylem took place with a consequent upward movement in the transpiration stream. Other evidence shows quite clearly that a diurnal alternation between a movement of mineral salts into leaves and a movement out of the leaves is of frequent occurrence. Phillis and

Mason (1942) have shown by chemical analyses that the magnesium, potassium, chlorine, phosphorus, and nitrogen content of cotton leaves regularly increases during the daylight hours, and decreases at night.

The general picture of the movements of mineral salts in plants seems to be about as follows. After entering the roots from the soil most of the mineral salts move across the cells of the young roots to the xylem ducts in which they are carried upward into the leaves. Some of the molecules of which these elements are a part or of which they become a part as a result of chemical transformations in the leaf do not become immobilized in the leaves but move out of them *via* the phloem. Once in the stem such solutes move both upward and downward in the phloem, entering other leaves, and probably also reaching the apical growing regions of roots and stems. Some of the mineral compounds in the phloem also migrate laterally and enter the transpiration stream in which they are transported upward at a usually rapid rate. In a sense, therefore, we may speak of a "circulation" of mineral compounds within the various organs of a plant.

Lateral Translocation of Solutes. Lateral translocation of water in a tangential direction readily occurs in woody stems. This does not appear to be true of many solutes. In straight-grained trees the sugars from the leaves on one side of the tree are translocated to the roots directly below them and, if nitrates are added to the soil on one side of a tree, increase in nitrogen occurs principally in the leaves and branches above the roots on that side. A similar lack of lateral translocation of solutes has been shown in certain species of herbaceous plants. On the other hand, radial transfer of solutes from xylem to phloem or *vice versa* appears to occur readily. Some of the radial movement of solutes probably occurs along the vascular rays.

In perennial woody plants, however, there is evidence that reorientation of conductive tissues may occur in such a way as to offset the effects of a lack of lateral translocation of solutes. If all branches on one side of a tree are removed or destroyed, for example, the effect of diminishing growth of trunk and roots on that side usually does not persist for more than one year. Subsequently developed conductive tissues are usually oriented in such a fashion as to permit translocation to the trunk and roots on the side of the tree which no longer bears branches.

Mechanism of translocation of solutes in the xylem. Dissolved mineral salts and organic solutes, when present, are carried along with the ascending streams of water in the xylem ducts which are pulled up through the plant according to the mechanism. During upward translocation some of the solutes are lost by lateral movement into living cells of the stem adjacent to the xylem conduits. In the leaves the solutes of the xylem vessels migrate into the living cells of the mesophyll. The rates at which solutes are translocated upward through the xylem of the stem will correspond with the rates of translocation of water.

Basic considerations regarding mechanism of translocation of solutes in the phloem. Although several theories have been advanced to account for the mechanism of the movement of solutes in phloem tissues, no one theory has ever received general acceptance. As a background for considering the principal theories that have been proposed, it will be helpful to review briefly some of the pertinent facts regarding the problem of movement of solutes in the phloem.

1. *Living cells are essentail.* Unlike movements of water and solutes in the xylem, translocation in the conducting elements of the phloem ceases when the cells are killed. Evidence for this fact is presented earlier in the discussion.

2. *The movement may be bidirectional.* There is no doubt that movement of solutes in the phloem can sometimes occur in one direction and sometimes in the other. The question of whether or not simultaneous bidirectional movement can occur through the same region of phloem has an important bearing on concepts of the mechanism. Experiments already discussed show that both carbohydrates and phosphates move both up and down in the phloem of a stem from the point at which

they enter it from the petiole of a leaf. This phenomenon does not, however, involve bidirectional movement through the same phloem at the same time.

Indications that two different kinds of solutes—sugar and the dye flurorescein—could move simultaneously through the same phloem region in opposite directions were obtained by Palmquist (1938). Simultaneous movement of phosphate and carbohydrate, labeled with radioactive phosphorus (P^{32}) and radioactive carbon (C^{14}) respectively, in opposite directions through the phloem tissue of geranium stems has been shown to occur, at least over short distances and at relatively slow rates. It was further shown that phosphate could move up in the phloem while carbohydrate was moving down in the same phloem, and *vice versa*.

Chemical potential of ions

Since a substance spontaneously tends to move to regions where its chemical potential is lower this quantity is a useful parameter for analyzing passive movements of solutes. Using a linear combination of the various contributors, the chemical potential of any species *j* is given by Equation as follows : $\mu_j = \mu_j^* + RT \ln a_j + \overline{V}_j P + z_j FE + m_j gh$. Since water is uncharged ($z_w = 0$), the electrical term does not enter into its chemical potential. For ions, however, $z_j FE$ becomes important. In fact, for charged solutes under most conditions of biological interest, differences in the electrical term are usually far larger than are changes in the terms for the hydrostatic pressure or gravity. In particular, for movements across membranes, Δh is zero and so we will omit the gravitational term in this chapter, while changes in the $\overline{V}_j P$ term are also generally ignored because they are relatively small (as we will indicate below). Therefore the chemical potential generally used when dealing with ions, $\mu_j^* + RT \ln a_j + z_j FE$, is commonly referred to as the *electrochemical* potential. Although this emphasizes the role played by electrical potentials, "chemical potential" is only a shorthand expression for the sum of *all* the different contributors affecting a particular species, and it is not really necessary to use a special term like "electrochemical" potential when dealing with ions.

Electrical potential. The difference in electrical potential E between two locations is a measure of the amount of work involved in moving a charge from one location to the other. When the charge is expressed in coulombs and the potential difference in volts, then the amount of work in joules equals the product of the two. The zero level for electrical potential is actually arbitrary. Since both the initial and the final electrical potential must be expressed relative to the same zero level, the arbitrariness is of no consequence when the electrical potential difference (final minus initial potential) is determined.

The charge carried by an ion of species j is a positive or negative integer z_j (the charge *number*) times the charge on a proton. For instance, z_j is $+1$ for the potassium ion and -2 for sulphate (SO_4^{--}). The electrical charge carried by a proton is commonly called the *electronic charge*, since it is the same in magnitude as the charge on an electron (but opposite in sign, of course). One unit for charge is the coulomb ; a proton has a charge of 1.6021×10^{-19} coulomb. Thus, a mole (Avegadro's number) of protons would have a charge equal to $(6.02252 \times 10^{23}$ protons/mole) $(1.6021 \times 10^{-19}$ coulomb/proton), or 96,487 coulombs/mole. Such a unit consisting of Avogadro's number of electronic charges (*i.e.*, one mole of single, positive charges) is called the Faraday, F. This quantity appearing in the electrical term of the chemical potential equals 96,487 coulombs/mole or 96,487 joules/mole-volt—a coulomb-volt is a joule—which can also be expressed as 23,060 cal/mole-volt.

To illustrate the rather large contribution the electrical term can make to differences in the chemical potential of a charged substance compared with the effect of changes in hydrostatic pressure, let us consider a small monovalent cation ($z_j = +1$) that has a partial molal volume ($\bar{V}_j$) equal to 30 cm^3/mole. What electrical potential difference (ΔE) in volts (v) corresponds to the same contribution to the chemical potential of such an ion as a hydrostatic pressure drop (ΔP) of one bar ?

In other words, what is ΔE so that $z_j F \Delta E$ equals $\bar{V}_j \Delta P$ under these conditions ? One bar is 10^6 dynes/cm^2, or 0.1 joule/cm^3. Thus, ΔE need be only (30 cm^3/mole)(0.1 joule/cm^3)/[(1)(96,487 joules/mole-volt)], that is, 3.1×10^{-5} v (0.031 mv). This is an extremely small electrical potential difference ! For comparison, the drop in electrical potential across a biological membrane is often near 100 mv, which is (100)/(0.031) or 3200 times larger than the potential difference calculated here. Stated another way, 100 mv across a membrane would make the same contribution to the chemical potential difference as a drop of 3200 bars of hydrostatic pressure from one side of the barrier to the other for the above case of a monovalent cation having a $\bar{V}_j$ of 30 cm^3/mole. Most hydrostatic pressures encountered in plant or cell physiology are only a few bars. Thus, hydrostatic pressure drops across membranes are usually negligible in comparison with the electrical contributions to the chemical potential differences of ions from one side of the membranes to the other.

Electroneutrality and membrane capacitance. Another consequence of the relatively large magnitude of electrical effects is the general occurrence of *electroneutrality*. Thus, in most aqueous regions that are large compared with atomic dimensions the total electrical charge carried by the cations is essentially equal in magnitude to that carried by the anions. What is the situation for plant cells ? Do we ever have an excess of negative or positive charges in cells or organelles ? If there were a net charge in some region, such as near a membrane, an electrical potential difference would exist from one part of the region to another. Can we relate the size of the electrical potential drop to the net charge ?

In order to relate charges and ΔE*s*, we need to introduce a new term, *capacitance*. Electrical capacitance is simply a coefficient of proportionality between a net charge and the resulting electrical potential difference. A high capacitance means that the region has the *capacity* to have many uncompensated charges without at the same time developing a large electrical potential drop. Let us now express this symbolically. The magnitude of such an electrical potential difference,

ΔE, is related to the capacitance of the region (C) as follows ;*

$$Q = C\Delta E \qquad ...(1)$$

where Q is the net charge. The unit for capacitance is the *farad* (f), which equals 1 coulomb/volt. The capacitances of most biological membranes are approximately the same per unit area, about 1 microfarad (μf)/cm^2 (see Bull, Cole, Keynes in Agin, or Ruch and Patton).

For convenience in estimating the net charge inside a cell which would lead to a typical membrane potential, we will consider a spherical cell of radius r. Suppose that the uncanceled or net charges are uniformly distributed in space at a concentration, c. (For a conductor—a body in which electrical charges can freely move—such as an aqueous solution, the uncanceled or net charges would not remain uniformly distributed. Rather, they repel each other and would therefore collect at the inner surface of the sphere. The quantity c is therefore the hypothetical concentration of the net charge if it were uniformly distributed throughout the interior of the sphere.) The charge Q within the sphere of radius r is then $(4/3)\pi r^3 cF$. (The concentration) c can be in moles/cm^3, and here refers to the net concentration of uncompensated singly charged particles ; F, which is in coulombs/mole, is necessary to convert from concentration to electrical charge.) The capacitance C of the sphere is $4\pi r^2 C'$, where $4\pi r^2$ is the surface area of a sphere and C' is the capacitance per unit area. After substituting these values of Q and C into Equation 1 and rearranging, we obtain the following expression for the electrical potential difference in the case of a spherical capacitor :

$$\Delta E = \frac{rcF}{3C'} \qquad ...(2)$$

Equation 2 gives the electrical potential difference from the center of the sphere to just outside its surface. For a conductor, the internal uncompensated charges are found near the surface. In that case, ΔE actually occurs across the bounding surface (such as a membrane) surrounding the spherical body under consideration. Equation 2 indicates that the electrical potential difference developed is directly proportional to the average concentration of net charge enclosed, and inversely proportional to the capacitance per unit area of the sphere.

To apply Equation 2 to a specific situation, let us take a spherical cell with a radius of 30 μm and an electrical potential difference across the membrane (inside negative) of —100 mv, a value close to that occurring for many cells. If the membrane capacitance per unit area has a typical value of 1 $\mu f/cm^2$, to what net charge concentration in the cell does this potential drop correspond ? Since a farad is a coulomb/volt, one $\mu f/cm^2$ is 10^{-6} coulomb/v-cm² while F equals 96,487 coulombs/mole (Apps. II and III). Using Equation 2, c equals $3C' \Delta E/rF$, hence the average concentration of net charge would be (3) $(10^{-6}$ coulomb/v-cm²$)(-0.1v)/[(30\times10^{-4}$ cm)(96,487 coulombs/mole)], or -1.0×10^{-9} mole/cm³, which is $-1.0-10^{-6}$ mole/liter (-1 μM). The sign of the charge is negative, indicating more internal anions than cations. The average concentration of the net (uncanceled) charge leading to a considerable electrical potential difference is rather small.

It is instructive to compare the net charge concentration averaged over the volume of the cell (c in Equation 2 with the total concentration of anions and cations in a plant cell. Specifically, since the positive and negative ions in plant cells can each have a total concentration near 0.1 M, the above calculated excess of —1 μM is only about one extra negative charge per (0.1 M)/(10^{-6} M), or 10^5 anions. Expressed another way, the total charge of the cations inside such a cell equals or compensates that of the anions to within about one part in 100,000. When cations are taken up by a cell to any appreciable extent, either anions must accompany them and/or cations must be released from inside the cell, Otherwise, if marked departures from electrical neutrality were allowed to occur in some region, sizable electrical potential differences would build up.

Activity coefficients of ions. We will now turn our attention from the electrical term in the chemical potential to the activity one, specifically to the chemical activity itself. The activity of species j, a_j, is the thermodynamically effective concentration.

For charged particles in an aqueous solution this activity can differ appreciably from the actual concentration, c_j—a point that has not always been adequately recognized in dealing with ions.

A quantitative description of the dependence of the activity coefficients of ions on the concentration of the various species in the solution was developed by Debye and Huckel in the 1920s. In a local region around a particular ion, the electrostatic forces describable by relations constrain the movement of other ions. As the concentration increases, the average distance between the ions becomes less, which facilitates ion-ion interactions. Equation [electrical force$=Q_1Q_2/(4\pi\epsilon_0 Dr^2)$] indicates, for example, that the electrostatic interaction between two charged particles varies inversely as the square of their separation, thus, electrical forces increase greatly as the ions get closer and closer together. When ions of opposite sign attract each other, the various other interactions of both ions are restricted, thus lowering their thermodynamically effective concentration or activity.

A simple and approximate form of the Debye-Huckel equation appropriate for estimating the values of activity coefficients of ions in relatively dilute aqueous solutions at 25° is

$$\log\gamma_{+}=\frac{0.509\, z_{+}z_{-}\sqrt{\frac{1}{2}\sum_{j} c_j z_j^2}}{1+\sqrt{\frac{1}{2}\sum_{j} c_j z_j^2}} \qquad \ldots(3)$$

where z_+ is the charge number of the cation, z_- that of the anion, concentrations are expressed in molarities, and the summations are over all charged species. Thermodynamic measurements cannot give the activity coefficient of a single ion, because we cannot have a solution of one type of ion by itself in which to measure γ_+ or γ_-. In fact, activity coefficients of ions always occur as the products of those of cations and anions, hence $\gamma_\pm$ in Equation 3 represents the mean activity coefficient of some cation-anion pair having charge numbers z_+ and z_-. Since z_- is negative, Equation 3 indicates that log $\gamma_\pm$ is also negative, and $\gamma_\pm$ is less than one. The activities of ions in dilute aqueous solutions are thus less than their concentrations, as we would indeed expect.

To estimate $\gamma_\pm$ under conditions approximating those that might occur in a plant or animal cell, let us consider an aqueous solution containing 0.100 M of both monovalent cations and anions and 0.025 M of both divalent cations and anions. For this solution, $\frac{1}{2}\sum_j c_j z_j^2$ (known as the *ionic strength*) is equal to $\frac{1}{2}[(0.100)(1)^2+(0.100)(-1)^2+(0.025)(2)^2+(0.025)(-2)^2]$, or 0.200 M. From Equation 3, log $\gamma_\pm$ is then $(0.509)(1)(-1)(\sqrt{0.200})/(1+\sqrt{0.200})$, or -0.157 for the monovalent ions. This corresponds to a mean activity coefficient of only 0.70, a value considerably less than one.

The activity coefficient of a particular ionic species depends on all the ions in the solution, as is indicated by the ionic strength terms in Equation 3. Therefore, even when some particular ionic species is itself dilute, its activity coefficient can still be appreciably less than one because of the many electrostatic interactions with other ions. The departure from 1.00 for activity coefficients is even greater for divalent and trivalent ions than for monovalent ions, as the z_+z_- factor in Equation 3 indicates. Thus although activity coefficients of ions are often set equal to unity for convenience, this is obviously not always justified. A practical difficulty arising under most experimental situations is that c_j is much easier to determine than a_j, especially for compartments like the cytoplasm or the interior of a chloroplast. In those circumstances where a_j has been replaced by c_j, caution must be exercised in the interpretations or conclusions. The activity coefficients for nonelectrolytes and water are generally much closer to unity than are those for ions, hence the assumption involved in replacing a_j, by c_j for such neutral species is not as severe as for the charged substances (*e.g.*, $\gamma_{sucrose}$ is about 0.96 for 0.3 M sucrose). An all-inclusive theory for activity coefficients is rather complicated and beyond the scope of this text. Equation 3 is presented only to illustrate in a quantitative manner that activity coefficients of ions can be appreciably less than 1.00 under conditions that may occur in a plant cell.

Nernst potential. Having considered the electrical and activity terms in some detail, we should now turn to the role of these quantities in the chemical potential of ions. Specifically, we will consider the deceptively simple and yet extremely important relationship between the electrical potential difference across a membrane and the accompanying distribution of ions across it at equilibrium.

When ions of some species j are in equilibrium across a membrane, its chemical potential outside (o) is the same as inside (i), *i e.*, $\mu_j^{\circ}=\mu_j^{i}$. A difference in the Hydrostatic pressure term almost always makes a negligible contribution to the chemical potential differences of ions across membranes, therefore $\bar{V}_jP$ can be omitted from μ_j in the present case. With this approximation and the definition of chemical potential the condition for equilibrium of ionic species j across the membrane is given by :

$$\mu_j^{*}+RT\ln a_j^{\circ}+z_jFE^{\circ}=\mu_j^{*}+RT\ln a_j^{i}+z_jFE^{i} \qquad ...(4)$$

The term μ_j^{*} in equation 4 is a constant referring to the same standard state of species j in the aqueous solutions on both sides of the membrane, and can be canceled from the equation.

Upon solving Equation 4 for the electrical potential difference, $E^{i}-E^{\circ}$, across the membrane *at equilibrium*, we obtain the following important relationship :

$$E_{Nj}=E^{i}-E^{\circ}=\frac{RT}{z_jF}\ln\frac{a_j^{\circ}}{a_j^{i}}$$

$$=\frac{59.2}{z_j}\log\frac{a_j^{\circ}}{a_j^{i}}\ \text{mv at } 25^{\circ} \qquad ...(5)$$

The electrical potential difference E_{Nj} in Equation 5 is called the *Nernst potential* of species j, *i.e.*, $E^{i}-E^{\circ}=\Delta E=E_{Nj}$ in this case, where the subscript N stands for Nernst, who first derived this relation about 1900. We derived it by assuming equality of the chemical potentials of some charged species on two sides of a membrane, but the Nernst potential can be considered in a broader context (discussed below). For convenience of calculation, we can replace the natural logarithm (ln) by 2.303 log, where log is the common logarithm to the base 10. The quantity 2.303 RT/F equals 58.2 mv at 20°, 59.2 mv at 25° (cf. Eq. 5), and 60.2 mv at 30°. Using such numerical values, we can express the Nernst potential in the useful form presented in the second line of Equation 5.

Equation 5, the Nernst equation, is an equilibrium statement showing how the internal and external activities of ionic species *j* are related to the electrical potential difference across the membrane Fig. 4.4. At equilibrium, a tenfold difference in activity of a monovalent ion across some membrane is energetically equivalent to a 59 mv difference in electrical potential (at 25°). Hence, a relatively small electrical potential drop can balance a large difference in activity across a membrane. For some calculations, $\gamma_j^o\gamma_j^i$ is set equal to one (a less stringent assumption than setting both γ_j^o and γ_j^i equal to one). Under this condition, a_j^o/a_j^i in Equation 5 becomes the ratio of the concentrations, c_j^o/c_j^i ($a_j = \gamma_j c_j$. Such a substitution may be justified when the ionic strengths on the two sides of the membrane are approximately the same, but it can lead to errors when the outside solution is much more dilute than the internal one, as would occur for *Chara* or *Nitella* in pond water.

Throughout the rest of this book we will represent the *actual* electrical potential drop existing across a membrane, $E^i - E^o$, by E_M, where the subscript M refers to membrane. Hence, for both equilibrium and non-equilibrium situations, we have $E^i - E^o = \Delta E = E_M$. When a particular ionic species *j* is in equilibrium across some membrane, E_M equals E_{Nj}, the Nernst potential for that species. However, regardless of the actual electrical potential difference existing across a membrane (E_M), a Nernst potential for an individual ionic species *j* can always be *calculated* from Equation 5 by using the ratio of the outside to the inside activity (a_j^o/a_j^i) or concentration (c_j^o/c_j^i), *e.g.*, $E_{Nj} = (RT/z_jF) \ln (c_j^o/c_j^i)$. For many plant cells, the Nernst potential across the plasmalemma or the tonoplast calculated for potassium ions is close to the electrical potential differences measured across the membranes. If some ionic species cannot penetrate the membrane or is actively transported across it, E_{Nj} can markedly differ from E_M. In fact, the minimum amount of energy needed to transport ions across a membrane is proportional to the difference between E_{Nj} and E_M. The aqueous phases desiganated as inside and outside can have more than one membrane intervening between them. For instance, the vacuolar sap and an external solution—the regions often considered experimentally—have two membranes separating them, the plasmalemma and the tonoplast. The thermodynamic arguments remain the same in the case of multiple membranes, with E_M and E_{Nj} referring to the electrical potential differences between the two regions actually under consideration, regardless of how many membranes occur between them.

Example of E_{Nk}. Data obtained on potassium for the large internodal cells of the green freshwater alga *Chara australis* are convenient for illustrating the use of the Nernst equation (see Vorobiev). The potassium ion activity in the medium bathing the cells, $a_K{}^o$ (where the element symbol K as a subscript is the conventional shorthand notation for K^+) was 0.096 mM, while $a_K{}^i$ in the vacuole was measured as 48 mM. Thus, $a_K{}^o/a_K{}^i$ was (0.096 mM)/(48 mM), or 1/500. Using Equation 5 (with a factor of 58.2, since measurement was at 20°), the Nernst potential for potassium is (58.2 mv) log (1/500), or −157 mv. The measured electrical potential of the vacuole relative to the external solution. E_M, was −155 mv—very close to the calculated Nernst potential for potassium. Thus, potassium in this case may be in equilibrium between the external solution and the vacuole. The potassium concentration in the vacuole of these *Chara* cells was 60 mM. The activity coefficient fo potassium in the vacuole ($\gamma_K{}^i$) equals $a_K{}^i/c_K{}^i$ ($a_j=\gamma_j c_j$, therefore $\gamma_K{}^i$ was (48 mM)/(60 mM), or 0.80, while the potassium activity coefficient in the bathing solution ($\gamma_K{}^o$) was about 0.96. For this example, where there are large differences in the internal and the external concentrations, the ratio $\gamma_K{}^o/\gamma_K{}^i$ is (0.96)/(0.80), or 1.20, which differs appreciably from 1.00. If instead of activities concentrations had been used in Equation 5

(c_K^o/c_K^i=0.1/60=1/600), E_{N_K} would have been calculated as −162 mv, which is somewhat lower than the measured potential of −155 mv. Calculating from the concentration ratio, the suggestion that potassium is in equilibrium from the bathing solution to the vacuole could not have been made with much certainty, if at all.

Equilibrium does not require that the various forces acting on a substance are zero, rather, that they cancel each other. In the above example for *Chara* the factors that tend to cause potassium to move are the differences both in its activity (or concentration) and the electrical potential across the membranes. The activity of potassium ions was much higher in the vacuole than in the external solution (cf. Fig. 4.4). The activity term in the chemical potential therefore represents a driving force on potassium directed from the inside of the cell to the bathing solution. The electrical potential is lower inside the cell as is indicated in Fig. 4.4, hence the electrical driving force on the positively charged potassium ions tends to cause potassium entry into the cell. At equilibrium these two tendencies for movement in opposite directions are balanced and no net flux of potassium occurs. As indicated above, the electrical potential difference existing across a membrane when K^+ is in equilibrium is the Nernst potential of potassium, E_{N_K}. Generally, c_K^i and a_K^i for both plant and animal cells are much higher than c_K^o and a_K^o, and potassium is often close to equilibrium across the cellular membranes. From these observations, we can expect that the interiors of cells are usually at negative electrical potentials compared with the outside solutions, as is indeed the case. In general, the chemical potentials of ions are not equal in all regions of interest, hence passive movements toward lower μ_j occur—a topic we will turn to next.

Fluxes and diffusion potentials

Fluxes of many different solutes are continually occurring across biological membranes. Influxes take mineral nutrients into a growing cell while certain products of metabolism flow out of cells. The primary concern of this section will be the passive fluxes of ions toward lower chemical potentials. Although the mathematical expressions on the next few pages are rather formidable, the underlying approach is really quite straightforward. First, we indicate that the flux of some species is directly proportional to the driving force causing the movement. Next, the driving force is expressed in terms of the relevant components of the chemical potential. We then examine the consequences of having electroneutrality when there are simultaneous passive fluxes of more than one type of ion. This leads us to an expression describing the electrical potential difference across a membrane in terms of the properties of the ions penetrating it.

Before discussing the relation between fluxes and chemical potentials, we will briefly consider fluxes already mentioned or

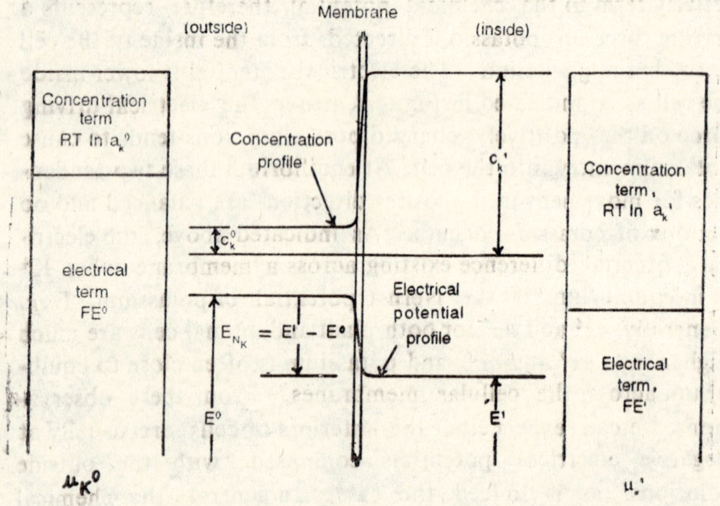

Fig. 4.4. Equilibrium of K^+ across a membrane. When potassium is in equilibrium across some membrane, the side with the higher concentration must be at a lower electrical potential in order for its chemical potential to be unchanged $(\mu_K{}^o = \mu_K{}^i)$ in crossing the membrane ($\mu_K{}^*$ is the same on the two sides of the membrane). The electrical potential difference across the membrane is then the Nernst potential for potassium, E_{NK}. Note that E^o and E^i must be expressed relative to some arbitrary baseline for electrical potentials.

which may be familiar from other contexts. In Fick's first law of diffusion, which says that the flux of (neutral) species j equals—$D_j \partial c_j/\partial x$, where we can consider that the force is the negative gradient of the concentration. We are all familiar with Ohm's law in the form ΔE equals IR, where I is the current and R is the resistance across which the electrical potential drop is ΔE. The current per unit area A is simply the flux of charge, J_c, which equals—$(1/\rho)\partial E/\partial x$ where ρ is the electrical resistivity and the negative gradient of the electrical potential represents the driving force. We usually replace —$\partial E/\partial x$ by $\Delta E/\Delta x$, which leads to $I/A=(1/\rho)\Delta E/\Delta x$. In order to make this conform with the usual expression of Ohm's law ($\Delta E=IR$), $\rho\Delta x/A$ must equal R, which indeed it does. The gravitational force, $-m_j g$, is simply the negative gradient of the potential energy in a gravitational field, *i.e.*, $-\partial m_j gh/\partial h = -m_j g$, where the minus sign indicates that the force is directed toward decreasing altitudes, *viz.*, toward the center of the earth. The gravitational force leads to the various forms of precipitation as well as to the percolation of water down through the soil.

We have considered examples of fluxes depending on each of the variable terms of the chemical potential ($\mu_j=\mu_j^* + RT \ln a_j + \bar{V}_j P + z_j FE + m_j gh$. In particular, the activity term ($RT \ln a_j$) leads to Fick's first law, the pressure term ($\bar{V}_j P$) accounts for Darcy's law and Poiseuille's law, the electrical term ($z_j FE$) yields Ohm's law, and the gravitational term ($m_j gh$) is responsible for fluxes caused by gravity. In each case, flux is found by experiment to be directly proportional to the appropriate driving force. We can actually generalize such relationships since nearly all transport phenomena can be represented by the statement: flux equals an appropriate force divided by some resistance (flux=force/resistance). Force is the negative gradient of a suitable potential, which we often conveniently take as the change in potential over some distance. But we have already shown that the chemical potential is an elegant way of summarizing all the factors which can contribute to the motion of a substance. It should not be surprising, therefore, that *in general* the flux of species j is proportional to the negative gradient of its chemical potential, $-\partial\mu_j/\partial x$.

Flux and mobility. Let us consider a charged substance, species *j*, which can cross a particular membrane. At equilibrium, the chemical potential of species *j*, μ_j, does not change with time or position—we are dealing with communicating regions—and there is no net flux of this solute across the membrane. When μ_j, changes with position but not with time, the situation is referred to as a *steady state*, a condition often used to approximate problems of biological interest. Actually, it is a matter of judgement whether μ_j is constant enough in time to warrant a description of the system as being in a steady state. (Similarly, constancy of μ_j for appropriate time and distance intervals is necessary before indicating that a system is in equilibrium.) Here we will let μ_j change with both position and time over regions where species *j* is free to move; later we will restrict our development to the steady state case.

When μ_j depends on position, a net passive movement or flux of species *j*, J_j, will tend to occur toward the region where the chemical potential is lower. The negative gradient of the chemical potential of species *j*, $-\partial\mu_j/\partial x$, acts as the driving force for this J_j. The greater $-\partial\mu_j/\partial x$ is, the larger will be the flux of species *j* in the *x*-direction. As a very useful approximation, we will assume that J_j is proportional to $-\partial\mu_j/\partial x$, where the minus sign means that a net positive flux occurs in the direction of decreasing chemical potential. The magnitude of a flux across some plane is also proportional to the local concentration of that species, c_j. That is, for a given driving force on species *j*, the amount of that substance which moves is proportional to how much of it is actually present—the more present, the greater the flux. Thus, for the one-dimensional case of crossing a plane perpendicular to the *x*-axis, J_j can be expressed as

$$J_j = u_j c_j \left(-\frac{\partial\mu_j}{\partial x}\right)$$
$$= \bar{v}_j c_j \qquad \ldots(6)$$

where u_j is a coefficient called the *mobility* of species *j*. A mobility is generally the ratio of some velocity to the force which causes the motion—we will return to a consideration of u_j shortly.

As we have already indicated, the top line of Equation 6 is a representative example from the large class of expressions relating various flows to their causative forces. In this particular case, J_j is the rate of flow of moles of species *j* across unit area of a plane, and can be expressed in moles/cm^2-sec. Such a molar flux of species *j* divided by its local concentration, c_j, gives the mean velocity, $\bar{v}_j$, with which this species moves across the plane—when c_j is in moles/cm^3, J_j/c_j can have units of (moles/cm^2-sec)/(moles/cm^3) or cm/sec. Perhaps this important point can be better appreciated by considering it in the following way. Let us suppose that the average velocity of species *j* moving perpendicularly toward area A of the plane of interest is $\bar{v}_j$, and we will consider a volume element of cross-sectional area A which extends back from the membrane for a distance $\bar{v}_j dt$. In a time *dt*, all molecules in the volume element $\bar{v}_j dt$A will cross area A, which means that the number of moles of species *j* crossing in this interval is $(\bar{v}_j dt\mathrm{A}) \times (c_j)$, where c_j is the number of moles/unit volume. Hence, the molar flux, J_j (which is simply the number of moles crossing unit area per unit time), would be $\bar{v}_j dt\, \mathrm{A}c_j/(\mathrm{A}dt)$, or $\bar{v}_j c_j$. In other words, the mean velocity of species *j* moving across the plane ($\bar{v}_j$) times the number of those molecules per unit volume which can move (c_j) equals the flux of that species (J_j), as is given by the bottom line of Equation 6. The upper line of Equation 6 indicates that this average velocity, J_j/c_j, equals the mobility of species *j*, u_j, times $-\partial\mu_j/\partial x$, the latter being the force on *j* that causes it to move. Thus, mobility is simply the proportionality factor between the mean velocity of motion ($\bar{v}_j = J_j c_j$) and the causative force $(-\partial\mu_j/\partial x)$. The greater of mobility of some species, the larger its velocity in response to a given force.

The particular form of the chemical potential of species *j*, μ_j, to be substituted into Equation 6 depends on the specific application that we have in mind. For charged particles moving across biological membranes, the appropriate μ_j is μ_j^*+RT ln a_j+z_jFE. (As we mentioned before, the $\overline{V}_j$P term makes only a relatively small contribution to the $\Delta\mu_j$ of an ion, therefore it is not included. For treating the one-dimensional case described by Equation 6, μ_j must be differentiated with respect to x, $\partial\mu_j/\partial x$, which leads to RT∂ ln $a_j/\partial x$+z_jF∂E$/\partial x$ when T is constant. The quantity ∂ ln $a_j/\partial x$ equals $(1/a_j)\partial a_j/\partial x$, which is $(1/\gamma_j c_j)\partial\gamma_j c_j/\partial x(a_j=\gamma_j c_j)$. Using the above form of μ_j appropriate for charged solutes and this expansion of the derivative, ∂ ln $a_j/\partial x$, the net flux of species j appearing in Equation 6 can be written

$$J_j=-\frac{u_j RT}{\gamma_j}\frac{\partial\gamma_j c_j}{\partial x}-u_j c_j z_j F\frac{\partial E}{\partial x} \qquad ...(7)$$

Equation 7 becomes simply$-u_j$RT$(\partial c_j/\partial x)$, *i e.*, it is proportional to the j either across a membrane or in a solution in terms of two components of the driving force : the gradients in activity and electrical potential.

Before proceeding, we should examine a little more closely the first term on the right side of Equation 7. When γ_j varies across the membrane, $\partial\gamma_j/\partial x$ can be considered to represent a driving force on species j In keeping the with common practice, we will ignore this possible force, *i.e.*, we will assume that γ_j can be treated as a constant—in any case $\partial\gamma_j/\partial x$ would be extremely difficult to measure. For constant γ_j, the first term on the right side of Equation 7 becomes simply $-u_j$RT$(\partial c_j/\partial x_j$, *i.e.*, it is proportional to the concentration gradient. In the absence of electrical potential gradients $(\partial E/\partial x=0)$, and for neutral solutes $(z_j=0)$, Equation 7 indicates that J_j equals $-u_j$RT$(\partial c_j/\partial x)$. But this is just the type of flux described by Fick's first law $(J_j=-D_j\partial c_j/\partial x)$, with u_jRT taking the place of the diffusion coefficient, D_j. In other words, we can replace u_jRT in Equation 7 by D_j. Since D_j equals u_jRT, diffusion coefficients must depend on temperature. Moreover, u_j is generally inversely proportional to the viscosity, which decreases as T increases, thus the dependence of D_j on T can be pronounced. Consequently, the temperature should be specified when the value of a diffusion coefficient is given.

Let us now return to a consideration of fluxes. In the absence of electrical effects and for constant γ_j, the net flux of species j given by Equation 7 equals $-D_j\partial c_j/\partial x$ when u_jRT is replaced by D_j. Thus, Fick's first law ($J_j=-D_j\partial c_j/\partial x$, is a special case of our general flux relation (Equation 6), where we ignored first the pressure and gravitational effects in order to obtain Equation 7 and then the electrical effects. But this eventual reduction of Equation 6 to Fick's first law is as we should expect, since in fact the only driving force considered when we presented Fick's first law was the concentration gradient. Such agreement between our present thermodynamic approach and the *seemingly* more empirical Fick's first law is quite important. It serves to justify the logarithmic term for activity in the chemical potential ($\mu_j=\mu_j^*+RT \ln a_j+\overline{V}_jP+z_jFE+m_jgh$. In other words, if the activity of species j appeared in the above Equation in a form other than $\ln a_j$. Equation 7 would not reduce to Fick's first law under the appropriate conditions. Since Fick's first law has been amply demonstrated experimentally, such a disagreement between theory and practice would necessitate some modification in the expression used to define chemical potential.

The terms in the chemical potential can be justified or "derived" by different methods. The forms of some terms in μ_j can be readily appreciated since they follow from familiar definitions of work, *e.g.*, the gravitational term and the electrical term. The above comparison with Fick's first law indicates that the $\ln a_j$ form is the appropriate way of handling the activity term. Some of these derivations incorporate conclusions from empirical observations. Moreover, the fact that the chemical potential can be expressed as a series of terms which can be added together agrees with experiment. Thus, a thermodynamic expression for the chemical potential can (1) summarize the results of previous observations (2) stand the test of experiments, and (3) lead to new and useful predictions.

In contrast to the case for a neutral solute, the flux of an ion also depends on an electrical driving force, represented in

Equation 7 by $-\partial E/\partial x$. A charged solute spontaneously tends to move in an electrical potential gradient—a cation moving in the direction of lower or decreasing electrical potential. Of course, the concentration gradient also affects charged particles. If a certain type of ion were present in some region but absent in an adjacent one, the ions would tend to diffuse into the latter region. A such charged particles diffuse toward regions of lower concentration, an electrical potential difference is created. This electrical potential difference is referred to as a "diffusion potential," which we will discuss in detail below. The interrelationship between concentration and electrical effects is extremely important in biology. Most membrane potentials can be treated as diffusion potentials resulting from different rates of movement of the various ions across a membrane.

Diffusion potential in a solution. We will now use Equation 7 to derive the electrial potential difference created by ions diffusing down a concentration gradient in a solution containing one type of cation and its accompanying anion. This case is relatively simple compared with the more biological one to follow, thus it may more clearly illustrate the relationship between concentration gradients and the accompanying electrical potential differences. We will assume that the cations and anions are initially placed at one side of the solution. In time, they will *diffuse* across the solution toward regions of lower concentration. In general, one ionic species will have a higher mobility, u_j, than the other. The more mobile ions will tend to diffuse faster than their oppositely charged partners, resulting in a microscopic charge separation. This slight charge separation sets up an electrical potential gradient leading to a *diffusion potential.* Using certain simplifying assumptions, we will calculate the magnitude of the electrical potential difference so created.

For convenience of analysis. let us consider a solution containing only a monovalent cation (+) and its monovalent anion (−). We will assume that their activity coefficients are constant. As the previous calculations on electrical effects indicated, solutions are essentially neutral in regions which are large compared with atomic dimensions. Thus, c_+ equals c_-, and the concentration of either species can be designated c. Furthermore, no charge imbalance will develop in time, which means that the flux of the cation across some plane in the solution, J_+, equals that of the anion across the same plane, J_-. (A very small charge imbalance does develop, which sets up the electrical potential gradient, but this uncompensated flux is transitory and in any case negligible compared with J_+ or J_-.) Both fluxes, J_+ and J_-, can be expressed by Equation 7 and then equated to each other, with gives

$$-u_+RT\frac{\partial c}{\partial x}-u_+cF\frac{\partial E}{\partial x}=-u_-RT\frac{\partial c}{\partial x}+u_-cF\frac{\partial E}{\partial x} \qquad ...(8)$$

where the plus sign on the right side occurs because the monovalent anion carries a negative charge ($z_-=-1$). Rearrangement of Equation 8 yields the following expression for the electrical potential gradient :

$$\frac{\partial E}{\partial x}=\frac{u_- - u_+}{u_+ + u_-}\,\frac{RT}{Fc}\,\frac{\partial c}{\partial x} \qquad ...(9)$$

Equation 9 indicates that a nonzero $\partial E/\partial x$ occurs when the mobility the cation differs from that of the anion and there is a concentration gradient. If u_- is greater than u_+, the anions move (diffuse) faster than the cations toward regions of lower concentration. As some individual anion moves ahead of its "partner" cation, an electric field is set up in such a direction as to speed up the cation and slow down the anion until they both move at the same speed, thus preserving electrical neutrality.

In order to obtain the difference in electrical potential produced by diffusion between planes of differing concentration, we must integrate Equation 9. At this point we will restrict our consideration to the *steady state* case where neither E nor c changes with time. In going along the x axis from region I to region II, the change in the electrical potential term in Equation 9 is $\int_{I}^{II} (\partial E/\partial x)dx$, which becomes $\int_{I}^{II} dE$ for the steady state condition and so equals $E^{II} - E^{I}$. The integral of the concentration term is $\int_{I}^{II} (1/c)(\partial c/\partial x)dx$, which becomes $\int_{I}^{II} dc/c$, or $\ln (c^{II}/c^{I})$. Using these two relations, integration of Equation 9 leads to

$$E^{II} - E^{I} = \frac{u_{-} - u_{+}}{u_{+} + u_{-}} \frac{RT}{F} \ln \frac{c^{II}}{c^{I}}$$

$$= 59.2 \frac{u_{-} - u_{+}}{u_{+} + u_{-}} \log \frac{c^{II}}{c^{I}} \text{ mv at } 25° \qquad \ldots(9)$$

where in the second line ln has been replaced by 2.303 log, and the value 59.2 mv at 25° has been substituted for 2.303 RT/F. In the general case, the anions have mobilities different from the cations. As the ions diffuse to regions of lower chemical potential in the solution, an electrical potential difference—given by Equation 9, and called a diffusion potential—is set up across the section where the concentration changes from c^{I} to c^{II}.

An example of a diffusion potential describable by Equation 9 occurs at the open end of the special micropipettes used for measuring electrical potential differences across membranes. (The "pulling," filling, and using of micropipettes—commonly referred to as microelectrodes or microsalt-bridges. The fine tip of the micropipette is open and provides an electrically conducting pathway into the cell or tissue. Ions diffusing from this fine tip give rise to a diffusion potential between the interior of the micropipette and the aqueous compartment into which the tip is inserted. To estimate the magnitude of this potential, we will assume that there is a concentration ratio of 30 across the micropipette tip, and then calculate the diffusion potentials for typical electrolytes, NaCl and KCl. The chloride mobility, u_{Cl}, is about 1.52 times u_{Na} and so $(u_{-} - u_{+})/(u_{+} + u_{-})$ is (0.52)/(2.52), or 0.21 for NaCl. When the concentration ratio is 30, the diffusion potential calculated by Equation 9 is

(59.2 mv)(0.21) log ($^1/_{30}$), or -18 mv. For KCl as the electrolyte, u_{Cl} is $1.04u_K$, and thus $(u_- - u_+)/(u_+ + u_-)$ is (0.04)/(2.04), or 002. This leads to a diffusion potential of (59.2)(0.02) log ($^1/_{30}$) or only -2 mv in going from the interior of the micropipette, where the concentration of KCl is. say, 3 M, to a region where it is 0.1 M. Thus, from the point of view of minimizing the diffusion potential across the fine tip, KCl is a much more suitable electrolyte for micropipettes than is NaCl. In fact, KCl is employed in nearly all micropipettes used for measuring membrane potentials (generally c_K inside cells is also higher than c_{Na}, and it does not vary as much from cell to cell as does c_{Na}). In any case, the closer u is to u_+, the smaller will be the diffusion potential predicted by Equation 9 for a given concentration ratio.

Membrane fluxes. As is the case for diffusion potentials in open solution, membrane potentials also depend on the different mobilities of the various ions and on their concentration gradients. In this case, however, the "solution" in which the diffusion takes place toward regions of lower chemical potential is actually the membrane itself. In this regard, we should mention were that a membrane is often the main barrier, or rate-limiting step for the diffusion of molecules into and out of cells or organelles. We would therefore except it to be the phase across which the diffusion potential is expressed. Under biological conditions a number of different types of ions are present, hence the situation is more complex than for the single cation-anion pair analyzed above. Furthermore, the quantities of interests such as $\partial E/\partial x$ and γ_j are those within the membrane, hence they are not accessible to experimental measurement. We will consider the ways of dealing with these difficulties in this section.

In order to calculate membrane diffusion potentials, we must make certain assumptions before an integration of fluxes or gradients can be made. As a start, we will assume that the electrical potential (E) varies linearly with distance across the membrane. This means that $\partial E/\partial x$ is a constant equal to $E_M/\Delta x$, where E_M is the electrical potential difference across the membrane—*i.e.*, $E^i - E^o = \triangle E = E_M$, and $\triangle x$ is the membrane thickness. This assumption of a constant electric field across the membrane - the electric field equals $-\partial E/\partial x$ in our one-dimensional case—was originally suggested by D.E. Goldman in 1943. It appreciably simplifies the integration conditions leading to an expression describing the electrical potential difference across biological membranes. As another very useful approximation, we will assume that the activity coefficient of species j, γ_j, is constant across the membrane (a less severe restriction than setting it equal to one). A partition coefficient is needed to describe concentrations within a membrane since the solvent properties of a membrane are different from those of the aqueous solutions on either side of it, where the concentrations are actually determined. Thus, c_j in Equation 7 should be replaced by $K_j c_j$, where K_j is the partition coefficient for species j. Incorporating these various simplifications and conditions, and after transferring the electrical term to the left side, we can rewrite Equation 7 as follows :

$$u_j K_j c_j z_j F \frac{E_M}{\Delta x} + J_j = -u_j RTK_j \frac{\partial c_j}{\partial x} \qquad \text{...(10)}$$

In order to cast Equation 10 into a form that can be integrated, it is advantageous to put all terms containing the concentration on the same side of the equation. Furthermore, since the two most convenient variables are c_j and x, their differentials should appear on opposite sides of the equation. Being guided by these two objectives, and after first multiplying each term by $dx/(u_j RT)$, we can rearrange Equation 10 to give the following expression :

$$\frac{z_j F E_M}{RT \Delta x}\, dx = -\frac{K_j dc_j}{K_j c_j + \dfrac{J_j \Delta x}{u_j z_j F E_M}} \qquad \text{...(11)}$$

where $(\partial c_j/\partial x)\, dx$ has been replaced by dc_j in anticipation of the restriction to steady-state conditions.

When the term $J_j \Delta x/(u_j z_j FF_M)$ is constant, Equation 11 can be readily integrated from one side of the membrane to the other. The factors Δx, z_j, F, and E_M are all fixed, and only J_j and u_j need be further considered here. When the flux of species j

does not change with time or position, then J_j across any plane parallel to and within the membrane is the same, and species j is not accumulating or being depleted in any of the regions of interest. Consequently, $\partial c_j/\partial t$ is zero, while of course $\partial c_j/\partial x$ is nonzero, which is the steady state condition. Our restriction to a steady state therefore means that J_j is constant and $(\partial c_j/\partial x)\, dx$ equals dc_j, a relation which we have already incorporated into Equation 11. For convenience, the mobility (u_j) of each species is also assumed to be constant within the membrane. With these restrictions, the quantity $J_j\Delta x/c_j z_j FE_M)$ becomes a constant, and we can integrate Equation 11 from one side of the membrane (the outside, o) to the other (the inside, i).

We know $\int dx/(x+b)$ equals $\ln(x+b)$: Hence,

$$-\int_{K_jc_j^o}^{K_jc_j^i} K_j dc_j/(K_jc_j+b)$$

is

$$-\ln(K_jc_j+b)\Big|_{K_jc_j^o}^{K_jc_j^i}$$

or $-\ln(K_jc_j^i+b)/(K_jc_j^o+b)$, which equals $\ln(K_jc_j^o+b)/(K_jc_j^i+b)$. Since $\int_{z^o}^{z^i} dx$ equals Δx, integration of Equation 11 gives

$$\frac{z_jFE_M}{RT}=\ln\frac{\left(K_jc_j^o+\dfrac{J_j\Delta x}{u_jz_jFE_M}\right)}{\left(K_jc_j^i+\dfrac{J_j\Delta x}{u_jz_jFE_M}\right)} \quad \text{.. (12)}$$

After taking exponentials of both sides of Equation 12 in order toput it into a more convenient form, and then multiplying by $K_jc_j^i+J_j\Delta x/(u_jz_jFE_M)$, Equation 12 becomes

$$K_jc_j^i\, e^{z_jFE_M/RT}+\frac{J_j\Delta x}{u_jz_jFE_M}\, e^{z_jFE_M/RT}$$
$$=K_jc_j^o+\frac{J_j\Delta x}{u_jz_jFE_M} \quad (13)$$

A quantity of considerable interest in Equation 13 is J_j, the net flux of species *j*. This equation can be readily solved for J_j, giving

$$J_j = J_j^{in} - J_j^{out} = \left(\frac{K_j u_j z_j F E_M}{\Delta x} \right) \left(\frac{1}{e^{z_j F E_M/RT} - 1} \right) \left(c_j^o - c_j^i \, e^{z_j F E_M/RT} \right) \quad \ldots(14)$$

where j_j^{in} is the influx of species *j*, J_j^{out} is its efflux, and their difference is the net flux. (The net flux J_j can represent either a net influx or a net efflux, depending on which of the unidirectional components, J_j^{in} or J_j^{out}, is larger.) Although the mathematical manipulations necessary to go from Equation 10 to Equation 14 are lengthy and cumbersome, the resulting expression is of extreme importance for understanding both membrane potentials and the passive fluxes of ions. Moreover, throughout this text all the steps involved in a particular derivation have been shown in order to avoid statements such as "it can *easily* be shown that such and such follows from so and so" —expressions which can be most frustrating and often are untrue. Equation 14 will be used to derive the Goldman equation—describing the diffusion potential across membranes—and will be invoked later in the derivation of the Ussing-Teorell equation, a relation, obeyed by certain passive fluxes.

Equation 14 shows how the passive flux of some charged species *j* depends on its internal and external concentrations and on the electrical potential difference across the membrane. For most cell membranes, E_M is negative, *i.e.*, the inside of the cell is at a lower electrical potential than is the outside. For a cation (z_j a positive integer) and a negative E_M, the terms in the first two parentheses on the right side on Equation 14 are both negative, hence their product is positive. On the other hand, for an anion (z_j a negative integer) and negative E_M, both parentheses are positive, and their product is still positive. (For ($E_M > 0$, the product of the first two parentheses is also positive for both anions and cations.) Thus, the sign of J_j depends on the value of c_j^o relative to that of $c_j^i \, e^{z_j F E_M/RT}$. When c_j^o is greater than $c_j^i \, e^{z_j F E_M/RT}$, the expression in the last parenthesis of Equation 14 is positive, and a net inward flux of species j occurs ($J_j > 0$). Such a condition may be contrasted with Equation [$J_j = P_j(c_j^o - c_j^i)$], where the net flux in the absense of electrical effects was inward when c_j^o was simply larger than c_j^i, as would adequately describe the situation for neutral solutes. However, knowledge of the concentration difference alone is not sufficient to predict the magnitude or even the direction of the flux of ions ; we must also consider the electrical potential difference between the two regions.

Membrane diffusion potential—Goldman equation. Passive fluxes of ions, which can be described bp Equation 14 and are caused by gradients in the chemical potentials of the various species, lead to an electrical potential difference (diffusion potential) across a membrane. We can determine the magnitude of this electrical potential difference by considering the contributions from all ionic fluxes across the membrane and the condition of electroneutrality. Certain assumptions are needed, however, to keep the equations manageable. Under usual biological situations not all anions and cations can easily move through the membranes. Many divalent cations do not readily enter or leave plant cells passively, which means that their mobility in the membranes is small. Such ions usually do not make a large enough contribution to the fluxes into or out of plant cells to influence markedly the diffusion potentials across the membranes. Thus, we will omit them in the present analysis, which is nevertheless rather complicated.

For many plant cells, the total ionic flux consists mainly of movements of K^+, Na^+, and Cl^-. These three ions generally have fairly high concentrations in and near plant cells, and we would therefore expect them to make substantial contributions to the total ionic flux. More specifically, a flux of species j depends on the product of its concentration and its mobility ($J_j = -u_j c_j \partial\mu_j / \partial x$, Equation 6, thus, ions having relatively high local concentrations or moving in the membrane fairly easily (high u_j) will tend to be the major contributors to the total ionic flux. In some cases there may be a sizable flux of H^+ or OH^- (which can have high u_js) as well as other ions, the restriction here to three ions being partially in the interest of algebraic simplicity. However, the real justification for considering only K^+, Na^+, and Cl^- is that the diffusion potentials calculated by using the passive fluxes of these three ions across biological membranes are often in good agreement with the measured electrical potential differences.

Our previous electrical calculations indicated that solutions are essentially neutral, a condition which we will again invoke here. In other words, because of the large effects resulting from small amounts of uncanceled charge, the net charge actually needed to cause the electrical potential difference across a membrane (E_M) is negligible compared with the existing concentrations of ions. Furthermore, the steady-state fluxes of the ions across the membrane will not change this condition of electrical neutrality, which means that no net charge is transported by the algebraic sum of the various charge movements across the membrane, *i.e.*, $\sum_j z_j J_j = 0$. When the bulk of the ionic flux consists of K^+, Na^+, and Cl^- movements, this important condition of electroneutrality can be described by equating the cationic fluxes ($J_K + J_{Na}$) to the anionic one (J_C), which leads to the following relation (as mentioned above, the various ions are indicated by using only their element symbols as subscripts):

$$J_K + J_{Na} - J_{Cl} = 0 \qquad ...(15)$$

Equation 15 describes the net ionic fluxes which lead to the electrical potential difference across the membrane. After substituting the expressions for the various J_js into Equation 15, we will solve the resulting equation for the diffusion potential across the membrane, E_M.

To obtain a useful expression for E_M in terms of measurabel parameters, it is convenient to introduce a permeability coefficient for species j, P_j. In such a permeability coefficient was defined as $D_j K_j / \Delta x$, where D_j is the diffusion coefficient of species j_j K_j is its partition coefficient, and Δx is the membrane thickness. Upon comparing Equation 7 with ($J_j = -D_j \partial c_j / \partial x$), we can see that $u_j RT$ takes the place of the diffusion coefficient of species j, D_j. The quantity $K_j u_j RT / \Delta x$ can thus be equated to $K_j D_j / \Delta x$, which is the permeability coefficient (P_j) of species j for some particular membrane. In this way, the unknown mobility of species j in a given membrane, the unknown thickness of the membrane, and the partition coefficient for the solute can be replaced by one parameter, which describes the permeability properties of that solute crossing the particular membrane under consideration.

With all the preliminaries out of the way, let us now determine the expression for the diffusion potential across a membrane for the case where most of the ionic flux is due to K^+, Na^+, and Cl^- movements. Using the permeability coefficients of the three ions and substituting in the net flux of each species as defined by Equation 14, Equation 15 becomes

$$P_K\left(\frac{1}{e^{FE_M/RT}-1}\right)(c_K{}^o-c_K{}^i e^{FE_M/RT})+P_{Na}\left(\frac{1}{e^{FE_M/RT}-1}\right)\times (c_{Na}{}^o-c_{Na}{}^i e^{FE_M/RT})+P_{Cl}\left(\frac{1}{e^{-FE_M/RT}-1}\right)(c_{Cl}{}^o-c_{Cl}{}^i e^{-FE_M/RT}) =0 \quad ...(16)$$

where z_K and z_{Na} have been replaced by 1 z_{Cl} by -1, and FE_M/RT has been canceled from each of the terms for the three net fluxes. To simplify this rather unwieldly expression, the quantity $1/(e^{FE_M/RT}-1)$ can be canceled from each of the three terms represented in Equation 16 $1/(e^{-FE_M/RT}-1)$ appearing in the last term is the same $-$as $e^{FE_M/RT}/\, e^{FE_M/RT}-1)$. Equation 16 then assumes a much more manageable form :

$$P_Kc_K{}^o-P_Kc_K{}^i\, e^{FE_M/RT}+P_{Na}c_{Na}{}^o-P_{Na}c_{Na}{}^i\, e^{FE_M/RT} -P_{Cl}c_{Cl}{}^o\, e^{FE_M/RT}+P_{Cl}c_{Cl}{}^i=0 \quad ...(17)$$

After solving Equation 17 for $e^{FE_M/RT}$ and taking logarithms, we obtain the following expression for the electrical potential difference across the membrane :

$$E_M=\frac{RT}{F}\ln\frac{(P_Kc_K{}^o+P_{Na}c_{Na}{}^o+P_{Cl}c_{Cl}{}^i)}{(P_Kc_K{}^i+P_{Na}c_{Na}{}^i+P_{Cl}c_{Cl}{}^o)} \quad ...(18)$$

Equation 18 is generally known as the Goldman, or constant field equation. The electric field equals $-\partial E/\partial x$, which Goldman in 1943 set equal to a constant (here $-E_M/\Delta x$) in order

Table 4.3. *Concentrations, potentials and fluxes of various ions for Nitella translucens in the light and the dark.* The superscrip to refers to concentrations in the external bathing solution, superscript *i* refers to the cytoplasm. The Nernst potentials (E_{N_j}) were calculated from Equation 5 by using concentration ratios and a numerical factor of 58.2 mv since the temperature was 10°. The potential across the plasmalemma (E_M) was −138 mv. The fluxe in licated for the dark refer to values soon after removing the illumination (about 1000 lux) from the cells. A picomole is 10^{-12} mole.

Ion	c_j^o (mM)	c_j^i (mM)	E_{N_j} (mv)	$c_j^o/c_j^i e^{z_j FE_M/RT}$	Light j_j^{in}	Light j_j^{out}	Dark j_j^{in}	Dark j_j^{out}
					(picomple cm²-sec)			
Na^+	1.0	14	−67	17	0.55	0.55	0.55	0.1
K^+	0.1	119	−179	0.20	0.85	0.85	0.2	(0.85)
Cl^-	1.3	65	99	0.000085	0.85	0.85	0.05	—

to facilitate the integration across a membrane. It turns out that the assumption of a constant electric field in the membrane is actually not essential for obtaining Equation 18, we could invoke Gauss's law and perform a more difficult integration. In 1949, Hodgkin and Katz applied the general equation derived by Goldman to the specific case of K^+, Na^+, and Cl^- diffusing across a membrane, and Equation 18 is often referred to as the Goldman-Hodgkin-Katz equation.

Equation 18 gives the diffusion potential existing across a membrane. We have derived it by assuming independent passive movements of K^+, Na^+, and Cl^- across a membrane in which $\partial E/\partial x$, γ_j, J_j, and u_j are all constant. We used the negative gradient of its chemical potential as the driving force for the net flux of each ion. Thus, Equation 18 gives the electrical potential difference arising from the different tendencies of K^+, Na^+, and Cl^- to *diffuse* across a membrane to regions of lower chemical potential. When other ions cross some particular membrane in appreciable amounts they will also make a contribution to its membrane potential. However, the inclusion of divalent and trivalent ions in the derivation of an expression for E_M complicates the algebra considerably (*e.g.*, if Ca^{++} is also considered Equation 17 has fourteen terms on the left side instead of six, and the equation becomes a quadratic in powers of $e^{FE_M/RT}$). The fluxes of such ions may be rather small, in which case Equation 18 may be adequate for describing the membrane potential.

Application of the Goldman equation. In certain cases, the terms appearing in Equation 18, *viz.*, the permeabilities and the internal and external concentrations of K^+, Na^+, and Cl^-, have all been measured. The validity of the Goldman equation can then be checked. In other words, we can compare the predicted diffusion potential with the actual electrical potential difference experimentally measured across the membrane.

As a specific example, we will use the Goldman equation to evaluate the membrane potential across the plasmalemma of *Nitella translucens*. The concentrations of K^+, Na^+, and Cl^-

in the external bathing solution and in the cytoplasm of *Nitella translucens* are given in Table 4.3. The ratio of the permeability of Na^+ to that of K^+, P_{Na}/P_K, is about 0.18 for *Nitella translucens*. The plasmalemma of *Nitella* is much less permeable to Cl^- than it is to K^+, probably only 0.1% to 1% as much. Thus, for purposes of calculation, we will let P_{Cl}/P_K be 0.0003. In terms of permeabilities relative to that for potassium, P_K for *Nitella* is then 1.00, P_{Na} is 0.18 and P_{Cl} is 0.003. Using these relative permeability coefficient and the concentrations given in Table 4.3 the argument of the logarithm appearing in Equation 18, $(P_K c_K{}^o + P_{Na} c_{Na}{}^o + P_{Cl} c_{Cl}{}^i)/(P_K c_K{}^i + P_{Na} c_{Na}{}^i + P_{Cl} c_{Cl}{}^o)$, has the numerical value of [(1.0)(0.1mM)+(0.18)(1.0 mM)+(0.003)(65mM)] divided by [(1.0)(119 mM)+(0.18)(14 mM)+(0.003)(1.3 mM], which is (0.475)/(121.5), or 0.00391. Since in is 2.303 log and 2.303 RT/F equals 58.2 mv at 20°, the Goldman equation predicts a membrane potential, E_M, of (58.2 mv) log (0.00391), or −140 mv. Thus, we expect the cytoplasm to be electrically negative with respect to the external bathing solution, as is indeed the case. In fact, the measured value of the electrical potential difference across the plasmalemma of *Nitella translucens* is −138 mv at 20° (Table 4.3 not yet supplied). This excellent agreement between the observed electrical potential difference and that calculated from the Goldman equation supports the contention that the membrane potential is in fact a diffusion potential. This point can be fairly easily checked by varying the external concentration of K^+, Na^+, and/or Cl^- and then seeing whether the membrane potential changes in accordance with Equation 18.

The different ionic concentrations on the two sides of a membrane help set up the passive ionic fluxes creating the diffusion potential, but the actual magnitude of the contribution of a particular ionic species to E_M also depends on the ease with which that type of ion can cross the membrane. Based on the relative permeabilities and concentrations, the major contribution to the electrical potential difference across the plasmalemma of *Nitella* comes from the K flux, with Na and Cl playing secondary roles. If the Cl terms are omitted from Equation 18 (*i.e.*, if P_{Cl} is set equal to zero), the calculated membrane potential is −153 mv, compared with −140 mv when Cl is included.

This rather small difference between the two potentials is a reflection of the relatively low permeability coefficient for chloride crossing the plasmalemma of *Nitella*, hence the Cl flux has less effect on E_M than does the K flux.

Changes in the amount of Ca^{++} or H^+ in the external medium cause some deviations from the predictions of Equation 18 for the electrical potential difference across the plasmalemma of *Nitella*. Thus, the diffusion potential is influenced by the properties of the particular membrane being considered, and ions other than K^+, Na^+, and Cl^- may have to be included in specific cases. To allow for the influence of the flux of H^+ on E_M, for example. we could include $P_H c_H{}^o$ in the numerator of the logarithm in the Goldman equation (Equation 18) and $P_H c_H{}^i$ in the denominator (note that a movement of H^+ in one direction has the same effect on E_M as movement of OH^- in the other). Nevertheless, restriction to the three ions indicated has proved to be adequate for treating the diffusion potential across many membranes, and the Goldman, or constant field expression in the form of Equation 18 has found widespread application.

Donnan potential. Another type of electrical potential difference encountered in biological systems is associated with immobile or fixed charges in some region adjacent to an aqueous phase containing small mobile ions, and is referred to as a *Donnan potential.* When a plant cell is placed in a KCl solution, for example, a Donnan potential arises between the interior of the cell wall and the bulk of the bathing fluid. The electrical potential difference arising from electrostatic interactions at such a solid-liquid interface can be regarded as a special type of diffusion potential, as we will show shortly.

Pectin and other macromolecules in the cell wall have a large number of immobile carboxyl groups (—COOH) from which hydrogen ions dissociate. This gives the cell wall a net negative charge, as indicated. Cations such as Ca^{++} are electrostatically attracted to the negatively charged cell wall, and the overall effect is an exchange of H^+ for Ca^{++} and other cations. The region containing the immobile charges—such as carboxyl groups in the case of the cell wall—is generally referred to as the *Donnan phase* (see Fig. 4.5). At equilibrium, a distribution of oppositely charged ions electrostatically attracted to these immobile charges occurs between the Donnan phase and the adjacent aqueous one (Fig. 4.5). This sets up an electrical potential gradient and a Donnan potential is thus created between the center of the Donnan phase and the bulk of the solution next to it. The sign of the electrical potential in the Donnan phase relative to the surrounding electrolyte solution is the same as the sign of the charge of the immobile ions. For example, because of the presence of dissociated carboxyl groups, the electrical potential in the cell wall is negative with respect to an external solution. Membranes also generally act as charged Donnan phases. In addition, Donnan phases occur in the cytoplasm where the immobile charges are due to proteins and other large polymers (RNA, DNA) that have many carboxyl and phosphate groups from which

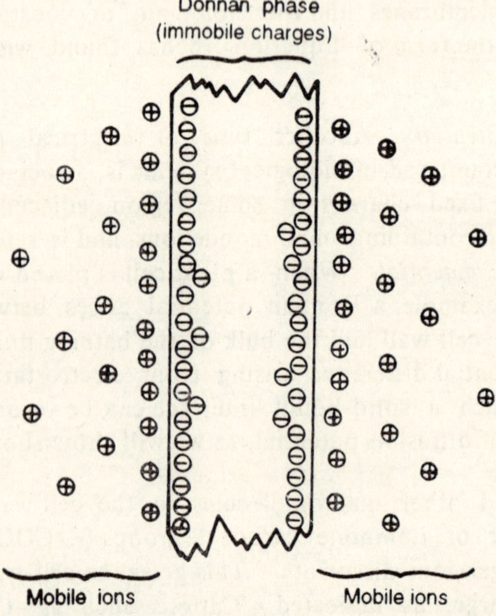

Fig. 4.5. Spatial distribution of positively charged mobile ions ⊕, occurring on either side of a Donnan phase in which are embedded immobile negative charges (⊖).

protons can dissociate, leaving the macromolecules with a net negative charge. Such polymers are fixed in the sense that they are unable to diffuse across either the plasmalemma or the tonoplast.

When the immobile or fixed ions are in some region, mobile ions of opposite sign can occur in adjacent layers on either side (Fig. 4.5). For example, layers containing cations are often formed in the aqueous solutions on each side of biological membranes, which generally act as Donnan phases having a net negative charge at usual pH's. At equilibrium no net movement of the ions occurs, so the chemical potentials of each of the mobile ions (*e.g.*, K^+, Na^+, Cl^-, Ca^{++}) have the same values up close to the Donnan phase as they do in the surrounding aqueous phase. The electrical potential difference (Donnan potential) can then be calculated by assuming constancy of the chemical potential. But this is exactly the same principle that we used in deriving the Nernst potential, $E_{N_j} = E^{II} - E^{I} = (RT/z_jF) \ln (a_j^{I} / a_j^{II})$, Equation 15, between two aqueous compartments. In fact, since the argument again hinges on the constancy of the chemical potential, the equilibrium distribution of any ion from the Donnan phase to the aqueous phase extending away from the barrier must satisfy the Nernst potential (E_{N_j}) for that particular ion.

The electrical potential difference known as the Donnan potential can also be regarded as a special case of a diffusion potential, as suggested above. To see why this is so, let us imagine that the mobile ions are tending to diffuse away from the fixed immobile ions of the opposite sign. We can assume that the mobile ions are initially in the same region as the immobile ones. In time, the mobile ions would tend to diffuse away. This tendency, based on thermal agitation, causes a slight charge separation and thus sets up an electrical potential difference between the Donnan phase and the bulk of the adjacent solution. For the case of a single species of mobile cations and the anions fixed in the membrane (both assumed to be monovalent here), the diffusion potential across that part of the aqueous phase next to the membrane can be described by

Equation 9, $E^{II} - E^{I} = [(u_{-}-u_{+})/(u_{+}+u_{-})](RT/F) \ln (c^{II} / c^{I})$, which we derived for diffusion toward regions of lower chemical potential in a solution. Fixed anions have zero mobility ($u_{-}=0$), hence $(u_{-}-u_{+})/(u_{+}+u_{-})$ here is $-u_{+}/u_{+}$, or -1. Equation 9 then becomes $E^{II} - E^{I}$ equals $-(RT/F) \ln (c^{II} / c^{I})$. which is simply the Nernst potential (Equation 5) for monovalent cations $[-\ln (c^{II} / c^{I}) = \ln (c^{I} / c^{II})]$. Thus, the Donnan potential can also be regarded as a type of diffusion potential occuring as the mobile ions tend to diffuse away from the charges of opposite sign, which remain fixed in the Donnan phase.

Active transport

The expression "active transport" implies that energy derived from metabolic processes is somehow involved in moving a solute across a membrane toward a region of higher chemical potential. There are three different aspects to this description of active transport, *viz.*, a supply of energy, movement, and an increase in chemical potential. Although not really necessary, the term "active transport" has conventionally been restricted to the case of movement in the energetically uphill direction, and we will follow this restriction here. By itself, a difference in chemical potential of a certain species across a membrane does not necessarily imply that active transport of that species is occurring. For example, if the solute does not penetrate the membrane, the solute is unable to attain equilibrium across it, and μ_j would not be expected to be the same on the two sides. For ions actually moving across some membrane, the ratio of the influx to the efflux of a particular ionic species provides information which helps to indicate whether or not active transport is taking place. Specifically, the Ussing-Teorell, or flux ratio equation is satisfied for that species if it moves passively and independently across the membrane. A very simple and often effective approach for determining whether fluxes are active or passive is to remove possible energy sources. For photosynthesizing plant tissue, this can mean comparing the fluxes in the light with those in the dark. (In addition, we should check whether the permeability of the membrane changes or other alterations occur upon illumination). Compounds or treatments which disrupt metabolism also can be useful for ascertaining whether metabolic energy is being used for the active transport of various solutes.

We will begin our discussion by indicating how active transport can directly affect membrane potentials. We will then compare the temperature dependencies of metabolic reactions with those for diffusion processes across a barrier to show that a marked enhancement of solute influx caused by increasing the temperature does not necessarily indicate that active transport is taking place. Next, we will consider a much more reliable criterion for deciding whether fluxes are passive or not,*viz.*, the Ussing-Teorell, or flux ratio equation. After examining a specific case in which active transport is involved, we will discuss a relationship that in often invoked for describing the concentration dependence of ion uptake.

Electrogenicity. One of the possible consequences of actively transporting a certain ionic species into a cell or organelle is the development of an excess of electrical charge inside the membrane-surrounded entity. If for some ionic species the active transport process involves an accompanying ion of opposite charge or an equal release of a similarly charged ion, the total charge in the cell or organelle is unaffected. However, if the charge of the actively transported ion is not directly compensated for, the process is *electrogenic*, *i.e.*, it tends to generate an electrical potential difference across the membrane. An electrogenic uptake of an ion which gives a net transport of charge into the cell or organelle will thus affect the membrane potential (see Higinbotham ; Higinbotham *et al.* ; Hope ; as well as Spanswick). We can appreciate this effect by referring to Equation 1 ($Q=C\Delta E$), which indicates that the drop in potential across the membrane, ΔE, equals Q/C, where Q is the net charge enclosed within the cell or organelle and C is its capacitance. The initial movement of net charge across the membrane by active transport leads to a fairly rapid charging or discharging of the membrane capacitance.

To be specific, we will consider a spherical cell into which there is an electrogenic influx of chloride caused by active transport, $J_{a.t.cl}^{in}$. The amount of charge transported in time t across the surface of the sphere (area$=4\pi r^2$) is $J_{a.t.cl}^{in}\ 4\pi r^2 t$, where r is the radius. This active uptake of Cl^- increases the internal concentration of negative charge by the amount moved in divided by the cellular volume, or $J_{a.t.cl}^{in}\ 4\pi r^2 t/(4\pi r^3/3)$, which is $3J_{a.t.cl}^{in}t/r$. Let us suppose that $J_{a.t.cl}^{in}$ is 1.0 picomole/cm²-sec (a picomole $=10^{-12}$ mole, and that the cell has a radius of 30 μm. In 1 sec the concentration of Cl^- actively transported in would then be $(3)(10^{-12}$ mole/cm²-sec)(1 sec)/(30×10^{-4} cm), or 1.0×10^{-9} mole/cm³. Assuming a membrane capacitance of 1 μf/cm², we saw that such a cell has 1.0×10^{-9} mole of uncompensated negative charge/cm³ when the interior is 100 mv negative with respect to the external solution. If no change were to take place in the other ionic fluxes, the electrogenic uptake of Cl^- into this cell would cause its interior to become more negative at the rate of 100 mv/sec. The charging of the membrane capacitance is indeed a rapid process.

The initiation of an electrogenic process causes an adjustment of the passive fluxes across the membrane. In particular, the net charge actively brought in is electrically compensated for very soon by appropriate passive movements of other ions into or out of the cell. The actual electrical potential difference across the membrane then results from the diffusion potential caused by these new passive fluxes plus a contribution from the electrogenic process involving the active transport of some charged species. We can represent the electrical potential drop generated by the active transport of species j, $E_{a.t.j}$, as follows :

$$E_{a.t\ j}=z_j F J_{a.t.j} R^{memb} \qquad \ldots(19)$$

where z_j is the charge number of species j, F is the Faraday, $J_{a.t.j}$ is the net flux of species j caused by active transport, and R^{memb} is the membrane resistance. $FJ_{a.t.j}$ in Equation 19—which is really a form of Ohm's law— has units of (coulombs/mole)$\times$(moles/cm²-sec) or coulombs/cm²-sec, which is amperes/cm². Hence, R^{memb} must be expressed in ohms-cm², so that

$z_j F J_{a.t.j} R^{memb}$ in Equation 19 can have units of ampere-ohms, *i.e.*, volts the proper unit for electrical potentials. For most plant cells, R^{memb} is usually from 1000 to 8000 ohm-cm^2.

Let us now see how large $E_{a.t.j}$ given by Equation 19 might be for representative electrogenic "pumps." The electrical potential drop created by active transport of 1 picomole of a monovalent cation/cm^2-sec across a membrane resistance of 3000 ohm-cm^2 would be $(+1)$(96,487 coulombs/mole)(10^{-12} mole/cm^2-sec)(3000 ohms-cm^2), or 0.3×10^{-3} volt, which is a rather small potential difference. When the membrane-resistance is higher or the net flux by the electrogenic pump is greater, the potential created by the active transport process would of course be larger. For certain plant cells and some animal cells, R^{memb} is about 10,000 ohm-cm^2 and $J_{a.t.j}$ may be 20 picomoles/cm^2-sec. Equation 19 indicates that the potential drop created by such an electrogentic pump for a monovalent cation would be approximately $(+1)(96{,}487)20 \times 10^{-12})(10{,}000)$, or 0.019 v (19 mv). In any case, the actual electrical potential difference across a membrane is obtained by adding the potential drop caused by active transport of uncompensated charge (Equation 19) and that for passive fluxes, the latter being predicted by the Goldman equation (Equation 8).

Q_{10}*, *a temperature coefficient. In general, most metabolic reactions are markedly influenced by temperature, while processes like light absorption are insensitive to temperature. What temperature dependence should we except for diffusion ? Can we decide whether the movement of some solute into a cell is by active transport or by passive diffusion once we know how the fluxes depend on temperature ? To answer such questions, we need an expression describing the distribution of energy among the molecules as a function of temperature in order to determine what fraction of the molecules has the requisite energy for a particular process. In aqueous solutions, the relevant energy is generally the kinetic energy of motion of the species involved. Hence, we will first relate the distribution of kinetic energy among molecules to temperature. At the end of this section we will present the related concept of *activation energy*—activation energies are used to discuss reactions which have an energy barrier that must be surmounted before the process can take place. The topics that we will introduce here are important for a basic understanding of many aspects of biology—all the way from biochemical reactions to the consequences of light absorption.

We begin by noting that very few molecules possess extremely large kinetic energies. In fact, the probability that a molecule has a given kinetic energy E decreases exponentially as E increases. The precise statement of this is known as the *Boltzmann energy distribution*, which describes the frequency with which specific kinetic energies are possessed by molecules at equilibrium at a given temperature. A convenient form for the Boltzmann energy distribution at temperature T is as follows :

$$n(\mathrm{E}) = n_{total}\, e^{-\mathrm{E}/k\mathrm{T}} \qquad \ldots(20)$$

where $n(\mathrm{E})$ is the number of molecules possessing an energy of E or more out of the total number of molecules, n_{total}, and k is Boltzmann's constant $k = 8.617 \times 10^{-5}$ electron volts (ev)/molecule-°K, where 1 ev is the energy change for a particle of unit electronic charge when it moves through an electrical potential difference of 1 volt. The quantity $e^{-\mathrm{E}/k\mathrm{T}}$, which equals $n(\mathrm{E})/n_{total}$ by Equation 20 is often referred to as the *Boltzmann factor*. Equation 20 indicates that the number of molecules with an energy of zero or greater, $n(0)$, equals $n_{total}e^{-0/k\mathrm{T}}$ or $n_{total}e^{-0}$, which is simply n_{total}, the total number of molecules present.

Due to collisions based on thermal motion, energy is continually being gained or lost by individual molecules in a random fashion. Hence a wide range of kinetic energies is possible, although very high energies are less probable, cf. Equation 20. As the temperature is raised, not only is the average energy per molecule higher, but the number of molecules in the "high energy tail" of the exponential Boltzmann distribution also increases substantially.

For diffusion across a membrane, the appropriate Boltzmann energy distribution indicates that the number of molecules with a kinetic energy of U or greater per molecule resulting from velocities in some particular direction is proportional to $\sqrt{T}e^{-U/kT}$ (see Davson and Danielli). A minimum kinetic energy (U_B) is often necessary to diffuse past some barrier or to cause some specific reaction. Thus, any molecule with a kinetic energy of U_B or greater has sufficient energy for the particular process. For the Boltzmann energy distribution appropriate to this case, the number of such molecules is proportional to $\sqrt{T}\,e^{-U_B/kT}$. (These expressions having the factor $\sqrt{T}$ actually only apply to one-dimensional cases, *e.g.*, for molecules diffusing across a membrane). At a temperature 10° higher, the number is proportional to $\sqrt{(T+10)}\,e^{-U_B/[k(T+10)]}$. The ratio of these two quantities is called the Q_{10}, or temperature coefficient of the process :

$$Q_{10}=\sqrt{\frac{T+10}{T}}\;e^{\frac{10U_B}{kT(T+10)}} \qquad \text{...(21)}$$

To obtain the form of the exponential given in Equation 21, we not that $U_B/[k(T+10)]+U_B/(kT)$ is equal to $-TU_B/[kT(T+10)]+(T+10)U_B/[kT(T+10)]$, which is $10U_B/[kT(T+10)]$.

A Q_{10} near one is characteristic of some passive processes which have no energy barrier to surmount, *i.e.*, where U_B equals 0. On the other hand, most enzymatic reactions take place only when the reactants have a considerable kinetic energy, so that such processes tend to be quite sensitive to temperature. A value of 2 or greater for Q_{10} is often considered to indicate the involvement of metabolism, such as occurs for active transport of a solute into a cell or organelle. However, Equation 21 indicates that *any* process having an appreciable energy barrier (U_B) can have a large temperature coefficient.

The value of the temperature coefficient Q_{10} for a particular process can be indicative of the minimum kinetic energy required (U_B), and vice versa. A membrane often represents an appreciable energy barrier for the diffusion of charged solutes ; let us say that the U_B for passive ion movement across it is 0.50 ev/molecule. Using Equation 21, this would lead to a temperature coefficient for T equaling 20° of

$\sqrt{(303/293)}\, e^{(10)(0.50)/((8.617\times10^{-5})(293)(303))}$, or $1.02\, e^{0.654}$, which equals 1.96. In other words, the uptake of this particular ion would approximately double with only a 10° increase in temperature. This example indicates that a passive process can have a rather high Q_{10} if there is an appreciable energy barrier, indicating that a large temperature coefficient for ion uptake does not necessarily imply active transport.

A kinetic energy of 0.50 ev or greater is possessed by only a relatively few molecules in the Boltzmann energy distribution—$n(E)=n_{total}e^{-E/kT}$, Equation 20. For instance, at 20° the Boltzmann factor $e^{-E/kT}$ is $e^{-(0.50\ ev)/(0.0253\ ev)}$, or only 2.5×10^{-9} (kT =0.0253 ev at 20°. As T is raised, the number of such molecules in the high energy part of the Boltzmann distribution increases greatly. Many more molecules thus have the requisite kinetic energy, U_B, and consequently can take part in the process being considered. In particular, at 30° the Boltzmann factor in Equation 20 becomes $e^{-(0.50)/(0.0261)}$, or 4.6×10^{-9} for a kinetic energy of 0.50 ev/molecule (kT=0.0261 ev at 30°). Hence, the Boltzmann factor for a U_B of 0.50 ev/molecule essentially doubles for a 10° rise in temperature, consistent with our Q_{10} calculation for this case.

So far in this section we have calculated on the basis of energy *per molecule*, while in many cases it is more convenient to consider energy *per mole*. In order to change from a molecule to a mole basis, we multiply Boltzmann's constant k (energy/molecule-°K) by Avogadro's number N (molecules/mole), which gives us the gas constant R (energy/mole-°K) *i.e.*, R equals kN. If U and U_B are expressed as energy per mole, we simply replace k in the Boltzmann energy distribution (Equation 20) and Equation 21 by R. Finally, as indicates, 1 ev/molecule is equal to 23.06 kcal/mole.

The existence of an energy barrier of height U_B is related to the concept of *activation energy*, which refers to the minimum amount of energy necessary for some reaction to take place. In the case of a membrane, we can evaluate U_B by determining how the number of molecules diffusing across the membrane varies with temperature, *e.g.*, by invoking Equation 21. In fact, the kinetic energy (U_B) corresponds to the activation energy for crossing the barrier. For a chemical reaction, we can also experimentally determine how the process is influenced by temperature. If we represent the activation energy per mole by A, the rate constant for such a reaction should vary with temperature as follows :

$$\text{rate constant} = Be^{-A/RT}$$

where B is essentially a constant.

Equation 22 is referred to as the *Arrhenius equation.* It was originally proposed on experimental grounds by Arrhenius at the end of the 19th century and only subsequently interpreted theoretically. A plot of the logarithm of the rate constant versus 1/T is commonly known as an *Arrhenius plot.* Hence, the slope of an Arrhenius plot is proportional to the activation energy by Equation 22—ln (rate constant) = ln B − A/RT. Many reactions of importance in biochemistry have large values for A and are therefore extremely sensitive to temperature. Moreover, an enzyme greatly increases the rate of the reaction which it catalyzes by reducing the value of the activation energy needed (Equation 22). Thus, the related concepts of activation energy and energy barrier have many applications in biology.

Let us finally consider the activation energies for certain case s of diffusion. For diffusion in water at 20°, A is about 4 kcal/mole for K^+ and 5 kcal/mole for mannitol (see Stein). After replacing U_B by A and k by R in Equation 21, we can calculate that the appropriate Q_{10} for such diffusion would be 1.3. The activation energy for the passive efflux of K^+ from many cells is about 15 kcal/mole near 20°, which corresponds to a Q_{10} of 2.5. Again, we conclude that a purely passive process, such as diffusion across membranes, can have a marked temperature dependence.

Ussing-Teorell equation. We will now consider ways of distinguishing between active and passive fluxes which are more reliable than determining Q_{10}s. One of the most useful physicochemical criteria for deciding whether a particular ionic movement across a membrane is active or passive is the application of the *Ussing-Teorell*, or *flux ratio* equation. For ions moving passively, this expression shows how the ratio of the influx to the efflux depends on the internal and external concentrations of that species as well as on the electrical potential difference across the membrane. If the Ussing-Teorell equation is satisfied, active transport of the ions need not be invoked. We can readily derive this expression by considering how the influx and the efflux could each be determined experimentally, as the following arguments indicate.

For measuring the unidirectional inward component, or influx of a certain ion J_j^{in}, the plant cell or tissue can be placed in a solution containing a radioactive isotope of species *j*. Initially, none of the radioisotope is inside the cells, hence the internal specific activity for this isotope equals zero at the beginning of the experiment. (As for any radioisotope study, only some of the molecules of species *j* are radioactive. The particular fraction is known as the *specific activity*, and must be determined for both c_j^o and c_j^i in the present experiment.) Since none is inside, the initial unidirectional outward component or efflux of the radioisotope, J_j^{out}, is zero, therefore, the initial net flux (J_j) of the isotope reflects J_j^{in}. From Equation 14, this influx of the radioisotope can be represented by $(K_j u_j z_j F E_M/\Delta x)\ [1/(e^{z_j F E_M/RT} - 1)] c_j^o$. After the isotope has entered the cell, some of it will start coming out. Therefore, only the initial flux will give an accurate measure of the influx of the radioisotope of species *j*.

Once the radioactivity has built up inside to a substantial level, we remove the radioisotope from the outside solution. The flux of the isotope will then be from inside the cells to the external solution. In this case, the specific activity for c_j^o equals zero, and c_j^i determines the net flux of the radioisotope. From Equation 14, this efflux of the isotope of species *j* differs in magnitude from the initial influx only by having the factor c_j^o replaced by $c_j^i e^{z_j F E_M/RT}$, while the quantities in the first two parentheses remain the same. The ratio of these two fluxes—each of which can be separately measured—takes on the following relatively simple form :

$$\frac{J_j^{in}}{J_j^{out}} = \frac{c_j^o}{c_j^i e^{z_j F E_M / RT}} \qquad \ldots(23)$$

Equation 23 was independently derived by both Ussing and Teorell in 1949, and is known either as the Ussing-Teorell equation or the flux ratio equation. It is strictly valid only for ions moving passively without interacting with other substances that may also be moving across the membrane. The present derivation makes use of Equation 14, which gives the passive flux of some charged species across a membrane in response to differences in the chemical potential of that species across the barrier. Equation 14 was derived by considering only one species at a time, hence, possible interactions between the fluxes of different species are not included in Equation 23. The Ussing-Teorell equation can thus be used to determine whether the observed influxes and effluxes are passive (*i.e.*, responses to the chemical potentials of the ions on the two sides of a membrane), or whether additional factors such as interactions between species or active transport must be invoked. For example, when active transport of species j into the cell is taking place, the actual J_j^{in} would be the passive unidirectional flux (*i.e.*, the one predicted by Equation 23 *plus* the influx due to active transport. After one further comment about Equation 23, we will consider a specific example of the use of the Ussing-Teorell equation to evaluate membrane fluxes.

The ratio of the influx of species j to its efflux, as given by Equation 23, can readily be related to the difference in its chemical potential on the two sides of the membrane, $\mu_j^o - \mu_j^i$. This difference causes the flux ratio to differ from unity, and we will return to it to estimate the minimum amount of energy needed to actively transport that ionic species across the membrane. After taking logarithms of both sides of the Ussing-Teorell equation (Equation 23) and multiplying by RT, we obtain the following equalities :

$$\begin{aligned} RT \ln \frac{J_j^{in}}{J_j^{out}} &= RT \ln \frac{c_j^o}{c_j^i} - z_j F E_M \\ &= RT \ln a_j^o + z_j F E^o - RT \ln a_j^i - z_j F E^i \\ &= \mu_j^o - \mu_j^i \end{aligned} \qquad \ldots(24)$$

where the membrane potential (E_M) has been replaced by $E^i - E^o$ in keeping with our previous convention. The derivation is restricted to the case of constant $\gamma_j(\gamma_j^o = \gamma_j^i)$, which means that ln (c_j^o/c_j^i) equals ln $(\gamma_j^o c_j^o/\gamma_j^i c_j^i)$, or ln (a_j^o/a_j^i), which is ln $a_j^o - \ln a_j^i$. Finally, the $\overline{V}_j P$ term in the chemical potential is ignored for the charged species—actually, we need only assume that $\overline{V}_j P^o$ equals $\overline{V}_j P^i$—so that μ_j then equals $\mu_j^* + RT \ln a_j + z_j FE$, where μ_j^* has the same value on the two sides of the membrane. Thus, $\mu_j^o - \mu_j^i$ is $RT \ln a_j^o + z_j FE^o - RT \ln a_j^i - z_j FE^i$, as is indicated both in Equation 24 and the caption to Figure 4.6.

A difference in chemical potential of species *j* across a membrane would cause the flux ratio of the passive fluxes to differ from unity, a conclusion that follows directly from Equation 24. When μ_j^o equals μ_j^i, the influx would balance the efflux, hence there would be no net passive flux of species *j* across the membrane ($J_j = J_j^{in} - J_j^{out}$ by Equation 14). This condition ($\mu_j^o = \mu_j^i$) is also described by Equation 4, used to derive the Nernst equation. In fact, in such a case of equality of the chemical potentials, the electrical potential difference across the membrane is the Nernst potential, E_N, given by Equation 7, $E_{Nj} = (RT/z_jF) \ln (a_j^o/a_j^i)$. Thus, when E_M equals E_{Nj} for some species J_j^{in} equals J_j^{out}, and no net passive flux of that ion is expected across the membrane nor is any energy expended in moving the ion from one side of the membrane to the other. When Equations 23 and 24 are not satisfied for some species, such ions are not moving across the membrane passively, or perhaps not moving independently from other fluxes. One way this may occur is for the various fluxes to be interdependent, a condition describable by irreversible thermodynamics. Another way is through active transport of the ions, whereby energy derived from metabolism is used to move solutes to regions of higher chemical potential.

Example of active transport. The above principles for deciding whether or not active transport of certain ions is taking place can be illustrated by using data obtained with the large internodal cells of *Nitella translucens* (Table 4.4). All the parameters appearing in the Ussing-Teorell equation have been measured for Na^+, K^+, and Cl^- using *Nitella*. For experimental

Table 4.4. *Concentrations, potentials, and fluxes of various ions for Nitella translucens in the light and the dark.* The superscript to refers to concentrations in the external bathing solution, superscript *i* refers to the cytoplasm. The Nernst potentials (E_{Nj}) were calculated from Equation 7 by using concentration ratios and a numerical factor of 58.2 mv since the temperature was 20°. The potential across the plasmalemma (E_M) was −138 mv. The fluxes indicated for the dark refer to values soon after removing the illumination (about 1000 lux) from the cells. A picomole is 10^{-12} mole.

					Light		Dark	
Ion	c_j^o (*mM*)	c_j^i (*mM*)	E_{Nj} (*mv*)	$c_j^o/c_j^i e^{z_j F E_M/RT}$)	J_j^{in}	J_j^{out}	J_j^{in}	J_j^{out}
					(*picomole/cm²-sec*)			
Na^+	1.0	14	−67	17	0.55	0.55	0.55	0.1
K^+	0.1	119	−179	0.20	0.85	0.85	0.2	(0.85)
Cl^-	1.3	65	99	0.000085	0.85	0.85	0.05	—

Source: E. A. C. Mac Robbie, *Journal of General Physiology* 45 : 861-878 (1962) and R. M. Spanswick and E. J. Williams, *Journal of Experimental Botany* 15 : 193-200 (1964). Data reprinted by permission.

purposes, this fresh water alga is often placed in a dilute aqueous solution containing 1 mM NaCl, 0.1 mM KCl, plus 0.1 mM $CaCl_2$—this solution is not unlike the pond water in which *Nitella* grows, and is generally referred to as "artificial pond water"—which indicates the values for all three c_j^o's. The concentrations of Na^+, K^+, and Cl^- measured in the cytoplasm, the c_j^iz, are given in Table 4.4, column 3. Assuming that activities can be replaced by concentrations, we can calculate the Nernst potential across the plasmalemma from these concentrations by using Equation 7. $E_{N_j}=(58.2/z_j) \log (c_j^o/c_j^i)$ in mv (the numerical factor is 58.2 since the measurements were at 20°). E_{N_j} for Na^+ is $[(58.2 \text{ mv})/(1)] \log [(1.0 \text{ mM})/(14 \text{ mM})]$, or −67 mv ; for K^+ it is −179 mv ; and for Cl^- the Nernst

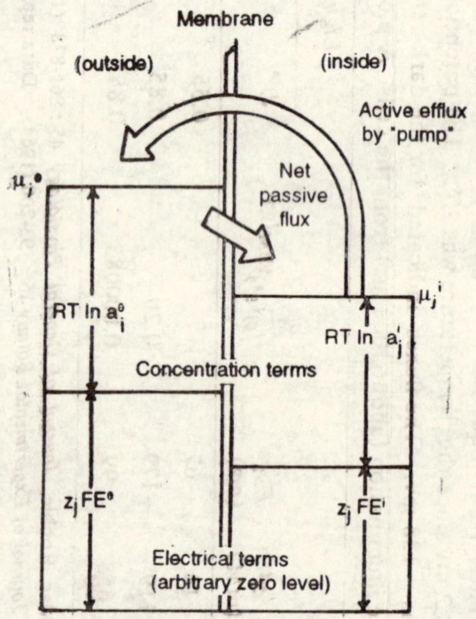

Fig. 4.6. Diagram illustrating the situation for an ion not in equilibrium across a membrane. Since μ_j^o is greater than μ_j^i there is a net passive flux into the cell. In the steady state, this net inward flux is balanced by an equal efflux caused by active transport of species j out of the cell. We note that $\mu_j^o - \mu_j^i$ equals $z_jF(E_{N_j} - E_M)$, where $z_jFE_{N_j}$ is RT ln a_j^o—RT ln a_j^i and $-z_jFE_M$ is $z_jFE^o - z_jFE^i$.

potential is 99 mv (column 4). A direct measurement of the electrical potential drop across the plasmalemma (E_M) gives -138 mv, as was indicated earlier in discussing the Goldman equation. Since E_{Nj} differs from E_M in all three cases, none of these ions is in equilibrium across the plasmalemma of *Nitella*.

A difference between E_M and E_{Nj} for a particular species indicates departure from equilibrium for that ionic species ; it also tells in what compartment μ_j is higher. Specifically, if the actual membrane potential is algebraically more negative than the calculated Nernst potential, the chemical potential in the inner aqueous phase (here the cytoplasm) is lower for a cation (Fig. 4.6) but higher for an anion, compared to the values in the external solution—cf. the effect of z_j in the electrical term in the chemical potential, z_jFE. Since E_M (-138 mv) is more negative

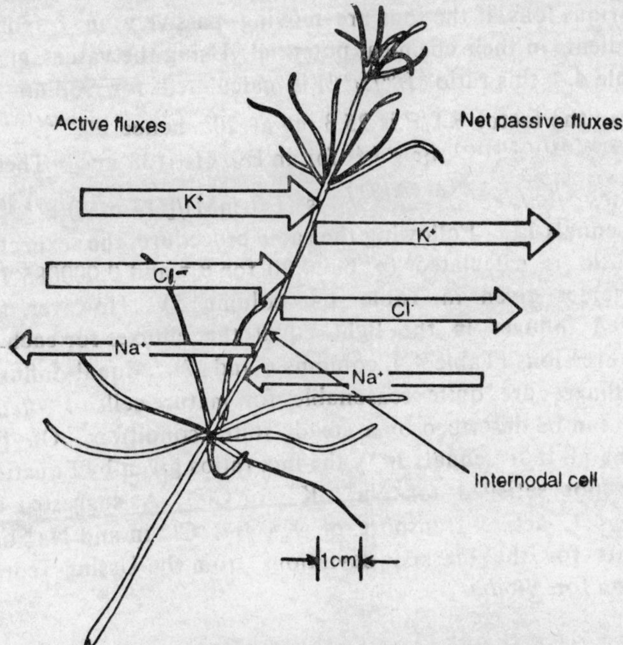

Fig. 4.7. In the steady state condition in the light, the three active fluxes across the plasmalemma of the large internodal cell of *Nitella* are balanced by the net passive effluxes of potassium and chloride and the net passive influx of sodium.

than $E_{N_{Na}}$ (−67 mv), Na^+ is at a lower chemical potential in the cytoplasm than outside in the external solution. Reasoning in an analogous manner, we find that K^+ ($E_{N_K} = -179$ mv) has a higher chemical potential inside, while Cl^- ($E_{N_{Cl}} = 99$ mv) is at a much higher chemical potential inside. If these ions can move across the plasmalemma, this suggests an active transport of K^+ and Cl^- into the cell and an active extrusion of Na^+ from the cell, as is schematically indicated in Fig. 4.7.

We can also consider the movement of the various ions into and out of *Nitella* in terms of the Ussing-Teorell equation to help determine whether active transport need be invoked to explain the fluxes. The Ussing-Teorell equation predicts that the quantity appearing on the right side of Equation 23 ($c_j^o / c_j^i e^{z_j FE_M/RT}$) should equal the ratio of influx to efflux of the various ions, if the ions are moving passively in response to gradients in their chemical potential. Using the values given in Table 4.4, this ratio (J_j^{in}/J_j^{out}) is calculated for sodium as follows : the factor RT/F is 25.3 mv at 20° hence $e^{z_j FE_M/RT}$ is $e^{(1)(-138\ mv)/(25.3\ mv)}$, or $e^{-5.45}$ for an E_M of −138 mv. Therefore, $c_{Na}^o/(c_{Na}^i e^{-z_{Na} FE_M/RT})$ is (1.0 mM)/[(14 mM)($e^{-5.45}$)], which equals 17. Following the same procedure, the expected flux ratio is calculated to be 0.20 for K^+ and 0.000085 for Cl^- (values given in Table 4.4, column 5). However, the observed influxes in the light equal the effluxes for each of these three ions (Table 4.4, columns 6 and 7). Equal influxes and effluxes are quite reasonable for mature cells of *Nitella*, which can be described by a steady-state condition. On the other hand, if J_j^{in} equals J_j^{out}, the flux ratios given by Equation 23 are not satisfied for Na^+, K^+, or Cl^-. As suggested by Figure 4.7, active transport of K^+ and Cl^- in and Na^+ out accounts for the marked deviations from the Ussing-Teorell equation for *Nitella*.

As we mentioned earlier, another approach for studying active transport is to remove the supply of energy. In the case of *Nitella*, cessation of illumination causes an appreciable decrease in the Na efflux, in the K influx, and in the Cl influx

(Table 4.4, columns 8 and 9). But these are the three fluxes which are toward regions of higher chemical potential for the particular ions involved, and thus we may reasonably expect all three to be active. On the other hand, some fluxes remain essentially unchanged by placing the cells in the dark (values in Table 4.4, columns 8 and 9, refer to the fluxes soon after extinguishing the light, not the steady-state fluxes). For instance, the Na influx and K efflux are initially unchanged when the *Nitella* cells are transferred from the light to the dark, *i.e.*, these unidirectional fluxes toward lower chemical potentials do not depend on energy derived from photosynthesis. (For reasons which are as yet unclear, the energetically downhill efflux of Cl^- apparently increases in the dark. This J_{Cl}^{out} is not included in Table 4.4).

The passive diffusion of the ions toward regions of lower chemical potential helps create the electrical potential difference across a membrane, while active transport is instrumental in maintaining the asymmetrical ionic distributions which sustain the passive fluxes. Thus, the passive and the active fluxes are interdependent in the ionic relations of cells and both are crucial for the generation of the observed diffusion potentials. Moreover, active and passive fluxes can occur simultaneously in the same direction. For example, we calculated that J_{Na}^{in} should equal 17 times J_{Na}^{out}, if both fluxes were passive ones obeying the Ussing-Teorell equation. Since J_{Na}^{in} is passive and equal to 0.55 picomole/cm^2-sec, we expect a passive efflux of Na^+ equaling (0.55)/(17), or 0.03 picomole/cm^2-sec. Thus, the active component of the Na efflux in the light may be 0.55−0.03, or 0.52 picomole/cm^2-sec. At cessation of illumination, J_{Na}^{out} decreases from 0.55 to 0.1 picomole/cm^2-sec (Table 4.4). Extinguishing the light removes photosynthesis as a possible enery source for active transport, but respiration could still supply energy in the dark. This helps explain why J_{Na}^{out} in the dark does not decrease all the way to 0.03 picomole/cm^2-sec, the value predicted for the passive efflux.

Suppose that the chemical potential some species is higher outside than inside a cell, as is illustrated in Figure 4.6. The minimum amount of energy needed to transport a mole of that species from the internal aqueous phase on one side of some membrane to the external solution on the other is the difference in chemical potential of that solute across the membrane, $\mu_j{}^o - \mu_j{}^i$ there, $\mu_j{}^o > \mu_j{}^i$). As we noted in considering Equation 24, the quantity $\mu_j{}^o - \mu^i$ for ions is RT ln $(a_j{}^o/a_j{}^i) - z_j FE_M$. Since the Nernst potential, E_{Nj}, is (RT/z_jF) ln $(a_j{}^o/a_j{}^i)$ (Equation 7), we can express the difference in chemical potential across the membrane as follows :

$$\mu_j{}^o - \mu_j{}^i = z_j F(E_{Nj} - E_M)$$

or

$$\mu_j{}^i - \mu_j{}^o = z_j F(E_M - E_{Nj}) \qquad ...(25)$$

We will use the bottom line of Equation 25 to discuss the active transport of ions *into* cells (*i.e.*, when $\mu_j{}^i > \mu_j{}^o$).

Using the Nernst potentials of Na^+, K^+, and Cl^- for *Nitella translucens* and the value of E_M given in Table 4.4, we can calculate $z_j(E_M - E_{Nj})$ in Equation 25 for transporting these ions across the plasmalemma of this alga. Such a quantity is $(+1)[(-138 \text{ mv}) - (-67 \text{ mv})]$, or -71 mv for Na^+, $+41$ mv for K^+, and $+237$ mv for Cl^-. These values for $z_j(E_M - E_{Nj})$ mean that Na^+ is at a higher chemical potential in the external bathing solution while K^+ and Cl^- are at higher chemical potentials inside the cell, as we concluded above (also, see Figs. 4.6 and 4.7. From Equation 25 the minimum energy required to actively transport, or "pump" Na^+ out across the plasmalemma of the *Nitella* cell equals F times the energy difference in millivolts, which is (0.02306 kcal/mole-mv)(71 mv), or 1.6 kcal/mole in the case of Na^+ (F is 0.02306 kcal/mole-mv). Potassium requires (0,02306)(41), or 0.9 kcal/mole to be pumped inward. The active extrusion of Na^+ from certain algal cells may be linked to the active uptake of K^+, ATP being implicated as the energy source for this coupled exchange process. The hydyolysis of ATP under biological conditions usually releases at least 10 kcal of energy/mole. For the case of a *Nitella* cell, this is sufficient energy per mole of ATP hydrolyzed to pump a mole of Na^+ out (1.6 kcal/mole) and one of K^+ in (0.9 kcal/mole). The transport Cl^- inward takes a minimum of (0.02306)(237), or 5.5 kcal/mole, according to Equation 25 which is a fairly large amount of energy. Neither the form of the energy used nor the actual mechanism involved in actively transporting chloride into *Nitella* or other plant cells is fully understood at the present time, although exchanges with OH^- appear to be involved.

The active uptake of K^+ and Cl^- together with an active extrusion of Na^+ in the case of *Nitella* are actually fairly widespread phenomena among plant cells. We might ask, why does a cell actively transport K^+ and Cl^- in and Na^+ out ? Although no completely satisfactory answer can be given to such a question, we shall *speculate* on possible reasons, based on the principles we have been considering.

Let us imagine that a membrane-bounded cell containing negatively charged proteins is placed in an NaCl solution, the latter possibly reflecting primeval conditions when life originated. When Na^+ and Cl^- are both in equilibrium across the membrane, E_M equals $E_{N_{Cl}}$ and $E_{N_{Na}}$. Using concentrations in Equation 3.5, log $(c_{Na}{}^o/c_{Na}{}^i)$ then equals $-$log $(c_{Cl}{}^o/c_{Cl}{}^i)$, or $c_{Na}{}^o/c_{Na}{}^i = c_{Cl}{}^i/c_{Cl}{}^o$, hence $c_{Na}{}^o c_{Cl}{}^o = c_{Na}{}^i c_{Cl}{}^i$. For electroneutrality in an external solution containing only NaCl, $c_{Na}{}^o$ equals $c_{Cl}{}^o$. Since $a^2 = bc$ implies that $2a \leqslant b+c$, we conclude that $c_{Na}{}^o + c_{Cl}{}^o \leqslant c_{Na}{}^i + c_{Cl}{}^i$. But the proteins which cannot diffuse across the membrane, also make a contribution to the internal osmotic pressure, so that π^o would be less than π^i. Water tends to flow of course toward higher osmotic pressures. When placed in an NaCl solution, water would therefore tend to enter such a membrane-bounded cell containing proteins causing it to swell without limit. Thus, an outwardly directed active transport of Na^+ might have originated to lower $c_{Na}{}^i$ and thus to prevent excessive swelling of primitive cells.

An energy-dependent uptake into a plant cell tends to increase π^i, which leads to a rise in P^i. This higher internal hydrostatic pressure favours cell enlargement and consequently cell growth. For a plant cell surrounded by a cell wall we might therefore expect an active transport of some species into the cell, *e.g.*, chloride. (Animal cells do not have to push against a cell wall in order to enlarge, and do not generally have an active uptake of Cl^-.) Enzymes have evolved that operate efficiently when exposed to relatively high concentrations of K^+ and Cl^-. In fact, many actually require K^+ for their activity, so that an inwardly directed K pump is probably necessary for biochemistry as we know it. The presence of a substantial concentration of such ions insures that electrostatic effects adjacent to a Donnan phase could in large part be screened out—otherwise, a negatively charged substrate, *e g.*, an organic acid or a phosphorylated sugar, might be electrostatically repelled from the catalytic site on an enzymes (proteins are generally negatively charged at cytoplasmic pH's). Once active transport has set up certain concentration differences across the membrane, the membrane potential is an inevitable consequence of the tendency of such ions to passively diffuse toward regions of lower chemical potential. Whether such diffusion potentials actually benefit the general plant cell is an open question. They are essential however for the transmission of electrical impulses in excitable cells in animals and in certain plants.

Carriers, solute uptake, and the Michaelis-Menten formalism. Although active transport is clearly of common occurrence, its actual mechanism—including the means whereby metabolic energy is used--remains uncertain. The possible involvement of a "carrier" molecule in such active transport of solutes across plant cell membranes was first suggested by W. J. V. Osterhout in the early 1930s. This carrier was proposed to bind selectively certain molecules and then to carry, or ferry them across the membrane. Carriers can provide a cell with the specificity or selectivity needed to control the entry and exit of the various types of solutes encountered. Thus, certain metabolites can be specifically taken into the cell, while photosynthetic and waste products can be selectively moved out across the membranes. Also, the active transport of certain inorganic nutrients into epidermal cells in the root allows a plant to obtain and accumulate these solutes from the soil. Even though the all-important mechanism for physically binding and moving the solutes through the membrane is still unknown, the carrier concept has found widespread application in the interpretation of experimental observations.

Based primarily on results from competition studies, certain solutes are described as being bound to, or associated with a particular carrier. When an ion of some species is attached to a carrier, another similar ion (of the same or a different species) competing for the same binding site cannot also be bound. For example, the similar monovalent cations K^+ and Rb^+ appear to bind in a competitive fashion to the same carries. For some cells, one and the same carrier might transport Na^+ out of the cell and K^+ in—the so-called sodium-potassium pump alluded to above. Calcium and strontium may compete with each other for binding sites on another common carrier. Two other divalent cations, Mg^{++} and Mn^{++}, are apparently transported by a single carrier that is different from the one for Ca^{++} and Sr^{++}. The halides (Cl^-, I^-, and Br^-) may also be transported by a single carrier. Although ions have been shown to bind specifically to certain proteins isolated from cellular membranes, much remains to be learned about carriers.

One of the most important variables in the study of carrier mediated uptake is the external concentration. As the external concentration of a solute increases, the rate of uptake generally reaches an upper-limiting value. We may then presume that all binding sites on the carriers for that particular ion or neutral solute have become filled or saturated. In particular, the rate of active uptake of species j, J_j^{in}, is often proportional to the external concentration of that solute (c_j^o) over the lower range of concentrations, leveling off as c_j^o is raised, and eventually approaching a maximum rate for the influx ($J_j^{in}{}_{max}$). We can describe this kind of behaviour by the following equation :

$$J_j^{in} = \frac{J_j^{in}{}_{max}\, c_j^o}{K_j + c_j^o} \qquad \text{(26)}$$

where K_j is a constant characteristic of species j crossing a particular membrane, expressed in the units of concentration. For the uptake of many ions into roots and other plant tissues, $J_j^{in}{}_{max}$ is 1 to 10 μ moles/g fresh weight-hr. Often, two different K_js are observed for the uptake of one and the same ion into a root. The lower one is generally between 6 and 100μM (which is in the concentration range of many ions in the soil water), while the other K_j can be above 10 mM.

Equation 26 is similar in appearance to the Michaelis-Menten equation used for describing enzyme kinetics in biochemistry. The substrate concentration in the latter relation is analogous to c_j^o in Equation 26 and the enzyme reaction velocity to J_j^{in}. The term in the Michaelis-Menten equation equivalent to K_j in Equation 26 is the substrate concentration for half-maximal velocity of the reaction, K_M (the Michaelis constant). We can press the analogy between enzyme kinetics and active uptake even further. The lower the K_M, the greater is the affinity of the enzyme for the substrate. Likewise. a low value for K_j indicates that the ion or other solute is more readily bound to some substance in the membrane such as a "carrier" molecule. The physiological consequence of a low K_j is that species *j* can be efficiently bound, even when its external concentration is relatively low; such a species is actually favored or selected for active transport into the cell.

The two most common ways of graphing data on solute uptake which fit Equation 26 are illustrated in Figure 4.8 a shows that when the external concentration of species *j*, c_j^o, is equal to K_j, J_j^{in} equals $\frac{1}{2}J_{j\ max}^{in}$, as we can see directly from Equation 26 *i.e.*, J_j^{in} is then $J_{j\ max}^{in}k/(K_j+K_j)$, or $\frac{1}{2}J_{j\ max}^{in}$. Thus, K_j is the external concentration at which the rate of

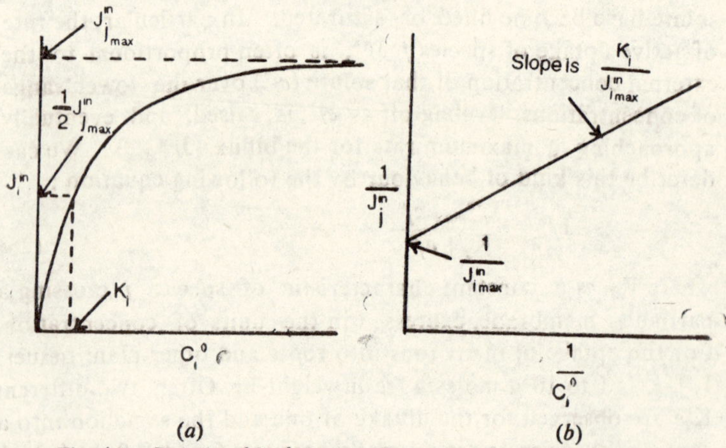

Fig. 4.8. Relationship between the external solute concentration (c_j^o) and the rate of influx (J_j^{in}) for active uptake according to a Michaelis-Menten type of kinetics, as given by Equation 26: (*a*) lineay plot (*b*) reciprocal plot.

active uptake is half-maximal—in fact, the observed values of K_j have proved to be convenient parameters for describing the uptake of various solutes. If two different solutes compete for the same site on some carrier, then J_j^{in} for species j will be decreased by the presence of the second solute, which is known as *competitive inhibition* of species j. As the concentration c_j^o is increased, the uptake continuously rises and eventually approaches an upper limit, $J_j^{in}{}_{max}$, which is the asymptotic value for the active influx (Fig. 4.8*a*). In the case of competitive inhibition, $J_j^{in}{}_{max}$ is not affected, since in principle we can raise c_j^o high enough so that the same maximum rate for the active uptake of species j is eventually obtained. However, the half-maximum rate occurs at a higher concentration, which is another way of saying that K_j is raised if a competing solute is present.

For many purposes, it is advantageous to plot the experimental data in such a way that a linear relationship is obtained when Equation 26 is satisfied for the active uptake of species j. Taking reciprocals of both sides of Equation 26 we note that $1/J_j^{in}$ equals $(K_j+c_j^o)/(J_j^{in}{}_{max}\,c_j^o)$, which is $K_j/J_j^{in}{}_{max}\,c_j^o)$ $+1/J_j^{in}{}_{max}$. Hence, when $1/J_j^{in}$ is plotted versus $1/c_j^o$ (Fig. 4.8*b*), Equation 26 takes the form of a straight line of slope $K_j/J_j^{in}{}_{max}$, with an intercept on the ordinate of $1/J_j^{in}{}_{max}$. This latter method of treating the experimental results has proved to be rather convenient, and has found widespread application in studies of solute uptake by plant tissues, especially into roots. Since K_j increases while $J_j^{in}{}_{max}$ remains the same for competitive inhibition, the presence of a competing species causes the slope of the line in Figure 4.8*b* (*i.e.*, $K_j/J_j^{in}{}_{max}$) to be greater, while the intercept on the y-axis ($1/J_j^{in}{}_{max}$) is unchanged. Certain compounds that are structurally unrelated to species j can inhibit the functioning of the carriers. Such a *noncompetitive* inhibitor does not bind to the site used for transporting species j across the membrane, and K_j is unaffected. Since a noncompetitive inhibitor lowers the maximum rate of active influx ($J_j^{in}{}_{max}$), both the slope and the y-axis intercept in Figure 4.8*b* are changed.

The basis of the mathematical form of Equation 26 for describing uptake is the competitive binding of solutes to specific sites on a limited number of carriers. In other words, active processes involving metabolic energy do not necessarily have to be invoked. If a solute were to diffuse across the membrane only when bound to a carrier, the expression for the influx could also be Equation 26. This passive entry of a solute mediated by a carrier is termed *facilitated diffusion.*

Since facilitated diffusion is so important in biology and yet is often misunderstood, we will briefly elaborate upon it here. Certain molecules passively enter cells more readily than would be expected from consideration of their molecular structure or observations with analogous substances (*e.g.*, by comparing P_js). It would appear that some mechanism is *facilitating* their entry. Moreover, the net flux is still toward lower chemical potentials and hence would be in the same direction as ordinary diffusion. To help explain facilitated diffusion. carriers are proposed to act as shuttles for a net passive movement of the specific molecules across the membrane towards regions of lover energy. Instead of the usual diffusion across the barrier—based on random thermal motion of the solutes—carriers select out and bind certain molecules. The solute-carrier complex then diffuses across the membrane—or perhaps the solute-carrier complex simply rotates 180° in the membrane—and, finally, the solutes are somehow released on the other side of the membrane without the intervention of metabolism. Such facilitation of entry by carriers may also be regarded as a special means of lowering the activation energy needed for the solute to cross the energy barrier represented by the membrane. Thus, carriers facilitate the influx of solutes in the same way that enzymes facilitate biochemical reactions.

There are certain general characteristics of a facilitated diffusion system. As already mentioned, the net flux is toward lower chemical potentials. (According to the usual definition, active transport is in the energetically uphill direction and, conceivably, it can employ the same carriers used in facilitated diffusion.) Movement by facilitated diffusion is more rapid than that expected for ordinary diffusion. Also, the carriers can exhibit a high degree of selectivity, *i.e.*, they can be specific for certain types of molecules, while not binding closely related species. In addition, carriers in facilitated diffusion become saturated when the external concentration of the species transported is raised sufficiently, a behaviour consistent with Equation 26. Finally, since carriers can exhibit competition phenomena, the flux of a species entering the cell by facilitated diffusion can be reduced when structurally analogous molecules are added to the external solution. (Such molecules would compete for sites on the carrier and thereby reduce the binding and the subsequent transfer of the original species into the cell.)

For convenience, we have been discussing facilitated diffusion *into* a cell, while exactly the same principles apply for exit. Let us assume that a carrier for potassium exists in the membrane of a certain cell and that it is being used as a shuttle for facilitated diffusion. Not only would the carrier lead to an enhanced net flux toward the side with the lower chemical potential, but also both the unidirectional fluxes J_K^{in} and K_K^{out} could be increased over the values predicted for ordinary diffusion. This increase in the unidirectional fluxes by a carrier has often been called "exchange diffusion." In such a case the molecules are interacting with a membrane component, *viz*, the carrier, hence the Using-Teorell equation—Equation 23 $J_j^{in}/J_j^{out}=c_j^o/(c_j^i e^{z_jFE_M/RT})$—would not be obeyed, since it presupposes that there are no interactions with other substances. In fact, observation of departures from predictions of the Using Teorell equation is often the way cases of exchange diffusion are actually discovered.

Both active and passive fluxes across the cellular membranes can occur concomitantly; these movements depend on concentrations in rather different ways. For passive diffusion, the unidirectional component J_j^{in} is proportional to c_j^o for neutral solules $[J_j=P_j(c_j^o-c_j^i)]$, and by Equation 14 for ions. This proportionality strictly applies only over the range of external concentrations for which the permeability coefficient is essentially independent of concentration, and the membrane potential must not change in the case of charged solutes.

Nevertheless, ordinary passive influxes do tend to be propor tional to the the external concentration, while an active influx or the special passive influx known as facilitated diffusion—either of which can be described by a Michaelis-Menten type of formalism—shows saturation effects at the higher concentrations. Moreover, facilitated diffusion and active transport exhibit both a high degree of selectively and competition phenomena, while ordinary diffusion does not.

Principles of irreversible thermodynamics

So far, we have been using classical thermodynamics—though it may have been noticed that we have often done so somewhat illegitimately. For example, let us consider Equation ($J_{V_w} = L_w \Delta\psi$). Since there is a difference in chemical potential given by a change in water potential, we expect a net (and irreversible) flow of water from one region to another, which is obviously not an equilibrium situation. Strictly speaking, classical thermodynamics is concerned solely with equilibria, not with movement—classical thermodynamics might have been better-named "thermostatics". Thus, we have frequently been concerned with a kind of hybrid—appealing to classical thermodynamics for the driving forces, and using nonthermodynamic arguments and analogies to discuss fluxes. One of the objectives of irreversible thermodynamics is to help legalize the arguments. But, as well see, legalizing them brings in new ideas and considerations.

Irreversible thermodynamics uses the same parameters as classical thermodynamics, *viz.*, temperature, pressure, free energy, activity, and so on. But these quantities have been strictly defined for macroscopic amounts of matter in equilibrium situations only. How can we use them to discuss processes *not* in equilibrium, the domain of irreversible thermodynamics ? This dilemma immediately circumscribes the range of validity of the theory of irreversible thermodynamics it can only deal with "slow" processes or situations which are not very far from equilibrium, for only in such circumstances can equilibrium-related concepts such as temperature and free energy retain their validity—at least approximately. We have to assume from the outset that we can talk about and use classical thermodynamic parameters even in nonequilibrium situations.

Another refinement that we should introduce into our theory is to recognize that the movement of one species may affect the flux of a second species. A particular flow of some solute may interact with another flux by way of collisions, each species flowing under the influence of its own force. For example, water, ions, and other solutes moving through a membrane toward regions of lower values for their respective chemical potentials can exert a frictional drag on each other. In this case, the magnitude of the flus of a solute may depend on whether water is also flowing. In this way, the fluxes of various species across a membrane become interdependent. Stated more formally, the flux of a solute is not only dependent on the negative gradient of its own chemical potential—which is the sole driving force we have recognized up to now—but it may also be influenced by the gradient in the chemical potential of water. Again using Equation ($J_{V_w} = L_w \Delta \psi$) as an example, we have considered that the flow of water depends only on the difference in its own chemical potential between two locations, and have thus far ignored any coupling to concomitant fluxes of solutes.

A quantitative description of interdependent fluxes and forces is given by irreversible thermodynamics, a subject which treats nonequilibrium situations such as those actually occurring under biological conditions. In this brief introduction to irreversible thermodynamics we will emphasize certain underlying principles and then derive the *reflection coefficient*, a parameter important in plant and cell physiology. In order to keep the analysis manageable, we will restrict our attention to isothermal conditions, which approximate many biological situations where fluxes of water and solutes are considered.

Fluxes, forces, and Onsager coefficients. In our previous discussion of fluxes, the driving force leading to the flux of species j, J_j, was the negative gradient in its chemical potential, $-\partial\mu_j/\partial x$. Irreversible thermodynamics takes a more general view, namely, that the flux of species j depends not only on—

$\partial\mu_j/\partial x$, but can be affected by *any* other force occurring in the system, such as the negative gradient in the chemical potential of some other species. A particular force, X_k, can likewise influence the flux of any species. Thus, the various fluxes become interdependent, or coupled, since they can respond to changes in any of the forces. Another premise in irreversible thermodynamics is that J_j is linearly dependent on the various forces. This means that in practice we will treat only those cases which are not too far from equilibrium. Even with such a simplification the algebra often becomes rather cumbersome, owing to the coupling of the various forces and fluxes.

Using a linear combination of all the forces, we can represent the flux of species j by

$$J_j = \sum_k L_{jk}X_k = L_{j1}X_1 + L_{j2}X_2 + \ldots + L_{jj}X_j + \ldots \qquad \ldots(27)$$

where the summation, $\sum_k$, is over all forces (all X_k), and the L_{jk} are referred to as the *Onsager coefficients*, or the phenomenological coefficients, in this case for conductivity. The first subscript on these coefficients, *e.g.*, j on L_{jk}, identifies the flux; the second subscript (*i.e.*, k on L_{jk}) designates the force. Each term, $L_{jk}X_k$, is thus the partial flux of species j due to the particular force X_k. The individual Onsager coefficients in Equation 27 are therefore the proportionality factors indicating what contribution the force X_k makes to the flux of species j, J_j Equation 27 is sometimes referred to as the *phenomenological equation.* Phenomenological equations are used to describe relations between observable quantities, without regard for explanations in terms of atoms or molecules. For instance, Ohm's law and Fick's laws are also phenomenological equations, likewise they assume linear relations between forces and fluxes.

Are there any relations between the various phenomenological coefficients ? In fact, L_{jk} equals L_{kj}, which is known as the *reciprocity relation.* Such an equality of cross-coefficients was derived in 1931 by Onsager—awarded the Nobel Prize in chemistry in 1968—from statistical considerations which utilize the principle of "detailed balancing"—an argument involving microscopic reversibility, *i.e.*, for local equilibrium, any molecular process and its reverse will be taking place at the same average rate in that region. In discursive language, the Onsager reciprocity relation means that the proportionality coefficient relating the flux of species k caused by the force on species j equals the proportionality coefficient giving the flux of j due to the force on λ. (Strictly speaking, conjugate forces and fluxes must be used, as will be the case below. The fact that L_{jk} equals L_{kj} can be further appreciated by considering Newton's third law; action and reaction must be equal. For example, the frictional drag exerted by a moving solvent on the solute is equal to the drag exerted by the moving solute on the solvent. The pairwise equality of cross-coefficients given by the Onsager reciprocity relation reduces the number of coefficients needed to describe the interdependence of forces and fluxes in irreversible thermodynamics and consequently leads to a simplification in solving the sets of simultaneous equations.

Water and solute flow. As a specific application of the principles just introduced, we will consider in some detail the important coupling of water and solute flow. The driving forces for the fluxes are the negative gradients in chemical potential, which we will assume are proportional to the differences in chemical potential across some barrier, here considered to be a membrane. In particular, we will represent $-\partial\mu_j/\partial x$ by $\Delta\mu_j/\Delta x$, which is $(\mu_j^o-\mu_j^i)/\Delta x$ in the present case (For convenience, the thickness of the barrier, Δx, will be incorporated into the coefficient multiplying $\Delta\mu_j$.) To help keep the algebra relatively simple, the development will be for a single nonelectrolyte. The fluxes are across a membrane permeable to both water (w) and the single solute (s). From Equation 27, we can represent the fluxes as the following linear combination of the differences in chemical potential:

$$J_w=L_{ww}\Delta\mu_w+L_{ws}\Delta\mu_s \qquad ...(28)$$

$$J_s=L_{sw}\Delta\mu_w+L_{ss}\Delta\mu_s \qquad ...(29)$$

Equations 28 and 29 allow for the possibility that each of the fluxes may depend on the differences in both chemical potentials, $\Delta\mu_w$ and $\Delta\mu_s$. Four phenomenological coefficients are used in these two equations. By the Onsager reciprocity relation, L_{ws} equals L_{sw}. Thus, three different coefficients (L_{ww}, L_{ws}, and L_{ss}) are needed to describe the relationship of these two fluxes to the two driving forces. This is in constant to Equation $J_j = u_j c_j\,(-\partial\mu_j/\partial x)$, where each flux depends on but one force; accordingly, only two coefficients would then be involved in describing J_w and J_s.

To obtain more convenient formulations for the fluxes of water and the solute, we generally express $\Delta\mu_w$ and $\Delta\mu_s$ in terms of the differences in the osmotic and the hydrostatic pressures, $\Delta\pi$ and ΔP—it is usually easier to measure ΔP and $\Delta\pi$ than it is to determine $\Delta\mu_w$ and $\Delta\mu_s$. This is a straightforward matter for $\Delta\mu_w$; the only possible ambiguity is in deciding the algebraic sign. For the present case, $\Delta\mu_w$ is the chemical potential of water on the outside minus that on the inside, $\mu_w{}^o - \mu_w{}^i$. From equation $(\mu_w = \mu_w - \bar{V}_w\pi + \bar{V}_w P + m_w gh)$, $\Delta\mu_w$ is given by

$$\Delta\mu_w = -\bar{V}_w\Delta\pi + \bar{V}_w\Delta P \qquad \ldots(30)$$

where $\Delta\pi$ here equals $\pi^o - \pi^i$ and ΔP is $P^o - P^i$ ($\Delta h = 0$ across a membrane).

To express $\Delta\mu_s$ in terms of $\Delta\pi$ and ΔP, we must first consider the activity term, RT ln a_s, in a little detail. We will begin with the differential RT $d(\ln a_s)$, which equals RT da_s/a_s, or RT $d(\gamma_s c_s)/\gamma_s c_s$. When γ_s is constant, this latter quantity becomes simply RT dc_s/c_s, From Equation $(\pi_s = RT\sum_j c_j)$, RTdc_s equals $d\pi$ for a dilute solution of a single solute. Hence, RT $d(\ln a_s)$ can be replaced by $d\pi/c_s$ as a useful approximation. In expressing the difference in chemical potential across a membrane we are interested in macroscopic changes, not in the infinitesimal changes given by differentials. To go from differentials to differences (Δ), RT$d(\ln a_s)$ becomes RTΔ ln a_s and $d\pi/c_s$ can be replaced by $\Delta\pi/\bar{c}_s$, where $\bar{c}_s$ is essentially the mean concentration of solute *s*, in this case across the membrane. (Alternatively, we can simply define c_s as that concentration for which RTΔ ln a_s exactly equals $\Delta\pi/\bar{c}_s$.) Therefore, we can replace the term RTΔ ln a_s in $\Delta\mu_s$ by the equivalent term, $\Delta\pi/\bar{c}_s$. From Equation, μ_s equals μ_s + RT ln a_s

$+\bar{V}_s P$ for a neutral species, and the difference in chemical potential of the neutral solute across the membrane ($\Delta\mu_s$) becomes

$$\Delta\mu_s = \frac{1}{c_s}\Delta\pi + \bar{V}_s\Delta P \qquad \text{...(31)}$$

The two expressions representing the driving forces, $\Delta\mu_w$ (Equation 30) and $\Delta\mu_s$ (Equation 31), are expressed as functions of the same two pressure differences, $\Delta\pi$ and ΔP, the latter being rather convenient from the point of view of experiment.

Volume flow, L_P, and σ. Now that we have appropriately expressed the chemical potential differences of water and the solute, we will direct our attention to the fluxes. In our usual units, the fluxes J_w and J_s are the moles of water and solute, respectively, moving across one cm² of membrane surface in a second. A quantity of considerable interest in plant and cell physiology is the volume flow, J_V, which is the rate of movement of the total *volume* of both water and solute across unit area of the membrane. J_V has the units of volume per unit area per unit time, *e.g.*, cm³/(cm²-sec), which is cm/sec.

The molar flux of some species j (J_j) in moles/cm²-sec times the volume occupied by each mole of it ($\bar{V}_j$) in cm³/mole gives the volume flow for that component (J_{Vj}) in cm/sec. For solute and water both moving across the membrane, the total volume flow J_V is the volume flow of water plus that of solute, which in the case of a single solute can be represented as follows :

$$J_V = V_w J_w + V_s J_s \qquad \text{...(32)}$$

It is generally simpler and more convenient to measure the total volume flux (such as that given in Equation 32 than one of the component volume fluxes ($J_{Vj} = \bar{V}_j J_j$). For instance, we can often determine the volumes of cells or organelles under different conditions and relate any changes in volume to J_V. The volume flux of water, J_{V_w} is simply $\bar{V}_w J_w$.

However straightforward it may be, the algebraic substitution necessary to incorporate the various forces and fluxes into the volume flow given by Equation 32 leads to a rather cumbersome expression. Nevertheless, it provides—albeit in symbols—insight into the origin of the reflection coefficient, an important quantity that we will consider in detail below. When the fluxes of water and solute given by Equations 28 and 29, respectively, are substituted into the expression for the volume flow (Equation 32), we obtain the following relation :

$$J_V=(V_w L_{ww}+V_s L_{sw})\Delta\mu_w+(V_w L_{ws}+V_s L_{ss})\Delta\mu_s \qquad \ldots(33)$$

As we mentioned above, L_{sw} equals L_{ws}, so that only the symbol L_{ws} need be retained here. Next, we will incorporate the chemical potential differences across the membrane for water and the solute, as given by the Equation 30 and 31 respectively, into Equation 33. Upon rearrangement, J_V is then as follows :

$$J_V=(\bar{V}_w^2 L_{ww}+2\bar{V}_s \bar{V}_w L_{ws}+\bar{V}_s^2 L_{ss})\Delta P$$

$$-(\bar{V}_w^2 L_{ww}-\bar{V}_w\frac{1}{\bar{c}_s}L_{ws}-\bar{V}_s\frac{1}{\bar{c}_s}L_{ss}$$

$$+\bar{V}_s\bar{V}_w L_{ws})\Delta\pi \qquad \ldots(34)$$

Equation 34 gives the dependence of the total volume flow on the hydrostatic and osmotic pressure differences across some membrane permeable to both water and the solute.

The quantities multiplying ΔP and $\Delta\pi$ in Equation 34 are rather cumbersome, so we will find it convenient to define some new parameters. The factor multiplying the difference in hydrostatic pressure, ΔP, is called the *hydraulic conductivity coefficient*, L_P. From Equation 34, this hydraulic conductivity coefficient has the following value.

$$L_P=\bar{V}_w^2 L_{ww}+2\bar{V}_s\bar{V}_w L_{ws}+\bar{V}_s^2 L_{ss} \qquad \ldots(35)$$

Such a coefficient describes the membrane conductivity, since the larger L_P, the greater is J_V for a given hydrostatic pressure difference cross the membrane. The other unwieldly factor appearing in Equation 34, *viz.*, the parenthesis multiplying $\Delta\pi$, is incorporated into the reflection coefficient, σ. It is advantageous to define σ as the ratio of the coefficient of $\Delta\pi$ divided by the coefficient of ΔP in the expression for the volume flow (Equation 34). Using the definition of L_P (Equation 35), σ is therefore given by

$$\sigma=\frac{\bar{V}_w^2 L_{ww}-\bar{V}_w \frac{1}{\bar{c}_s} L_{ws}-\bar{V}_s \frac{1}{\bar{c}_s} L_{ss}+\bar{V}_s\bar{V}_w L_{ws}}{L_P} \quad ...(36)$$

The reflection coefficient defined by Equation 36 is perhaps the most important parameter from irreversible thermodynamics relevant to the fluxes generally considered in plant physiology. It was introduced by Staverman in 1948 and specifically applied to biological situations by Kedem and Katchalsky in 1958. Since 1958, numerous applications of irreversible thermodynamics, in general, and reflection coefficients, in particular, have been made in biology. We will shortly consider specific values of σ for solutes entering chloroplasts. We will also show that the maximum value of the reflection coefficient occurs when the solute is unable to penetrate the membrane—the molecules are then all *reflected* from the barrier. In the absence of a hydrostatic pressure difference across a membrane, the volume flow (J_V) equals—$L_P\sigma\Delta\pi$ by Equation 34. Thus, the magnitude of the dimensionless parameter σ determines the volume flow expected in response to a difference in osmotic pressure across a membrane—it is this use which is most pertinent in plant physiology. After further considering the expression for volume flow, we will discuss the properties of reflection coefficients (σ is derived in a more conventional).

Using L_P and σ as defined by Equations 35 and 36 respectively, we can rewrite the expression for the total volume flow (Equation 34 in the following convenient form :

$$J_V=L_P(\Delta P-\sigma\Delta\pi) \quad ...(37)$$

Equation 37 indicates that the osmotic contribution to the volume flow across the membrane is changed by the factor σ, which is less than one when the solute can across the membrane (as we will soon derive).

Many different solutes can penetrate a biological membrane under usual conditions. Each such species *j* can be characterized by its own reflection coefficient, σ_j, for that particular membrane. The volume flow given by Equation 37 can then be generalized as

$$J_V = L_P(\triangle P - \sum_j \sigma_j \triangle \pi_j) \qquad \text{...(38)}$$

where $\triangle \pi_j$ is the osmotic pressure difference across the membrane for species j (*e.g.*, $\triangle \pi_j = RT\triangle c_j$. Although interactions with water are still taken into account, the generalization represented by Equation 38 introduces the assumption that the solutes do not interact with each other as they penetrate the membrane. Moreover, J_j in Equation 38 refers only to the movement of neutral species—otherwise we would also need a current equation to describe the flow of charge. Nevertheless, Equation 38 is a useful approximation of the actual situation, describing the multicomponent solutions encountered by cells, and we with use it as the starting point for our general consideration of solute movement across membranes. Before discussing such movement, however, let us consider the range of value of reflection coefficients.

Values of reflection coefficients. A reflection coefficient characterizes some particular solute interacting with a specific membrane. In addition, σ_j depends on the solvent on either side of the membrane, as the subscripts w for water in Equation 36 clearly indicate—water is the only such solvent we will consider here. Two extreme conditions can describe the passage of solutes : impermeability, which leads to the maximum value of unity for the reflection coefficient, and nonselectivity, where σ_j is zero. A reflection coefficient of zero may describe the movement of a solute across a very coarse barrier (one with large pores) which does not distinguish or select between solute and solvent molecules, or it may refer to the passage through a membrane of a molecule very similar in size and structure to water itself.

Impermeability describes the limiting case where water can cross some membrane, but the solute cannot. When the solute does not penetrate the membrane at all. J_s must remain zero for any and all values of $\triangle\mu_w$ and $\triangle\mu_s$. Since J_s equals $L_{sw}\triangle\mu_w + L_{ss}\triangle\mu_s$ (Equation 29), the only way in which J_s will always be zero is for both L_{ss} and L_{ws} to be zero. Therefore, when the membrane is impermeable to the particular solute under consideration, both L_{ss} and L_{ws} are zero. Putting these zero values for L_{ss} and L_{ws} into Equations 35 and 36, we obtain the following two relations : $L_P = \overline{V}_w^2 L_{ww}$ (Equation 35) and $\sigma = \overline{V}_w^2 L_{ww}/L_P$ (Equation 35). Consequently, σ then equals $\overline{V}_w^2 L_{ww}/\overline{V}_w^2 L_{ww}$, which is one. A reflection coefficient of unity thus signifies that the solute is not crossing the membrane, *i.e.*, all of the solute is "reflected" from the membrane, and the reflection coefficient has its maximum value. When a solute can penetrate, L_{ss} takes on a positive value, *i.e.*, the force on solute *s* causes a flux of it across the membrane. The $-L_{ss}\overline{V}_s/c_s$ term then causes the numerator of the expression for σ (Equation 36) to decrease as L_{ss} increases, while the denominator (L_P, see Equation 35) tends to increase as the solute's ability to penetrate goes up. The reflection coefficient thus becomes less than one when the solute can cross the membrane. This last comment emphasizes the fact that the reflection coefficient defined by Equation 36 handles the general case of solute permeability. Moreover, the two terms containing L_{ws} in the numerator of the expression for the reflection coefficient (Equation 36) take into account the further possibility that solute and solvent may interact when crossing the membrane, *i.e.*, σ can also be affected by coupling or interdependence of forces and fluxes.

When a membrane or barrier is nonselective, both water and the solute move across it at the same velocity, *i.e.*, $v_s = v_w$. The solution is then the same on both sides of the barrier, so that $\Delta\pi$ across the barrier is zero, which means that $\Delta\mu_w$ equals $\overline{V}_w \Delta P$ and $\Delta\mu_s$ is $\overline{V}_s \Delta P$. (Admittedly, the idea of a solution containing a single solute is not realistic from a biological point of view, but it is convenient for illustrating the minimum value for σ). As we discussed above v_j is simply J_j/c_j. Using these various conditions as well as Equations 28 and 29, we obtain the following string of equalities : $v_w = J_w/c_w = L_{ww}\Delta\mu_w/c_w + L_{ws}\Delta\mu_s/c_w = (L_{ww}\overline{V}_w/c_w + L_{ws}\overline{V}_s/c_w)\Delta P = v_s = J_s/c_s = L_{ws}\Delta\mu_w/c_s + L_{ss}\Delta\mu_s/c_s = (L_{ws}\overline{V}_w/c_s + L_{ss}\overline{V}_s/c_s)\Delta P$. Next, we will let the number of moles of water per unit volume (c_w) times the volume per mole of water ($\overline{V}_w$) equal unity

$-c_w\bar{V}_w \cong 1$ is another way of stating the dilute solution approximation. Also, we will identify c_s with $\bar{c}_s$. Putting these last two relations into the factors multiplying ΔP in the above string of equalities yields $L_{ww}\bar{V}_w^2 + L_{ws}\bar{V}_s\bar{V}_w = L_{ws}\bar{V}_w/\bar{c}_s + L_{s},\bar{V}_s/\bar{c}_s$. Finally, we can use this desired relation to evaluate the numerator of Equation 36. Accordingly, σ equals $0/L_P$, which is zero, *i.e.*, for the case of nonselectivity ($v_s = v_w$), the reflection coefficient of the solute must be zero for that barrier.

Another relatively simple condition by which σ equals zero can be readily understood from Equation 37 $[J_v = L_P(\Delta P - \sigma\Delta\pi)]$. Let us consider the rather realistic situation where ΔP is zero across a membrane. The volume flow (J_v) is then simply $-L_P\sigma\Delta\pi$, by Equation 37. For a solute having a reflection coefficient equal to zero for that particular membrane, the volume flow would be zero. By Equation 32 ($J_V = \bar{V}_wJ_w + \bar{V}_sJ_s$), a zero J_V implies that $\bar{V}_wJ_w$ equals $-\bar{V}_sJ_s$. In words, the volume flow of water must be equal and opposite to the volume flow of the solute in order to get no net volume flow. It follows that when ΔP is zero but $\Delta\pi$ is nonzero, the absence of a net volume flow J_v across a membrane permeable to both water and the single solute indicates that the reflection coefficient for that solute is zero. This condition of σ equaling zero occurs when the volume of water flowing toward the side with the higher π is balanced by an equal volume of solute diffusing across the membrane in the opposite direction toward the side where the solute is less concentrated (lower π). Such a situation of zero volume flow anticipates the concept of a "stationary state," introduced in the next section.

Solutes interacting with biological membranes exhibit properties ranging from impermeability ($\sigma_j = 1$) all the way to nonselectivity ($\sigma_j = 0$). Substances which are retained in, or excluded from plant cells have reflection coefficients rather close to unity for the cellular membranes. For instance, the σ_js for sucrose and amino acids are usually near 1.0 for plant cells. Methanol and ethanol enter cells very readily and have reflection coefficients of approximately 0.3 for some *Chara* and

Nitella internodal cells. On the other hand, σ_j can essentially equal zero for solutes crossing porous barriers, such as those presented by cell walls, or for molecules that are penetrating very readily across membranes, such as D_2O. As is the case for a permeability coefficient, the reflection coefficient of some species is the same for traversal in either direction across a membrane.

For many small neutral solutes not interacting with carriers in the membrane, the reflection coefficients are correlated with the partition coefficients. For example, when K_j for nonelectrolytes is less than about 10^{-3}, σ_j is generally close to one. Thus compounds which do not readily enter the lipid phase of the membrane (low K_j) also do not cross the membrane easily (σ_j near 1). When the partition coefficient is one or greater, the solutes can enter the membrane in appreciable amounts, and σ_j is generally close to zero. Considering the intermediate case, the reflection coefficient of some species j can be near 0.5 for a small nonelectrolyte having a K_j of about 0.1 for the membrane lipids, although individual molecules differ depending on their molecular weight, branching, and atomic composition. Neglecting frictional effects with other solutes, we see that the intermolecular interactions affecting the partition coefficient are similar to those governing the value of the reflection coefficient, and there is somewhat of a correlation therefore between the permeability coefficient of a solute ($P_j = D_j K_j / \Delta x$) and its reflection coefficient for the same membrane. In fact, Fig. 4.9 illustrates that there is even a correlation between the reflection coefficients of a series of nonelectrolytes determined with animal membranes and the permeability coefficients of the same substances measured for plant membranes. Thus, as the permeability coefficient goes from very small to very large values, the reflection coefficient decreases from unity (describing relative impermeability) to zero (for the opposite extreme of nonselectivity). In the next section we will specifically consider some of the consequences of reflection coefficients differing from unity for solutes crossing cellular and organelle membranes.

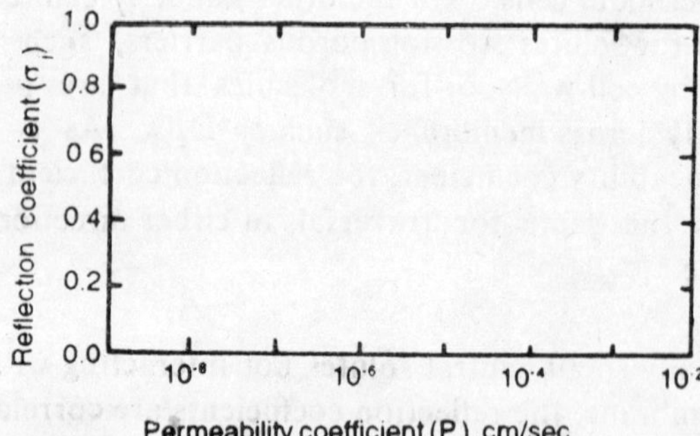

Fig. 4.9. Correlation between the reflection coefficients for a series of nonelectrolytes (determined using rabbit gallbladder epithelium) and the permeability coefficients for the same compounds (measured by R. Collander using *Nitella mucronata*). Most of the measurements are in the area indicated.

Solute movement across membranes

We can profitably reexamine certain aspects of the movement of solutes into and out of cells and organelles by using the more general equations developed from irreversible thermodynamics. One particularly important situation amenable to rather uncomplicated analysis is when the total volume flow J_V is zero, an example of a *stationary state*. This stationary state, in which the volume of the cell or organelle does not change over the time period of interest, can be brought about by having the net volume flow of water in one direction across the membrane equal the net volume flow of solutes in the opposite direction. A stationary state is therefore not be same as a steady state or an equilibrium condition for one cellular organelle; it represents a situation occurring only at a particular time or under some special experimental arrangements in this regard we should point out that μ_j can depend on both position and time for a stationary state but only on position form steady state, and on neither position nor into at equilibrium. When water is moving into a cell, $\Delta\mu_w$ is non-zero (we are not at equilibrium) and, in general μ_w will be increasing with time (we are not even in a steady state). If over some time interval the volume flow solutes out of the cell equals this J_V, in the cell will be in a *stationary state*, where its volume is not changing. One restriction here to cases of zero volume flow considerably simplifies the algebra and emphasizes the role played by reflection coefficients. Moreover, a stationary state of no volume change often characterizes the experimental situation under which the Boyle-Van't Hoff relation or the expression describing incipient plasmolysis.

What is the relationship between the internal and the external Ps and πs in the case of a stationary state ? In order to answer this we must first express the consequences of a stationary state in symbols. Our point of departure is Equation 38 $J_V = L_P \Delta P - \sum_j \sigma_j \tau_j)$, where the stationary state condition of zero net volume flow ($J_V = 0$) leads to the following equalities $0 = \Delta P - \sum_j \sigma_j \sigma \tau_j = P^o - P^i - \Sigma\sigma \quad (\tau_j^{\,o} - \tau_j^{\,i})$, Here, the osmotic pressures are for the effect of solutes on water activity, while in the general case both osmotic contributions from solutes (τ_i) atmospheric pressures resulting from the presence of interfaces (τ) might occur. Volume measurements for osmotic studies involving incipient plasmolysis or the Boyle-Van't Hoff relation are generally made when the external solution is at atmospheric pressure ($P^o = 0$) and when there are no external interfaces ($\tau^o = 0$). The stationary state condition of J_V equaling zero then leads to

$$\sum_j \sigma j \pi_j^{\,o} = \sum_j \sigma_j \pi_j^{\,i} + \tau^i - P^i \qquad \text{...(39)}$$

where the possibilities of interfacial interaction and hydrostatic pressures within the cell or organelle are explicitly recognized by the inclusion of τ^i and P^i. Equation 39 describes the stationary state of zero volume flow. It applies when the external solution is at atmospheric pressure (and $\tau^o = 0$) and when the solutes are capable of crossing the barrier, as occurs for molecules interacting with real, not idealized, biological membranes.

The influence of reflection coefficients on incipient plasmolysis. We used classical thermodynamics to derive the condition for incipient plasmolysis ($\pi^o_{plasmolysis}=\pi^i$), which occurs when the internal pressure P^i inside a plant cell just becomes zero. This derivation was made assuming equilibrium of water, *i.e.*, equal water potential, across a membrane impermeable to solutes. The assumptions of water equilibrium and impermeability are often not valid. Can we rectify the situation using an approach based on irreversible thermodynamics ?

Measurements of incipient plasmolysis can be made when there is zero volume flow ($J_V=0$) and for simple external solution ($\tau^o=0$) at atmospheric pressure ($P^o=0$). In this case, Equation 39 ($\sum_j \sigma_j\pi_j^o=\sum_j \sigma_j\pi_j^i+\tau^i-P^i$) would be the appropriate underlying expression from irreversible thermodynamics, instead of the less realistic condition of water equilibrium previously used. Using this stationary state condition, we obtain the following expression describing incipient plasmolysis ($P^i=0$) when the solutes can cross the cell membrane :

$$\sigma^o\pi^o_{plasmolysis}=\sum_j \sigma_j\pi_j^i+\tau^i \qquad ...(40)$$

where $\sum_j \sigma_j\pi_j$ has been replaced be $\sigma^o\pi^o$. In other words, it is often convenient to characterize the external solutes by an average reflection coefficient, σ^o, where $\sum_j \sigma_j\pi_j^o$ equals $\sigma^o\pi^o$ and π^o is the total external osmotic pressure, *i.e.*, $\pi^o=\sum_j \pi_j^o$.

Since the value of σ^o depends on the particular external solutes present, Equation 40 indicates that the external osmotic pressure (π^o) at incipient plasmolysis varies from solute to solute placed in the solution surrounding the plant cells. Let us suppose that solute *i* cannot penetrate the membrane, so that σ_i equals 1, a situation often true for sucrose. Another solute, *j*, can penetrate into the cells ($\sigma_j<1$), as is the case for many small nonelectrolytes. If we are at the point of incipient plasmolysis for each of these two solutes—taken in turn as the sole species in the external solution—$\sigma_i\pi_i^o{}_{plasmolysis}$ equals $\sigma_j\pi_j^o{}_{plasmolysis}$ by Equation 40. But solute *i* is unable to penetrate membrane ($\tau_i=1$). Hence we obtain the following relationships :

$$\sigma_j=\frac{\tau_i^o{}_{plasmolysis}}{\pi_j^o{}_{plasmolysis}}=\frac{\text{effective osmotic pressure of species } j}{\text{actual osmotic pressure of species } j} \qquad ...(41)$$

Equation 41 suggests a rather straightforward way of describing σ_j. Since (by supposition) species j can cross the cell membrane, σ_j must be less than one. Therefore, by Equation 41, $\pi_j{}^o{}_{plasmolysis}$ is greater than $\pi_i{}^o{}_{plasmolysis}$ where the latter refers to the osmotic pressure of the impremeable solute at the point of incipient plasmolysis. In words, a higher external osmotic pressure is needed to cause plasmolysis if that solute is able to enter the plant cell. The actual osmotic pressure indicated in Equation 41 is defined either by $\pi = -(RT/\bar{V}_w) \ln a_w$, where $\pi = \sum_j{}^o \pi_j$, or by $\pi_j = RT\gamma_j c_j$, when activity coefficient are included. When the membrane is impermeable to the solute, σ_j equals one, and the apparent (effective) osmotic pressure of species j equals the actual $\sigma_j{}^o$. Effective osmotic pressure takes into consideration the fact that most solutes can move across biological membranes ($\sigma_j < 1$), hence the π_j effective in leading to a net volume flow is reduced from its actual value—cf Equation 38, $J_V = L_P(\Delta P - \sum_j \sigma_j \Delta \pi_j)$. In summary, the reflection coefficient of species j indicates how effectively the osmotic pressure of that solute can be exerted across a particular membrane or other barrier.

We can also use the condition of incipient plasmolysis described by Equation 41 to evaluate specific reflection coefficients. It is difficult in experiments to replace one external solution by another with no changes in the tissue taking place, or with none of the previous solution adhering to the cell. Also, although it is easy in principle and involves only the use of a light microscope, determination of when the plasmalemma just begins to pull away from the cell wall is a rather subjective judgement. Nevertheless, the use of Equation 41 provides a simple way of considering individual reflection coefficients for various solutes entering plant cells.

Since osmotic pressures play a key role in plant physiology, σ_js are important parameters for quantitatively describing the solute and water relations of plants. In particular, the incorporatic of reflection coefficients allows the role of osmotic pressure to be precisely stated. In Poiseuille's law. $J_V = -(r^2/8n)$

$\theta P/\partial x$ which can adequentely describe movement in the xylem and possibly in the phloem, then flow is driven by the hydrostatic pressure gradient. An alternative view of the same situation is that σ equals zero for the solutes. In that case, osmotic pressures would have no direct effect on the movement described by Poiseuille's law. The real interest in reflection coefficients comes when σ_j is not at one of its two extremes of zero and unity. For such cases—intermediate between nonselectivity and impermeability—the volume flow does not depend on the full osmotic pressure difference across the barrier, but would be invalid to completely ignore the osmotic contribution of species *j*.

To illustrate some of the points we have been making concerning incipient plasmolysis ond reflection coefficients, let us consider Fig. 4.10. The cell in the upper left of the diagram is at the point of incipient plamolysis ($P^i=0$) for a nonpenetrating solute ($\sigma^o=1.0$) in the external solution (π^o=A, where A is a constant), *i.e.*, the plamalemma is just beginning to pull away from the cell wall. Equation 40 indicates that $\sigma_j\pi_j$ would then also equal A. If we place the cell in a second solution containing a penetrating solute ($\sigma^o<1$), Equation 41 indicates that the external solution must have a higher osmotic pressure in order for the cell to remain at the point of incipient plasmolysis. For instance, for a second solution with a σ_j of 0.5, the external osmotic pressure at the point of incipient plasmolysis would be (2A (cf Fig. 4.10*b*). Thus, when the external solute can penetrate, π^o is not as effective in balancing the internal osmotic pressure *c* in leading to a flow of water (cf. Eq. 41).

On the other hand if π^o were 2A for an impermeable external solute, extensive plasmolysis of the same cell would occur, as is illustrated in Figure 4.10*c*. Since $\sigma^o\pi^o$ is 2 A in this case, Equation 39 ($\sigma^o\pi^o=\sigma^i\pi^i-P^i$, when $\pi^i=0$) indicates that $\sigma^i\pi^i$ must also be 2A which means that essentially half of the internal water has left the cell. Finally if the reflective coefficient were 0.5 and the external osmotic pressure were A, $\sigma^o\pi^o$ would be $\frac{1}{2}$A, and we would not be at the point of

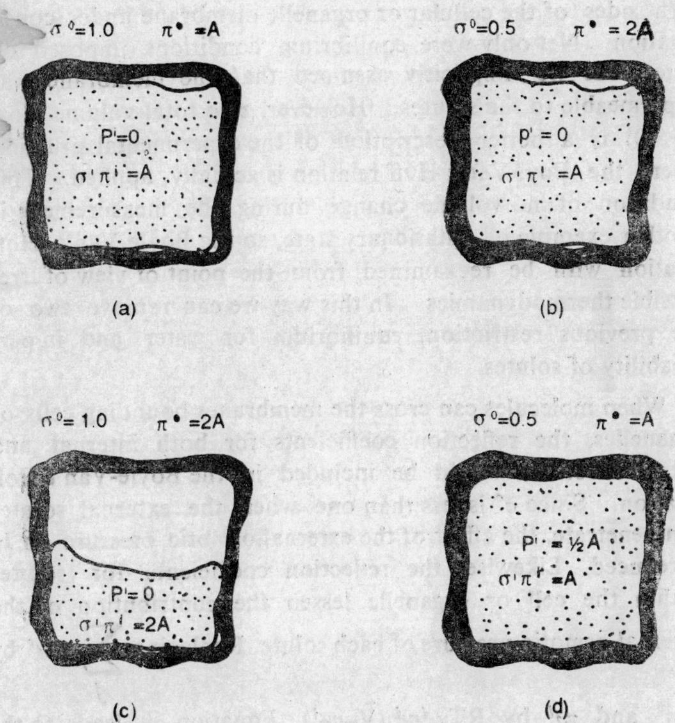

Fig. 4.10. Diagrammatic cell showing cell wall (shaded region) and a plasmalemma (thin line) for various external osmotic pressures : (*a*) point of incipient plasmolysis in the presence of an impermeable solute (*b*) point of incipient plasmolysis with a permeating solute (*c*) extensive plasmolysis (we are ignoring any effect of plasmodesmata) (*d*) cell under turgor.

incipient plasmolysis. In fact, the cell would be under turgor with an internal hydrostatic pressure equal to ½ A, at least until the concentration of the penetrating solute began to build up inside. We must therefore take into account the reflection coefficients of external (and internal) solutes in order to describe conditions at the point of incipient plasmolysis and, by extension, predict the direction and the magnitude of volume fluxes across membranes.

Extension of the Boyle-Van't Hoff relation. According to Boyle-Van't Hoff relation the water potential was the state on both sides of the cellular or organelle membrane under consideration. Net only were equilibrium conditions imposed on water, but we implicitly assumed that the membrane was impermeable to the solutes. However, zero total volume flow ($J_V=0$) is a better description of the experimental situation where the Boyl-Van't Hoff relation is actually applied. This condition of no volume change during the measurement is another example of a stationary state, so the Boyle-Van't Hoff relation with be reexamined from the point of view of irreversible thermodynamics. In this way we can remove two of the previous restriction, equilibrium for water and impermeability of solutes.

When molecules can cross the membranes bounding cells or organelles, the reflection coefficients for both internal and external solutes would be included in the Boyle-Van't Hoff relation. Since σ^o is less than one when the external solutes can penetrate, the effect of the external osmotic pressure on J_V is reduced. Likewise, the reflection coefficients for solutes within the cell or organelle lessen the contribution of the internal osmotic pressure of each solute. Replacing $\sum_j \sigma_j \pi_j^o$ by $\sigma^o \pi^o$ and π_j^i by $RT\gamma_j^i n_j^i/(\bar{V}_w n_w^i)$. Equation 39 leads to the following Boyle-Van't Hoff relation for the stationary state condition ($J_V=0$ in Eq. 38) :

$$\pi^o = RT \frac{\sum_j \sigma_j \gamma_j^i n_j^i}{\sigma^o \bar{V}_w n_w^i} + \frac{\tau^i - P^i}{\sigma^o} \qquad \text{...(42)}$$

The reflection coefficients of the membrane for both internal and external solutes enter into this extension of the expression relating volume and external osmotic pressure. The quantity $V-b$ in the conventional Boyle-Van't Hoff relation, $\pi^o(V-b) = RT \sum_j \varphi_j n_j$ can be identified with $\bar{V}_w n_w^i$. If we see Equation 42, it is evident that the osmotic coefficient of species j, φ_j, can be equated to $\sigma_j \gamma_j^i/\sigma^o$ as an explicit recognition of the activity coefficient of species j and the permeation properties of solutes, both internal and external. In fact, failure to recognize the effect of reflection coefficients on φ_j has led to misunderstandings on osmotic responses.

The volume of pea chloroplasts responds linearly to $1/\pi^o$ indicating that $\tau^i - P^i$ in such plastids may be negligible compared with the external osmotic pressures used. To analyze experimental observations, it is convenient to replace $\sigma^o\pi^o$ by $\sigma_x\pi_x{}^o + \alpha$, where $\pi_x{}^o$ is the contribution to the external osmotic pressure of solute x whose reflection coefficient (σ_x) is being considered, and α is the sum of $\sigma_j\pi_j{}^o$ for all the other external solutes. We can represent $RT \sum_j \sigma_j\gamma_j{}^i n_j{}^i$ by β, and replace $\bar{V}_w n_w{}^i$ by $V-b$. Making these substitutions into Equation 42, we obtain the following relatively simple form for the Boyle-Van't Hoff relation in the case penetrating solutes :

$$\sigma_x\pi_x{}^o + \alpha = \frac{\beta}{V-b} \qquad \ldots(43)$$

If we vary $\pi_x{}^o$ and measure V, we can then use Equation 43 to obtain the reflection coefficients for various nonelectrolytes in the external solution.

Reflection coefficients for chloroplasts. When the refinements introduced by reflection coefficients are taken into account, we can use osmotic responses of cells and organelles to quantitatively describe the permeability properties of their membranes. As a specific application of Equation 43, the progressive addition of hydroxymethyl groups in a series of polyhydroxy alcohols causes the reflection coefficients to steadily increase from 0.00 to 1.00 for pea chloroplasts (Table 4.4). In this regard, the lipid : water solubility ratio decreases in going from methanol to ethylene glycol to glycerol to erythritol to adonitol, *i.e.*, the partition coefficient K_x decreases. Since the permeability coefficient P_x equals $D_xK_x/\Delta x$ we expect a similar decrease in P_x as hydroxymethyl groups are added. Figure 4.9 clearly shows that as the permeability coefficient decreases the reflection coefficient generally increases. Consequently, the increase in σ_x of alcohols as $-$CHOH groups are added can be interpretal as simply a lowering of K_x. (As we go from methanol to adonitol, D_x also decreases, perhaps by a factor of 3, while K_x decreases about a thousandfold so that changes in K_x are the predominant influence on P_x and σ_x in this case). The reflection coefficients of six-carbon polyhydroxy alcohols, such as sorbitol and mannitol, are essentially unity for pea chloroplasts. This indication of relative impermeability suggests that these compounds would serve as suitable osmotica in which to suspend chloroplasts, as is indeed the case.

We may reasonably expect partition coefficients or the chloroplast membranes to be lower for amino acids than for adonitol. Since $\sigma_{adonetal}$ is 1.00 for pea chloroplasts, we would therefore anticipate that the reflection coefficients of amino acids would be near unity. However, this is not the case, as five of the amino acids have a σ_x of 0.04 or less for pea chloroplasts. L-isoleucine ($\sigma_x = 0.33$), L-valine (0.53), and L-leucine (0.56) have somewhat higher reflection coefficients, but these amino acids all have a branch methyl group on the side chain which apparently hinders their passage across the chloroplast membranes. In contrast, the σ_xs of these same eight amino acids are in all cases near unity for the cellular membranes which have been studied (for example, see Wright and Diamond). The fact that the reflection coefficients of amino acids are low for chloroplasts and high for cells has important physiological implications. Amino acids needed for protein synthesis in chloroplasts can then readily diffuse in from the cytoplasm. Also, amino acids produced as photosynthetic products can move out into the cytoplasm without the utilization of any energy for active transport. On the other hand, if amino acids could readily diffuse across cellular membranes, they would rapidly be lost from the cell.

Since the reflection coefficients of amino acids for pea chloroplasts are much less than predictions based on the partition coefficients, the permeation of amino acids into the chloroplasts may not be governed by the partition of these compounds between membrane lipids and water. Evidently, carriers passively shuttle the amino acids into and out of chloroplasts and thereby lower the effective reflection coefficients. When the concentration of an amino acid is raised above 100mM in the external solution bathing the chloroplasts, the reflection coefficients increase to values near unity. This is interpreted as a saturation of the carriers used for facilitated diffusion, and can be analyzed using a Michaelis-Menten type of formalism (Equation 26). If an aliphatic amino acid—L-alanine, L-isoleucine, L-valine, and L-leucine—is placed in the external solution with glycine, the chloroplast osmotic responses indicate competition for the same carrier. A second carrier for L-serine, L-threonine, and L-methionine is also demonstrated. In short, three criteria for the use of carriers in facilitated diffusion, *viz.*, specificity, saturation, and competition, can all be demonstrated using the osmotic responses of chloroplasts as interpreted by the modified from of the Boyle-Van't Hoff relation (Eq. 43). Hence, the rather esoteric concepts of irreversible thermodynamics can be applied in a relatively simple manner to gain insights into the physiological attributes of membranes.

5

The Water Relations of Plant Cells

The membrane of plant cells. The membranes of plant cells are much more complex in their structure and properties than the nonliving membranes. With very few exceptions the protoplasm of each plant cell is enclosed by a more or less rigid wall which acts as a membrane. Through this wall, which usually consists of several layers, each with a complex organization of its own, must pass all substances moving either into or out of cell. In addition to the structure of the wall itself, the role of the plasmodesms must be considered in accounting for the movement of substances from one cell to another. Lining the cell wall of all mature plant cells is a layer of cytoplasm. The cytoplasm, or parts thereof, constitutes a second layer through which substances entering or leaving the vacuole of a cell must penetrate. Both theoretical considerations and experimentally observed facts support the view that the two limiting layers of cytoplasm possess different physiochemical properties from the intervening cytoplasm and may be regarded as distinct membranes. The cytoplasmic membrane adjacent to the cell wall is called the *plasmalemma* ; that enclosing the vacuole the *tonoplast* or *vacuolar membrane*. Since the boundaries between the cytoplasm and the cell wall and the cytoplasm and the vacuole constitute interfaces, certain protoplasmic ingredients probably become adsorbed there in greater concentrations than in the interior of the cytoplasm. For this reason alone the postulation that the cytoplasmic membranes possess different properties from the interior cytoplasm seems justified.

Results from several types of experimental investigations support this hypothesis. The immiscibility of protoplasm with water, already discussed, appears to result at least partly from the fact that protoplasm is coated with a layer of material which is insoluble in water. By means of micropipettes certain nontoxic dyes can be injected into the body of the cytoplasm through which they will spread rapidly. They do not, however, pass through either the plasmalemma or the tonoplast. Neither will they pass into the interior of the cytoplasm from a solution bathing the cell, nor from the vacuole if the latter is injected with the dye. This is direct evidence that the constitution of the plasmalemma and the tonoplast is different from that of the intervening cytoplasm. The presence of cytoplasmic membranes can also be demonstrated by the manipulation of living protoplasm with fine glass needles. It has been found possible to withdraw a plasmolyzed protoplast from its cell wall and to strip off the protoplasm from the vacuole leaving the latter as a free-floating sac full of cell sap enclosed by a delicate membrane—the tonoplast (Fig. 5.1). Most evidence indicates that the tonoplast and probably also the plasmalemma are thin liquid films which are immiscible with water.

A complete picture of the permeability of a cell to different compounds can be drawn only in terms of the permeability of both the cell wall and the cytoplasmic membranes. In entering a cell a substance must first pass through the various layers of the cell wall, thence in turn through the plasmalemma, the interior cytoplasm, and the tonoplast before reaching the

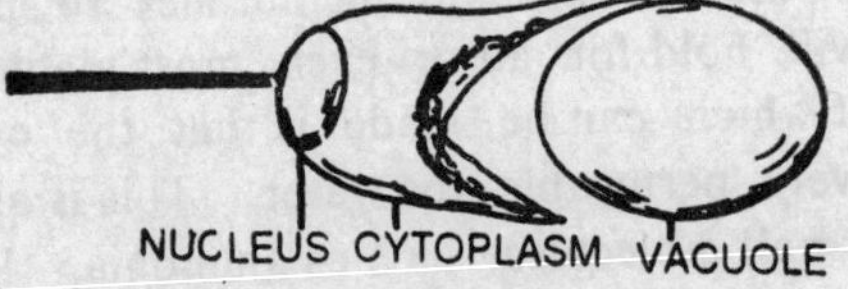

Fig. 5.1. **Removal of the outer cytoplasm from the tonoplast and vacuole of a plasmolyzed cell.**

vacuole. Although some investigators prefer to refer all permeability phenomena of plant cells to the cytoplasm as a whole, the evidence that the plasmalemma and the tonoplast exist as distinct membranes with the property of differential permeability must be regarded as very substantial.

Permeability of cell membranes. In general, cell walls composed of cellulose and pectic compounds, such as those in most parenchymatous tissues, are quite permeable to water and solutes. Water and solutes probably penetrate such walls through the hydrophilic intermicellar material. Lignified walls are also quite permeable to water and solutes, but cell walls in which suberin or autin are present in appreciable quantities, on the other hand, are less permeable both to water and solutes.

In contrast with the walls of most parenchymatous cells, the cytoplasmic membranes exhibit a high degree of differential permeability. The term "cytoplasmic membrane," is here employed in a loose sense to refer to the entire cytoplasmic system of membranes : plasmalemma plus interior cytoplasm plus tonoplast. Actually many substances entering plant cells are intercepted and utilized it one way or another in the cytoplasm, and do not diffuse in appreciable quantities through the tonoplast. Similarly compounds synthesized in the cytoplasm may pass out of the cell without crossing the tonoplast. Hence the plasmalemma is often regarded as the most important unit in the cytoplasmic system of membranes. It should be realized, however, that substances can move from one cell to another, through the plasmodesms, without crossing the plasmalemma.

Very few general statements can be made regarding the permeability of the cytoplasmic membranes to specific substances which will hold for all or even most plant cells. One such statement which can be made is that the cytoplasm is usually relatively permeable to water. This is also true for certain other small molecules, such as ammonia. Permeability of the cytoplasmic membranes to most solutes of physiological importance such as sugars, electrolytes, and amino acids, on the other hand, appears to be very variable. Studies of many kinds of cells, for example, indicate that their cytoplasmic membranes are relatively impermeable to sucrose and other sugars. The slowness with which many kinds of plant cells recover from plasmolysis in sugar solutions is also evidence of the relative impermeability of the cytoplasm to such compounds. Nevertheless there is much indirect evidence that sugars often move into and out of many kinds of cells at relatively rapid rates. Similarly, although many permeability studies indicate the cytoplasmic membranes to be relatively impermeable to electrolytes, they often penetrate relatively rapidly into plant cells, when in an actively metabolizing condition. The probable explanation of these apparent discrepancies is that most studies of permeability have been made upon mature cells which have lost most or all of their capacity for rapid absorption of electrolytes and sugars.

The permeability of the cytoplasmic membranes to specific substances is therefore not a fixed property, but one which differs from one kind of cell to another and varies in the same cell with the age of the cell, the environmental conditions to which it is subjected, and other factors. Furthermore, there are many indications that the cytoplasm is not merely a passive membrane like a sheet of collodion, but that it, at times at least, participates actively in the movement of molecules and ions into or out of cells, or from one part of a cell to another.

Plasmolysis. If a vacuolate plant cell which is at least partially distended is immersed in a hypertonic solution of some subtance to which the membranes are relatively impermeable, a characteristic series of changes takes place in the appearance of that cell. The first detectable occurrence is a gradual shrinkage in the volume of the entire cell as a result of outward osmosis. Consequently there is a reduction in the pressure exerted by the cell sap against the protoplasm and cell wall. This shrinkage in volume can be detected in many cells by measurement under a microscope, although there are some types of plant cells in which little change in volume occurs under such conditions. There is a lower limit to the elasticity of the cell wall, however, and when this is attained no further decrease in the volume of the cell will occur. Since the cell wall is quite freely permeable to the water and the solutes of the surrounding solution, and hence the hypertonic solution is in contact with the outer surface of the protoplast, the cell sap will continue to lose water by outward osmosis just as if no cell wall were present. Hence the protoplasm will continue to shrink in volume after the contraction of the cell wall has ceased. At this stage, therefore, the protoplasmic layer will begin to recede from the cell wall. If the hypertonic solution is strong enough, this separation of the protoplasm from the cell wall will become very pronounced. In some types of cells the protoplasm will "ball up" into a more or less spherical mass within the cell. More often the shrunken protoplasm assumes other configurations as shown in Fig. 5.2. This phenomenon is called *plasmolysis*. The pattern followed by the shrinkage of the protoplasm in plasmolysis is more or less typical for each kind of cell, although it may be modified somewhat depending upon physiochemical conditions within the protoplasm and the kind of solute used in the plasmolyzing solution. The space between the cell wall and the protoplasm

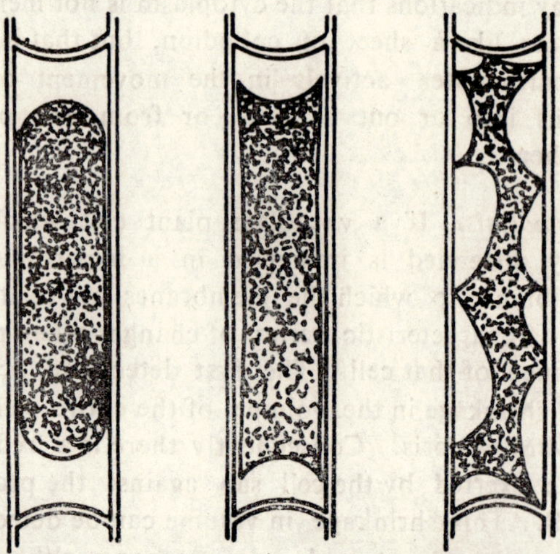

Fig. 5.2. Several common types of plasmolysis (diagrammatic).

will be filled, after the separation of the latter from the wall, with the external solution.

If a plasmolyzed cell is immersed in water, it will slowly recover and regain a turgid state as a result of osmotic movement of water into the vacuole. Similarly, if immersed in a solution hypotonic to the cell, sap recovery will also ensue, but the degree of targidity attained will be less than if the cell were immersed in pure water. Ripid "deplasmolysis," however, results in the death of many kinds of plant cells.

OSMOSIS AND OSMOTIC PRESSURE

Like gases and solutes, liquids exhibit diffusion phenomena. If, for example, water is brought into contact with another liquid such as ether, with which it is only slightly miscible, a slow diffusion of water molecules into the ether will occur. Simultaneously a slow diffusion of the molecules of the ether will take place into the water. Such diffusion will continue until the two liquids are mutually saturated.

Osmosis. This is by far the most familiar process involving the diffusion of liquids. An understanding of the dynamics of this process and of the significance of the physical quantity termed osmotic pressure is essential to an interpretation of the water relations of plant cells and tissues.

Let us first consider an experiment arranged as in Fig. 5.3. A sac-like membrane of collodion is completey filled with a

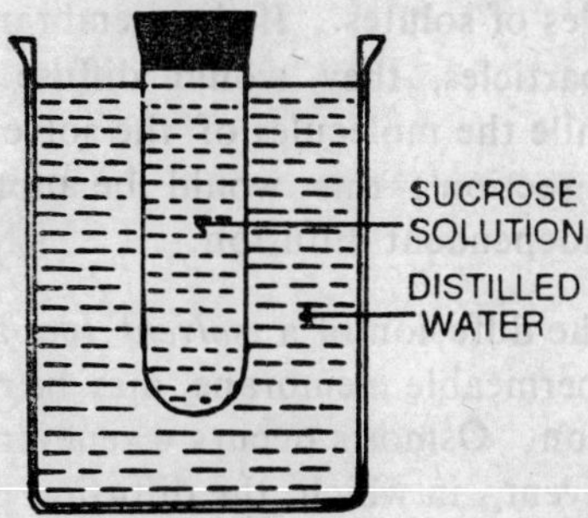

Fig. 5.3. Apparatus for the demonstration of osmosis through a collodion membrane.

strong sucrose solution and immersed in a beaker of water. The top of the sac is tightly plugged with a rubber stopper. The collodion membrane is prepared so that it is permeable to water, but impermeable or practically so to sucrose ; in other words, it is *differentially permeable*. It is also slightly elastic.

After a short time the originally limp sac becomes rigidly distended. This is a result of the diffusion of water through the collodion membrane into the interior of the sac Such a diffusion of water is an example of *osmosis*. The pressure developed as a result of the entrance of water prevails throughout the solution and is also exerted against the inside wall of the membrane. If the membrane is virtually impermeable to the solute, eventually the entire system will come to equilibrium, after which there will be no further increase in the volume of water inside of the membrane.

When the movement of the solvent is referred to in discussing osmosis it is always the net movement which is meant. Solvent molecules will always be moving across the membrane in both directions, but, except when an equilibrium has been attained, more molecules will move per unit of time in one direction than in the other. As will be shown in the later discussion this net movement is always from the region of the greater diffusion pressure of the solvent molecules to the region of their lesser diffusion pressure. The maintenance of the diffusion-pressure gradient of the liquid in most osmotic phenomena is made possible by the relative impermeability of the membrane to the molecules of solutes. If the membrane were permeable to the solute particles, they would diffuse outward across the membrane, while the molecules of the solvent were diffusing in the opposite direction ; this would be another example of the principle of independent diffusion.

Osmosis, the diffusion of a *solvent* (*not* a solution) across a differentially permeable membrane, may be regarded as a special case of diffusion. Osmosis occurs whenever two solutions with a common solvent, in which the diffusion pressures of the solvent are different, are separated by a membrane more permeable to the solvent than to the solutes. In living organisms water is the only important solvent which moves by osmosis, hence the following discussion of this process will be in terms of water and aqueous solutions.

Diffusion of the molecules or ions of a solute through a membrane is also a common phenomenon but occurs independently of the diffusion of the solvent molecules, in accordance with differences in the diffusion pressure of the solute, and seldom in amounts which are proportional to the diffusion of the solvent. The term "osmosis" is sometimes applied to such a diffusion of a solute across a membrane ; likewise it is sometimes applied to the diffusion of a gas across a membrane. Historical considerations, however, give priority to osmosis as a term for the diffusion of a solvent across a membrane. Furthermore, certain aspects of this phenomenon are so distinctive that it is convenient to have a distinguishing term for it.

Membranes and Permeability. The distinctive aspects of osmosis as compared with other diffusion processes result largely from the presence of the differentially permeable membrane. The concept of permeability is inseparable from the idea of a membrane. Permeability is a property of the membrane, not of the substance which diffuses through it. Thin layers or sheets of many different kinds of substances, such as rubber, parchment paper, collodion, cellophane, gelatin, and copper ferrocyanide, may serve as membranes.

Some membranes are impermeable to all, or virtually all, substances ; others allow all or most substances to diffuse through them with little impediment. Many biologically important membranes are of the type known as *differentially permeable.* Such membranes allow some substances to pass through them much more readily than others. They may be, and often are, impermeable or virtually so to some substances, while others diffuse through them quite freely. Alternative but less desirable terms for *differentially permeable* are *semipermeable* and *selectively permeable.*

The simplest kinds of membranes are essentially molecular sieves. The pores of a given membrane of this type are such that micelles, molecules and ions below a certain size can diffuse through them, while those exceeding this size cannot pass through. Membranes of collodion, parchment paper, and copper ferrocyanide owe their differential permeability largely to their sieve-like properties.

Some membranes are differentially permeable because certain substances are more soluble in them than others. The rubber balloon experiment described furnishes an example of such a membrane. Carbon dioxide gas is much more soluble in rubber than oxygen or nitrogen, hence a rubber membrane is much more permeable to carbon dioxide gas than to either of the other two.

Another example of a membrane of this type in operation can be seen in an experiment set up as in Fig. 5.4. A thin layer of water is introduced on top of a layer of chloroform in a test tube, the tube nearly filled with ether, and stoppered. A second tube is prepared in the same way except that xylene is substituted for the ether. After several days it will be observed that the

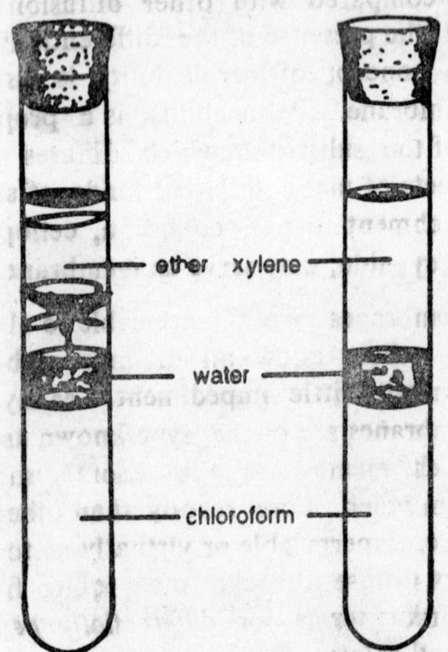

Fig. 5.4. Experimental arrangement for demonstrating the differential permeability of a water membrane.

level of the water layer in the first tube has risen, while in the second tube it has fallen, although the distance through which the layer moves is not as great in the second tube. In the first tube ether is osmosing through the water *membrane* more rapidly than the chloroform ; hence the volume of liquid below the membrane is increasing and the layer of water rises. In the second tube chloroform osmoses through the water membrane more rapidly than the xylene ; hence the water layer falls in the tube. Much less chloroform osmoses in a given time through the water layer in the second tube than does ether in the first tube. Of these three compounds, ether is the most soluble in water, chloroform next most soluble, and xylene least soluble. Permeability of the water membrane to these three compounds is clearly correlated with their solubility in water.

Diffusion Pressure of Liquids. Like a gas, water or any other liquid may be considered to possess a diffusion pressure, although the existence of such a physical quantity in liquids becomes apparent only under certain conditions as, for example, when a solvent and a solution are separated by a differentially permeable membrane. Diffusion pressure is the cause of diffusion not its result. Just as gases have a pressure, whether or not any diffusion is occurring, so do liquids and solutes have a diffusion pressure, whether or not any diffusion is occurring.

The only two factors which influence the diffusion pressure of pure water are pressure and temperature. If pressure is imposed on water, as by a piston in a closed system, or by confining effects of the walls of a closed osmometer or plant cell when the volume of the enclosed water is expanding as a result of osmosis, the diffusion pressure of the confined water increases by the amount of the imposed pressure. If the imposed pressure is 8 atm., for example, the diffusion pressure of the water increases by 8 atm. Correspondingly, if water is subjected to a "negative pressure" (tension), its diffusion pressure decreases by the amount of the negative pressure. The effect of the imposed pressure on the diffusion pressure of the liquid is quantitatively the same whether the liquid is pure, or whether solutes are present.

Although the diffusion pressure of a liquid is also theoretically affected by temperature, it is usually not necessary to consider this effect in evaluating the mechanism of the diffusion of water in plants.

In an aqueous solution, however, the diffusion pressure of the water is influenced by still another factor, the proportion of solute particles (ions plus molecules) to solvent molecules. The greater this proportion, the less the diffusion pressure of the water. Within a wide range of concentrations of the solute molecules or ions, the diminution in diffusion pressure is closely proportional to the ratio of solute particles to solvent molecules.

Ideally all diffusion phenomena including osmosis, should be interpreted on the basis of differences in diffusion pressures. Diffusion phenomena in gases can be readily analyzed in this way because gas pressures can usually be ascertained. The exact magnitudes of the diffusion pressures of liquids, on the other hand, cannot readily be measured. Hence use of this physical concept in the quantitative interpretation of the diffusion of liquids is not feasible. It is usually possible, however, to measure quantitatively the amount by which the diffusion pressure of the water in a given solution is less than that of pure water at the same temperature and under atmospheric pressure. This quantity is called the *diffusion-pressure deficit* or *pressure deficit* and can be used in the interpretation of osmosis and other diffusion phenomena in which water molecules participate, even if we do not know the exact magnitude of the diffusion pressure of water. In this indirect manner it is possible to use the concept of diffusion pressure in the analysis of diffusion phenomena in liquids.

Osmotic Pressure. This term was originally employed to designate the maximum pressure which develops in a solution enclosed within an osmometer under certain ideal conditions. An *osmometer* is an apparatus for measuring the magnitude of osmotic pressures. If a manometer or pressure gauge were inserted through the stopper of an apparatus set up as in Fig. 5.3 the arrangement would serve as a crude osmometer.

Most precise measurement of osmotic pressures have been made with osmometers constructed of cylindrically-shaped porous clay cups (Fig. 5.5) in the pores of which have been

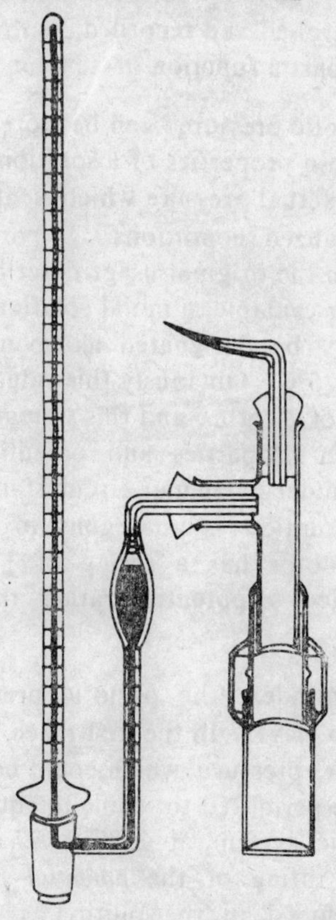

Fig. 5.5. Pfeffer's osmometer (shown in sectional view) with attached closed manometer for measuring the pressure developed. The lower part of the osmometer is a porous clay cup in the walls of which have been precipitated a differentially permeable membrane of copper ferrocyanide.

precipitated differentially permeable membranes of cupric ferrocyanide. The pressure developed a measured with a sensitive mercury manometer or in other ways. The necessary ideal conditions, only attained under rigorous experimental conditions, are that the membrane be permeable only to the solvent, that it be immersed in the pure solvent, and that the pressure equilibrium be attained without any appreciable dilution of the enclosed solution. The temperature of the experimental setup must also be controlled and recorded, as the osmotic pressure of a solution is in part a function of its temperature.

The term "osmotic pressure" can be more usefully employed as an index of certain properties of a solution, rather than as a designation for an actual pressure which is attained only under certain seldom realized conditions. In part such a usage is a derivative one from the original usage described in the preceding paragraph. For example, a molal solution of sucrose standing in a beaker may be designated as having an osmotic pressure of 27 atm. at 25°C. Obviously this solution is not exerting an actual pressure of 27 atm., and this statement merely implies that it has certain properties and potentialities which will become apparent under particular circumstances. Such a statement regarding a solution is analogous to the statement that an idle electric motor has a rating of ¼ horsepower ; both statements are indices of potential rather than existing capacities.

Used thus as an index, the osmotic pressure characterizes the solution in two ways. In the first place, it tells us that the maximum possible pressure which could be developed in that solution if it were permitted to come to equilibrium with pure water in an osmotic system at 25°C is 27 atm. The osmotic pressure is thus a rating of the *potential* maximum pressure which can be developed in the solution as a result of osmosis. In the second place, osmotic pressure is an index of the *diffusion-pressure deficit* of the water in a solution *insofar as this results from the presence of solutes*. Commonly used synonyms for osmotic pressure in this index sense are *osmotic concentration, osmotic value*, *osmotic potential*, and *osmotic power*.

Turgor pressure. *Turgor pressure* is the *actual* pressure which develops in a closed osmometer or plant cell as a result of osmosis or imbibition (see later). Use of the term "osmotic pressure" in this sense should be avoided as it only leads to confusion. A given solution has a unique osmotic pressure at a given temperature, but its turgor pressure is variable. Ordinarily the turgor pressure of a solution lies in the range between zero and its osmotic pressure, but under certain conditions it may exceed the osmotic pressure, and under certain other condition it may be negative in value. The turgor pressure prevailing within a solution is also exerted against the confining walls of the system.

Quantitative aspects of Osmosis. Let us suppose that a solution with an osmotic pressure of 20 atm. completely fills an essentially inelastic membrane which is permeable only to water and is arranged like that shown in Fig. 5.3. Let us further suppose this membrane to be immersed in pure water, which has, of course, a zero osmotic pressure. Both the solution within, and the water without, the membrane are initially under atmospheric pressure only, the effect of which (equivalent to a turgor pressure of 1 atm.) is conventionally disregarded when all parts of the system are under atmospheric pressure. From the previous discussion it should be clear that the diffusion-pressure deficit of the water in the enclosed solution is 20 atm.; that of the water outside of the sac, zero. Water, therefore, osmoses across the membrane into the solution. The entering water engenders a gradually increasing turgor pressure throughout the solution in the sac. Subjection of the water in this solution to such a pressure raises its diffusion pressure and hence lowers its diffusion-pressure deficit. An equivalent increase takes place in the diffusion pressure of the solutes present. A dynamic equilibrium will be attained when the turgor pressure of the internal solution reaches 20 atm. At this point the 20 atm. diffusion-pressure deficit which would otherwise result from the presence of solutes is offset by the 20 atm. turgor pressure and the diffusion-pressure deficit of the water in the solution is zero, which is the same as that of the pure water outside the membrane. When the diffusion-pressure deficit of the water on both sides of the membrane becomes equal, osmosis ceases, and the number of water molecules moving through the membrane in the inward direction will be the same as the number moving in the outward direction per unit of time. Since the membrane is assumed to be inelastic or nearly so, osmosis of only a very small amount of water into the solution is necessary to raise its turgor pressure to 20 atm. Since there is virtually no dilution of the solution, its final osmosis pressure, for all practical purposes, is the same as its initial osmotic pressure.

In living organisms water usually osmoses across a membrane from one solution to another, rather than from pure water into a solution. Let us consider, therefore, an analogous situation, exactly like that described above, except that the water on the outside of the membrane is replaced with a solution of 12 atm., osmotic pressure. Initially, before any turgor pressure develops, the diffusion-pressure deficit of the internal solution in this system is 20 atm. ; that of the external solution, 12 atm. Water therefore osmoses across the membrane into the solution within the sac. As the turgor pressure of the internal solution increases, the diffusion-pressure deficit of the water in that solution decreases. When a turgor pressure of 8 atm. has been attained, the diffusion-pressure deficit will have decreased to 12 atm. which is the same as that of the external solution. Dynamic equilibrium is thus attained in this particular system when the turgor pressure prevailing in the internal solution reaches 8 atm.

In the foregoing discussion an inelastic membrane has been assumed. Many membranes, however, especially in living organisms, are elastic within limits. When the membrane is appreciably elastic, the further complication of shifts in the osmotic pressure of the internal solution during osmosis is introduced. Let us assume an osmotic system identical with that described in the preceding paragraph, except that the membrane is elastic enough to permit a 25 per cent increase in the volume of the internal solution before a dynamic equilibrium is attained. As the internal solution increases in volume it becomes more dilute, *i.e.*, less concentrated. The osmotic pressure of a solution decreases with a diminution in its concentration. If we assume a proportionate decrease in osmotic pressure with decrease in concentration, which would be strictly true only in certain kinds of solutions, the final osmotic presure in the internal solution would be 16 atm. (20 : X=125 :

100). A turgor pressure of only 4 atm. would be required under these conditions to bring the water in the internal solution into dynamic equilibrium with that in the external solution which has an osmotic pressure of 12 atm. In such a system the turgor pressure of the internal solution is increasing while its osmotic pressure is decreasing. The only difference in end result when the membrane is elastic is that the turgor pressure and osmotic pressure at dynamic equilibrium are somewhat less than they would be otherwise. If the volume of the external solution is large relative to that of the internal solution, variations in the osmotic pressure of the former are usually negligible.

It should be evident from the foregoing discussion that the diffusion-pressure deficit of the water in a solution is always equal to the osmotic pressure of the solution less the turgor pressure which prevails in it. This relation can be conveniently expressed in the simple equation :

$$\text{D. P. D.} = \text{O. P.} - \text{T. P.}$$

As applied to the initial diffusion-pressure deficits in the second example given above these equations would work out as follows :

External solution : D. P. D. $= 12 - 0 = 12$ atm.
Internal solution : D. P. D. $= 20 - 0 = 20$ atm.

At dynamic equilibrium the diffusion-pressure deficit of this same external solution is unchanged ; that of the same internal solution is then calculated as follows :

$$\text{D. P. D.} = 20 - 8 = 12 \text{ atm.}$$

Magnitude of the osmotic pressures in plant cells. In many discussions of the osmotic pressure of plant cells no distinction is made between the "osmotic pressure of the cells at incipient plasmolysis" and their osmotic pressure under more or less turgid conditions. When the term osmotic pressure is employed without qualification it can be assumed to refer to the osmotic pressure of cells in their usual distended condition. It is generally considered that the results of determinations by the cryoscopic method represent approximately the osmotic pressure of the cells in this condition.

Different organs or tissues of the same plant may exhibit a wide range of osmotic pressures. Even similar organs on the same plant—leaves, for example—may vary considerably among themselves in the average osmotic pressure of their cells. Furthermore the tissues within the same organ usually show a considerable variation in osmotic pressures. The tatsophyll cells of most leaves, for example, show higher values than the epidermal cells.

In general, the osmotic pressures of the plant cells and tissues of the mesic species of North America vary in their extreme range from a fraction of an atmosphere to about 50 atm. Most of the values lie, of course, within a narrower part of this range, probably within the limits of a 4 to 30 atm. span of values. The osmotic pressure of a given plant cell or tissue is by no means a fixed quantity, but often fluctuates considerably. More or less regular daily or seasonal variations occur in the osmotic pressures of many kinds of plant cells or tissues

Factors influencing the osmotic pressure of plant cells. Any factor which influences either the water content of a plant cell or the solute content of its cell sap will have an effect on the magnitude of the osmotic pressure of that plant cell. The water content of the plant as a whole, and hence of its constituent cells, is controlled principally by the relative rates of transpiration and the absorption of water. The latter process is influenced very markedly by the water content and other conditions prevailing in the soil. Individuals of the same species invariably have a higher osmotic pressure when growing under drought conditions than when provided with a favourable water supply (Table 5.1). This is at least partially a result of the relatively lower water content of the leaves which obtains when the available soil water supply becomes low. Other factors which are also probably involved are a decrease in the growth rate of the plant which often permits an accumulation of mineral salts and soluble foods, and a shift of the starch⇄soluble carbohydrates equilibrium toward the side of the soluble carbohydrates.

Table 5.1. The effects of different soil water contents upon the osmotic pressure of maize plants.

Water content of soil, Per cent dry weight	*Osmotic pressure of tops*	*Osmotic pressare of roots*
31 per cent	22.06 atmos.	5.91 atmos.
25	23.08	7.23
16	24 36	7·79
14	25.04	9.24
13	25.47	11.34
11	26.48	11.98

The solute content of the cell sap is controlled by the specific metabolic processes of the plant and by the absorption of mineral salts by the plant from its environment. The rate of photosynthesis is an important factor in determining the osmotic pressure of plant cells, particularly those of leaf tissues. The influence of the inherent metabolic processes of any species upon the kinds and concentrations of various types of soluble organic compounds produced, such as simple carbohydrates, organic acids, and amino acids has an important effect on the magnitude of the osmotic pressure in any species. Metabolic conditions and their effects upon osmotic pressures may also be altered by environmental conditions. A well-known example of this is the difference in the osmotic pressures of sun and shade leaves on the same plant. The former almost invariably has the higher osmotic pressure, presumably at least in part as a result of their greater photosynthetic activity.

The mineral salts which contribute to the osmotic pressures of plant cells are all absorbed from the soil or in aquatic species from the water in which part or all of the plant is immersed. Different species of plants vary greatly in their toleration of high concentrations of mineral salts in the soil. All species can become adjusted, within limits, to a change in the mineral-salt content of the substratum. This adjustment takes the form of an increase in the osmotic pressure of the plant with an increase in the osmotic pressure of the medium from which it obtains its mineral salts (Table 5.2).

Other tissues of plants also show an increase in osmotic pressure with increase in the osmotic pressure of the soil solution, but such increases are generally most pronounced in the roots.

Species indigenous to saline soils usually have relatively high osmotic pressures. Such soils are rich in soluble salts, and the high osmotic pressures found in species native to such soils result from the absorption of relatively large quantities of mineral salts. In fact, the highest recorded osmotic pressure for any species of plant—202.5 atm.—was found in the saltbush (*Atriplex confertifolia*), a saline soil species.

Table 5.2. The effect of the osmotic pressure of the soil solution upon the osmotic pressure of the roots of maize.

Osmotic pressure of soil solution	*Osmotic pressure of sap from root cells*
1-21 atm.	4.59 atm.
1.99	5.48
3 38	6.61
4.96	7.51
7.22	8.19

Salt marsh plants also generally have relatively high osmotic pressures. This is also a correlation with the relatively high salt content of the substratum. Except in halophytes, however, a larger proportion of the osmotic pressure of most kinds of plant cells results from the presence of organic solutes than of mineral salts. Root cells are more likely to prove exceptions to this last statement than cells in the aerial organs.

The osmotic quantities of plant cells. Let us suppose three similar plant cells, each in a state of incipient plasmolysis (*i.e.*, with a zero turgor pressure), and each with a cell sap osmotic pressure of 10 atm., to be immersed, the first in pure water, the second in a solution with an osmotic pressure of 6 atm., and the third in a solution with an osmotic pressure of 10 atm. (Fig, 5.6). The volume of the liquid in which each cell is immersed is assumed to be very large in proportion to the volume of the cell.

In order to simplify the discussion, certain assumptions will be made regarding these cells and, unless stated to the contrary, regarding all other cells discussed in the remainder of this chapter. These assumptions are : (1) the cells are at the same temperature, (2) the cytoplasmic membranes are permeable only to water while the cell walls are permeable to both water and solutes, (3) the walls of all the cells are equally elastic, and (4) volume-changes in the cells are so small that any resulting effects upon the osmotic pressure of the cell sap are negligible and can be disregarded. A review of the discussion of the quantitative aspects of osmosis in physical systems will be helpful in following the ensuing closely analogous discussion with respect to cells.

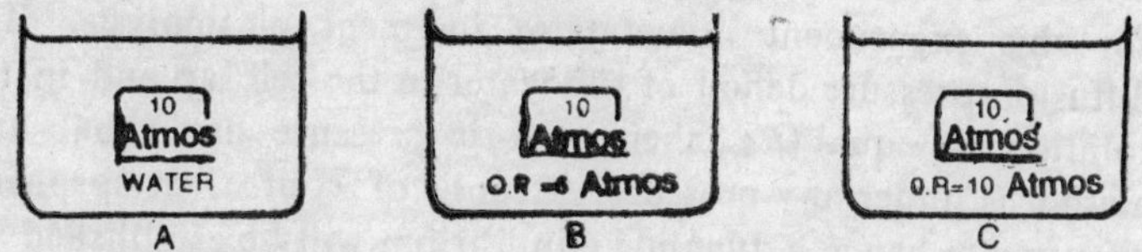

Fig. 5.6. Similar cells immersed in solutions of different osmotic pressures.

Osmosis of water into cell A begins immediately upon its immersion, because the diffusion-pressure deficit of the water in the cell sap is 10 atm. and that of the surrounding water is zero. The entrance of water into the cell results in a gradually increasing *turgor pressure* within the cell, just as it does in an osmometer. This pressure not only prevails within the water mass, but is also exerted against the protoplasmic layer and, in turn, against the cell wall. The increase in turgor pressure threfore results in a gradual distension of the cell. Imposition of a pressure of 10 atm. on the water in the cell sap will increase its diffusion pressure by 10 atm. Since the initial diffusion-pressure deficit of the water in the cell sap was 10 atm., imposition of a 10-atm. turgor pressure reduces its diffusion-pressure deficit to zero. This is also the diffusion-pressure deficit of the surrounding water, hence a dynamic equilibrium is attained when the turgor pressure in the cell sap reaches 10 atm.

The diffusion-pressure deficit of the water in the cell sap of cell B is 10 atm., that of the water in the surrounding solution 6 atm. Hence there will be a net inward movement of water into the cell. A turgor pressure thus develops within the cell, but at its maximum under these conditions it will attain a value of only 4 atm. Since the initial diffusion-pressure deficit of the water in the cell sap was 10 atm., imposition of a turgor pressure of 4 atm. reduces this to 6 atm. which is the diffusion-pressure deficit of the water in the surrounding solution. Hence the attainment of a turgor pressure of only 4 atm. results under these conditions in an osmotic equilibrium. Since the magnitude of the turgor pressure developed in cell B will be less than in cell A, the latter cell will be more distended than the former.

Cell C will remain, according to the conditions prescribed for this experiment, in a state of incipient plasmolysis. The diffusion-pressure deficit of the water in the cell sap and in the solution are equal (*i.e.*, their osmotic pressure are equal and neither is under any pressure, except, of course, atmospheric pressures) ; hence a dynamic equilibrium will be established as soon as the ce l is immersed in the solution. Since there is no net movement of water into the cell no turgor pressure will develop, a condition which will obtain in all plant cells in the condition of incipient plasmolysis.

In cells A and B a dynamic equilibrium has been attained by an adjustment of the diffusion-pressure deficit of the water in the cell sap until it is equal to that of the water in the surrounding liquid. This adjustment was attained in each of these cells by a shift in the magnitude of the turgor pressure, which is one of the two components determining the diffusion-pressure deficit of the water in the cell sap, the other being the osmotic pressure. In cell C no shift in the magnitude of the turgor pressure occurred, because the diffusion-pressure deficit was initially the same in the cell sap and in the solution. Unlike a solution exposed to the atmosphere, the diffusion-pressure deficit of a cell is not equal to its osmotic pressure except when the turgor pressure of the cell is zero. The term "diffusion-pressure deficit of a cell" should be considered as an abbreviation for "diffusion-pressure deficit of the water in the cell sap."

The interrelationships among the osmotic pressure, turgor pressure, and diffusion-pressure deficit of a plant cell should be further clarified by a study of Fig. 5.7. This diagram also takes into account the influence of volume changes in the cell upon these physical quantities. In the interests of simplicity the effects of such volume changes upon the osmotic quantities of the cell have been disregarded in the discussion up to this point. When the cell is completely flaccid (relative volume =1.0), its diffusion-pressure deficit is equal to its osmotic pressure (20 atm.) while its turgor pressure is zero. As the volume of the cell increases, as a result of influx of water, the osmotic pressure decreases because of the dilution of the cell sap. The pattern of the change in magnitude of the turgor pressure with increase in cell volume is not the same in all kinds of cells, but often follows a trend such as that shown in Fig. 5.7. The diffusion-pressure deficit, being equal to the difference between the osmotic and turgor pressures, shows a progressive decrease as the turgidity of the cell increases. When the cell attains a condition of maximum turgidity (relative volume=1.5), its turgor pressure is equal to its osmotic pressure while its diffusion-pressure deficit has fallen to zero.

The dotted extension of the line for turgor pressure to the left indicates that it sometimes has a negative value. Conditions under which this may occur are discussed shortly. The diffusion-pressure deficit of a cell in which the turgor pressure is negative is greater than its osmotic pressure.

In many types of cells volume changes are much less marked than indicated in this figure, and hence they can often be disregarded in generalized considerations of the water relations of plant cells. In some species many of the cell have walls which are virtually inelastic. This is especially true of the tissues of xerophytes and some water plants.

The fundamental interrelation between the osmotic pressure, turgor pressure, and diffusion-pressure deficit of a plant cell is

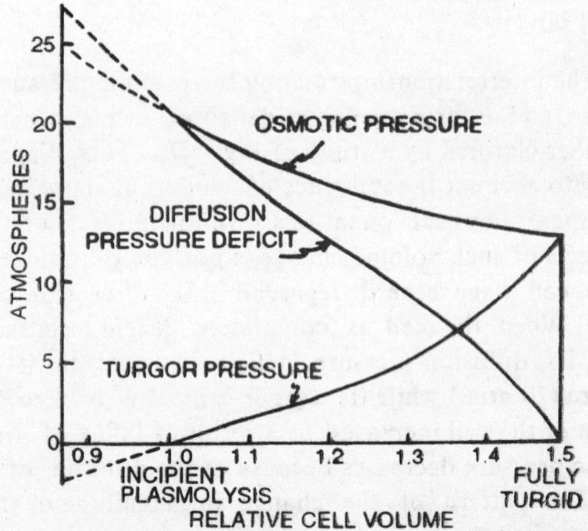

Fig. 5.7. **Interrelationships among the osmotic pressure, turgor pressure diffusion-pressure deficit, and cell volume of a plant cell. Based on data of Hofler (1920).**

the same as for an osmometer and is expressed by the same equation :

D. P. D.=O. P.−T. P.

The quantity to which the name "diffusion-pressure deficit" has been given in this discussion has been variously termed by writers in the field of plant physiology. The more commonly used synonyms for this term are "pressure deficit," "turgor deficit," "net osmotic pressure," "suction force," and "suction tension."

The osmotic pressure, diffusion-pressure deficit, and turgor pressure are collectively called the *osmotic quantities* of plant cells. As previous discussion has shown, a full evaluation of these quantities also requires a consideration of the volume changes of plant cells.

In plant tissues many of the cells are under a pressure imposed upon them by the surrounding cells. In addition to its own turgor pressure the water in such a cell is also sub-

jected to this pressure of external origin. This pressure is just as much a factor in determining the diffusion-pressure deficit of a cell as its own turgor pressure. The true diffusion-pressure deficit of such a cell is therefore less than that of the cell considered as an individual unit by the amount of this added pressure. Hence the equation representing the factors determining the diffusion-pressure deficit of a cell under such conditions becomes :

D. P. D.=O. P.—T. P.—Pressure exerted by surrounding cells

The water in the vessels and cells of plants frequently passes into a state of tension. Imposition of a pressure from an external source raises the diffusion pressure of water. Imposition of a tension, which is in effect a negative pressure, has precisely the opposite effect. In cells and vessels this can happen only if the enclosed water shrinks in volume to such a point that the encompassing protoplasm and cell walls are pulled inward as a result of adhesion between the walls and water. The resulting counter pull exerted by the walls on the water results in throwing it into a state of tension. Under such conditions the turgor pressure is negative in value (Fig. 5.7), and the tension (negative pressure) developed within the water will be equal to the negative turgor pressure. The equation for the diffusion-pressure deficit of a cell under such conditions becomes :

$$D.P.D. = O.P. - (-T.P.)$$

In other words, the diffusion-pressure deficit of the water in a cell or vessel when subjected to a tension is equal to the osmotic pressure plus the tension (negative pressure) imposed on the water.

Dynamics of the osmotic movement of water in plants. In order to simplify this part of the discussion, changes in the osmotic pressure of the cell sap resulting from volume changes in the cells, as shown in Fig. 5.7, will again be disregarded, as usually these are not great enough to modify seriously any generalized picture of the water relations of plant cells.

Let us imagine a certain cell (X) to have an osmotic pressure of 12 atm., and a turgor pressure of 6 atm., and a second cell (Z) to have an osmotic pressure of 10 atm. and a turgor pressure of 2 atm. Let us further suppose that these two cells can be brought into such intimate contact that osmotic movement of water can occur from one to the other (Fig. 5.8).

The water in the cell sap of X would have a diffusion-pressure deficit of 12 atm. were it under no pressure. The turgor pressure of 6 atm., however, reduces the diffusion-pressure deficit of the water in the cell sap to 6 atm. Similarly the water in the vacuole of cell Z would have a diffusion-pressure deficit of 10 atm., were it not also influenced by the turgor pressure of 2 atm. Hence the diffusion-pressure deficit of cell Z is 8 atm. The diffusion-pressure deficit of cell X is therefore less than that of cell Z. Water will, therefore, osmose from cell X to cell Z. Water will continue to show a net movement from X to Z until the diffusion-pressure deficits of the two cells are equal. In the movement of water from cell to cell in plants, it is the diffusion-pressure deficits and not the osmotic pressures which tend to equilibrate. This is only a special aspect of the fundamental tendency of the diffusion-pressure of water to attain a uniform value throughout any system. It is by no means impossible, therefore, for water to move from a cell of higher to one of lower osmotic pressure.

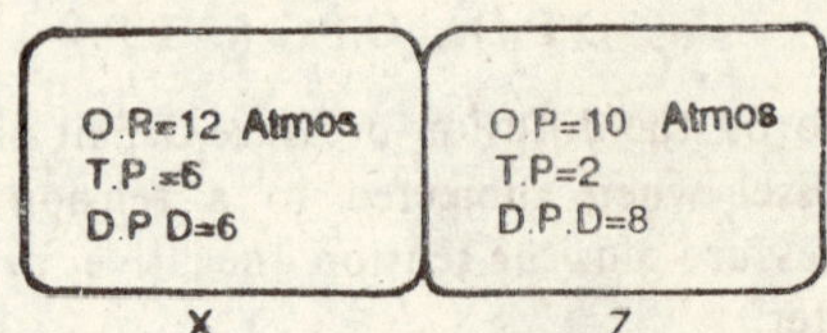

Fig. 5.8. Diagram of two adjacent cells used in explanation of the mechanism of the cell-to-cell movement of water.

After a dynamic equilibrium has been established between two cells their diffusion-pressure deficits will seldom be an exact average of their initial diffusion-pressure deficits. The only general statemnet which can be made is that, at equilibrium, the diffusion-pressure deficits of the two cells will be equal, and this value will lie somewhere between the two original values of the cells.

Whenever the diffusion-pressure deficits of two adjacent cells are dissimilar, a *diffusion-pressure-deficit gradient* exists between them. Movement of water from one cell to another can occur only when such a gradient exists. Other conditions being equal, the "steeper" this gradient, *i.e.*, the greater the difference in diffusion-pressure deficits, the more rapidly one cell will gain water from the other. The term "diffusion-pressure deficit gradient" can also be applied to a chain of cells in which the diffusion-pressure deficit increases serially from cell to cell. Several examples of such gradients are discussed in subsequent chapters.

The diffusion-pressure of water is influenced by temperature, although for small temperature differences the magnitude of this effect is not great. If two adjacent plant cells were to be maintained at different temperatures, theoretically this would have an effect on the rate or direction of osmosis. Actually, however, under the thermal conditions ordinarily prevailing in plants, this effect is so slight that it can be disregarded.

Methods of measuring the diffusion-pressure deficit of plant cells. Most methods of measuring this quantity are based on the principle, previously discussed, that if a cell is immersed in a solution with an osmotic pressure equal to the diffusion-pressure deficit of the cell, a dynamic equilibrium is immediately established, and no change will occur in the volume of the cell because no net movement of water takes place. In solutions with a higher osmotic pressure than the diffusion-pressure deficit of the cell it will decrease in volume; in solutions of lower osmotic pressure than its diffusion-pressure deficit it will increase in volume.

The "simplified method" of measuring diffusion-pressure deficits, introduced by Ursprung (1923), has been one of the most widely used. In this method narrow strips of tissues are cut from such structures as thin leaves or petals. The length of each strip is immediately measured under the microscope, usually while mounted in paraffin oil. Several strips are then immersed in each of a graded series of sucrose solutions in which they are allowed to remain until an equilibrium has been attained in each solution, after which their lengths are measured. The osmotic pressure of the solution in which no change in the length of the strips occurs is considered to equal the average diffusion-pressure deficit of the cells in the strip.

For the measurement of the average diffusion-pressure deficit of the cells in bulky tissues such as potato tubers, masses of tissue such as cylinders of approximately equal dimensions can be employed. The equilibrium point can be determined by measuring changes either in the weight or the volume of the cylinder. The solution in which the cylinder neither gains nor loses weight (or volume) is considered to have an osmotic pressure equal to the average diffusion-pressure deficit of the cells in the cylinder.

Determinations of diffusion-pressure deficits by the methods described above should not be confused with plasmolytic determinations of the osmotic pressures of plant cells. In the latter type of determination the critical measurement is the osmotic pressure of the solution with which the cells come to an equilibrium without any turgor pressure, *i.e.*, at incipient plasmolysis. In diffusion-pressure-deficit determinations the critical measurement is the osmotic pressure of the solution with which the cells come to equilibrium without any change in their turgor pressure, *i.e.*, without any change in the volume of the cells. Only when the cell is initially in a completely flaccid condition will determinations of its osmotic pressure and its diffusion-pressure deficit result in the same values.

All of the methods of measuring the diffusion-pressure deficits of plant cells are subject to most of the flaws which are also inherent in plasmolytic determinations of osmotic pressure, as well as some other serious errors. Except possibly for a few types of cells which are especially well adapted to such measurements, it is doubtful if the values obtained in determinations of diffusion-pressure deficits are ever more than fair approximations of the true values which obtain in plant cells.

The turgor pressure of a cell cannot be determined directly. However, if the osmotic pressure of a cell or group of cells has been determined, and likewise the diffusion-pressure deficit, the turgor pressure can be calculated from the equations given earlier in the chapter.

The imbibitional mechanisms of the movement of water. In the preceding discussion, plant cells have been considered purely as osmotic systems. There is no doubt that the osmotic mechanism of the cell-to-cell traslocation of water exists and operates in most if not all kinds of plant cells. Certain of the movements of water which occur in plants, however, result from the operation of an imbibitional rather than an osmotic mechanism. Whenever the diffusion-pressure deficit of water is greater in an imbibant than in a contiguous solution, water moves from the solution into the imbibant and *vice versa*.

Not only does water pass into dry, mature seeds by imbibition as the first step in germination, but there is evidence that water moves by this process into ovules which are ripening into seeds in the ovulary. During the latter stages of the development of seeds in the cotton boll, for example, their diffusion-pressure deficits exceed their osmotic pressures. This indicates that an osmotic mechanism does not adequately, account for the movement of water into the maturing seeds during this period. Since no evidence could be found for the existence of a metabolically activated mechanism of water absorption, the most probable explanation of these results is that the diffusion-pressure deficits of maturing cotton seeds are largely imbibitional in origin. It seems likely that imbibition plays a proportionately large role in the movement of water into many other young, actively growing tissues. Other examples of the imbibitional absorption of water can be seen in certain species native to semi-desert regions, such as the resurrection plant (*Selaginella lepidophylla*) or in species such as many kinds of lichens which are indigenous to locally dry habitats such as sun-swept cliffs. Such species may become air-dry during periods of drought, but rapidly imbibe water and resume their physiological activities upon the advent of rain. At least some species of lichens readily imbibe water-vapour as well as liquid water.

Within a plant cell the cell sap, the water in the protoplasm, and the water in the cell wall may each be regarded as possessing a diffusion-pressure deficit of its own. The diffusion-pressure deficit of the cell wall is largely of imbibitional origin, and that of the protoplasm partly so. If, for example, the diffusion-pressure deficit of the walls of a mesophyll cell is increased as a result of evaporation during transpirational water loss, water diffuses into the walls from the protoplasm and, in turn, from the cell sap into the protoplasm as long as the diffusion-pressure deficit of the wall exceeds that of the cell sap. In cells in which vacuoles are very small or lacking, or in which the interior of the cell is occupied by hydrophilic colloidal substances, imbibitional phenomena undoubtedly play a greater proportionate role in the movement of water than in cells with prominent vacuoles.

Possible metabolic mechanisms of the movement of water in plants. The osmotic and imbibitional mechanisms may not be the only ones which play a role in the movement of water from cell to cell in plants or from the external environment into plant cells. A number of investigators have suggested that metabolic mechanisms may also be involved, it usually being implied that the energy of respiration is utilized in the transport of water molecules across cell membranes.

Much of the edivence which has been advanced in support of such an hypothesis is highly indirect or otherwise unconvincing. Kelly (1947), however, has shown that a close correlation exists between the rate of respiration and rate of water intake in the oat coleoptile. It has also been shown that certain inhibitors of respiratory enzymes retard water intake by potato tuber tissue. The demonstration that auxins accelerate water absorption by potato tuber and other tissues when immersed in pure water has also been considered to be evidence for the existence of such a mechanism. Auxins are one type of plant hormone and are known to influence respiration rates. Hence their effect on water absorption might be evidence that respiration plays a role in this process. However, Levitt (1948) failed to find a correlation between the rates of auxin-induced water absorption and respiration in potato tuber tissue, and considers an increased plasticity of the cell walls to be a more likely explanation of this phenomenon.

That correlations do exist between rates of respiration and rates of water movement or absorption in at least some kinds of plant tissues is undeniably true. Whether or not this relationship is a very direct one, however, is not clear. In meristems, for example, passage of water into cells may be a resultant of their increased solute intake, increased fabrication of protoplasm, or expansion in cell-wall area, all processes which appear to be dependent upon respiration as a source of energy. The influeuce of respiration on water movement may thus be principally or entirely an indirect one.

While respiration undoubtedly exerts at least indirect effects on the movement of water in plants, there is almost no positive evidence that the energy of this process is utilized directly in the transport of water molecules. Furthermore, according to calculation of Levitt (1947), it appears improbable that a metabolic mechanism of movement of water of any appreciable magnitude could be maintained in most kinds of plant cells at the expense of respiratory energy.

The pathway of water through the root. Water enters the roots principally through the walls of the root hairs and epidermal cells of the root tips. Absorption of water by individual root hairs has been demonstrated experimentally with micropotometers. From the epidermal cells the water passes through successive rows of thin-walled cortical cells, and then through the cells of the endodermis.

The structure of the walls of the endodermal cells is peculiar. Two main types of such cells have been recognized. In one type (Fig. 5.9) the inner tangential and radial walls, or sometimes the entire wall, is thickened. These thickened walls are suberized and sometimes partly lignified. In a second type (Fig. 5.9B and C) a thickened strip is present on the inner surface of the radial and transverse walls, this thickening often being suberized. The width and general configuration of these *Casparian strips*, as they are called, vary with the species. Regardless of type, most thickened endodermal walls are pitted. In some species having the thick-walled type of endodermis, there are present, opposite the outer end of each area of xylem tissue (as seen in cross section) isolated thin-walled endodermal

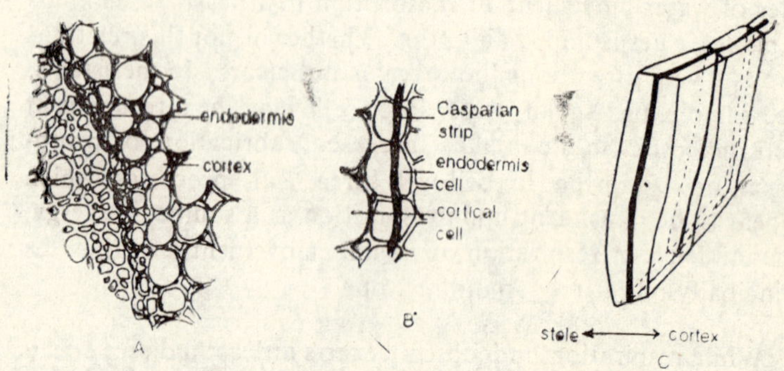

Fig. 5.9. (A) Thickened tangential walls of endodermis of a maize root, as seen in cross section. (B) Casparian strip in stem endodermis of *Piper macrophylla*, as seen in cross section. (C) Perspective diagram showing position of Casparian strip in an endodermal cell.

cells called *passage cells*. These are supposed to facilitate the movement of water and dissolved salts through the endodermis. Passage of water and mineral salts through the endodermis is probably also facilitated by the presence of lateral root initials which usually have developed by the time the endoderms is mature.

After passing through the endodermis water moves into the xylem ducts, in most species after traversing a few intervening layer of pericyclic cells.

The relation between roots and soil water. From the standpoint of the absorption of water by plants, a clear distinction should be made between conditions under which capillary movement of water occurs readily in a soil and those under which such movement of water is slow or non-existent. If capillary movement occurs readily, traslocation of water may take place toward the young roots whenever they are absorbing water. There are two principal conditions under which capillary translocation of water can occur in soils at appreciable rates : (1) in any zone of soil which is not more than a few feet above a water table, and (2) in the upper layers of any soil after a heavy rain or irrigation, but before the water content of the soil has decreased to its field capacity. As the water in the films surrounding the soil particles with which the root tips are in contact becomes depleted, more water moves toward those particles by capillarity. The actual rate of such capillary movement of water through the soil may become a factor influencing the rate of absorption. Root systems, however, are not static, but are more or less continually growing through the soil. The rate of root growth of most species decreases, as a general rule, with increasing wetness of the soil above the field capacity, because of the corresponding reduction in soil aeration. Hence in soils in which capillary movement of water occurs—which are necessarily relatively wet—the rate of elongation of roots is generally less than in otherwise similar but somewhat dryer soils. This continued growth of root tips through the soil brings them into contact with other portions of the soil water, so that, even if capillary movement to certain parts of the soil ceases, capillary movement of water to the roots may be re-established by the extension of the root tips themselves into zones of the soil that have not yet been depleted of water which can move by capillarity.

Many plants, much of the time, grow in soils at water contents between the permanent wilting percentage and the field capacity. In this range of soil-water contents capillary movement of water is very slow. Once most of the film water present on the soil particles with which the root tips are in contact has been absorbed it cannot be replaced in any significant quantity by capillary movement from adjacent regions of soil if the soil water content is below the field capacity. Neither does water move in vapour form through soils toward the absorbing regions of roots at appreciable rates. Under such conditions the absorbing region of every rootlet often becomes surrounded with a narrow cylindrical zone of soil which has been depleted to a water content much below that of the surrounding soil.

Since, in soils at a water content below the field capacity, movement of water toward the roots is very slow, the principal method by which the roots come in contact with additional increments of water is by continually growing through the soil. Mature root systems of many species of plants bear millions of root tips. Each of these numerous root tips may be pictured as progressing through the soil and absorbing most of the water present in the smaller interstices between the soil particles with which they come in contact. Relatively large quantities of water can be absorbed in this way, at least by some species of plants. For example, the total length of all the roots on a four-months-old rye plant was found by Dittmer (1937) to be 387 miles. On the average, therefore, the aggregate daily increase in the length of the roots on this plant was more than 3 miles. In addition, nearly 55 miles of new root hairs were formed, on the average, per day. Kramer and Coile (1940) calculated that such a rate of root extension would permit the absorption by such a plant of about 1.6 liters of water daily from a sandy loam, or about 2.9 liters from a heavy clay loam, when both soils were at the field capacity.

This picture of the role of root elongation in the absorption of water (and mineral salts) from the soil under certain conditions is probably an accurate one for many and perhaps most species of plants. The root systems of some species of plants, however, are of a sparse, sparingly branched type which would suggest that the quantities of water which they can absorb in the manner just described would be relatively small. Mycorrhizae may aid in the absorption of water by some species. Elongation of roots would also be less effective in contributing to the absorption of water by terrestrial species in which the roots bear few or no root hairs than in species on which root hairs develop in abundance.

The root-tip population of any plant is usually so large that often not all of the root tips are subjected to the same soil-water conditions. Some may be located in lower soil horizons which, under certain conditions, contain more water than the upper layers of the soil. Under other conditions the reverse situation may prevail After light rainfalls on a comparatively dry soil, for example, the root tips closer to the surface may be in contact with soil at a higher water content than those at greater depths. Hence the quantity of water absorbed in a given interval of time may differ greatly from one root tip to another as they advance through the soil. In general however, roots do not grow appreciably in soils which have a water content less than that of the permanent wilting percentage.

Mechanism of the absorption water. The development of a diffusion-pressure deficit in the mesophyll cells of leaves causes the water in the xylem vessels or tracheids to pass into a state of tension which results in an increase in the diffusion-pressure deficit of the water in the xylem ducts by the amount of the imposed tension. As soon as the diffusion-pressure deficit of the water in the xylem ducts in the absorbing region of the root exceeds that in contiguous cells, a gradient of diffusion-pressure deficits is established across the root, increasing consistently from cell to cell from its epidermal layer to the xylem conduits. Many authorities believe that the water in the cells in absorbing regions of roots often passes into a state of tension under conditions such as those just described. If this occurs greater diffusion-pressure deficits could develop in the peripheral cell walls of young roots than otherwise would be possible. However, the mechanism as just described will operate even if the water in the root cells never passes into a state of tension. The osmotic pressure of the root epidermal and root-hair cells of most species for which measurements are available is about 3-5 atm., although higher values undoubtedly occur in some species. Hence diffusion-pressure deficits of this magnitude can develop in the peripheral cells of roots even if the water in them is never under tension.

Whenever the diffusion-pressure deficit of the water in the peripheral walls of the young root cells exceeds that of the water in the soil, water will move from the soil into the root. Since the osmotic pressure of the soil solution in most soils is only a fraction of an atmosphere, the diffusion-pressure deficit of the absorbing cells of a root does not have to be very great before water will enter them from any soil with a water content equal to or greater than the field capacity.

The absorption process which has just been described is often called "passive absorption" because the entry of water into the roots is brought about by conditions which originate in the top of the plant and the root cells seemingly play only a subsidiary role. Although the general picture of this mechanism of absorption which has just been presented is probably correct in its essentials, it is almost certainly oversimplified. The influence of certain environmental factors upon absorption, particularly temperature and oxygen, suggests that the metabolic activities of living cells in the absorption zone of roots also play at least an indirect part in this process.

The mechanism of absorption just described undoubtedly accounts for the intake of most of the water which enters the roots of plants, but it is not the only mechanism of absorption which is known to operate in plants. In many species an internal pressure known as root pressure often develops in the xylem. The occurrence of sap exudation resulting from root pressure can be strikingly demonstrated with some species by immersing the root system of a decapitated plant in a potometer. After a time a dilute sap will begin to ooze from the cut stem, and the absorption of water will be indicated by the movement of the meniscus on the capillary arm of the potometer. If the volume of water exuded is measured it will be found to be not sensibly different from the volume absorbed. In other words, water is being absorbed and is moving in an upward direction through the plant as a result of processess which take place in the root cells. This type of absorption, in which the mechanism involved is localized within the root system, is often called "active absorption." Root pressure, or guttation, and active absortion are usually considered to be different aspects of the same phenomenon.

There is good evidence for the existence of a relatively simple osmotic mechanism of active absorption, but it is by no means certain that this is the only mechanism responsible for root pressure, xylem sap exudation, and guttation. The essentials of an osmotic theory of active absorption were first early suggested by Atkins (1916). Priestley (1920, 1922) has also advocated a similar hypothesis. In a number of species it has been shown that, although the osmotic pressure of the sap in the xylem ducts is relatively low, seldom exceeding 2 atm., it is higher than the diffusion-pressure deficit of most soils at the field capacity or a higher water content. Osmotic movement of water from the soil to the xylem ducts could therefore occur through the "multicellular membrane" of the intervening root cells in spite of the fact that such cells have a higher osmotic pressure than either the soil solution or the xylem sap. The mechanism of such a movement of water can be interpreted in terms of the establishment of a consistently increasing gradient of diffusion-pressure deficits from cell to cell across the root from the epidermis to the xylem ducts. Kramer (1932) has shown that in an analogous situation water will move across the cells of the petiole of the tropical papaw (*Carica papaya*).

Eaton (1943) found an almost proportional relation between the difference in osmotic pressures of the xylem sap and the external solution and the rate of exudation from decapitated young cotton plants. His results also indicated that exudation would cease when the value of this osmotic differential fell to zero suggesting that, under the conditions of these experiments at least, the osmotic mechanism was the only one in operation. The sensitivity of the osmotic mechanism is very great ; reversals occur in less than a minute from exudation to cessation of exudation and back to exudation upon transfer of the root systems of young plants from water to a dilute sucrose solution and back to water.

Although there is little doubt of the existence of an osmotic mechanism of active absorption, the results of certain experiments are difficult to reconcile with the view that this process can be accounted for entirely by such a mechanism. The findings of Grossenbacher (1938, 1939) and Skoog *et. al.* (1938) that an autonomous, more or less regular diurnal fluctuation occurs in the rate of exudation of xylem sap from detopped seedlings, with the maximum about noon and the minimum about midnight, suggests a more complex mechanism than the one described above. A retarding effect of deprivation of oxygen on absorption of water by isolated onion roots, and by root hairs of radish, has been demonstrated by refined experiments with potometers, suggesting that respiration plays at least an indirect part in the active absorption mechanism. The enhancing effect of auxin on xylem sap exudation rates also suggests that respiration plays a part in active absorption, since auxins are known to influence respiration rates. It has therefore been postulated that a second mechanism of active absorption exists which is dependent for its operation upon the metabolic activity of the root cells.

The influence of metabolic processess on active absorption of water may, however, be largely or entirely and indirect one. The absorption of mineral salts by cells is markedly influenced by the metabolic conditions prevailing within them, especially their rate of aerobic respiration. Furthermore, there are good reasons for believing that the passage of salts into the xylem ducts is largely controlled by the metabolic conditions in adjacent living cells. Metabolic conditions, therefore, may influence the steepness of the osmotic gradient between the xylem sap and the soil because of their effect on the absorption and cell-to-cell movement of mineral sales ; and their influence on the active absorption of water may be exerted largely or entirely in this or in other indirect ways.

The relation of transpiration to the movement of water through the plant. In most discussions of the cohesion of water theory of the rise of water in plants, the relation of transpiration to this process is so greatly emphasized that it has come to be almost indelibly associated with this concept. Transpiration is not, however, the fundamental cause of the upward movement of water in plants. Water ascends in the xylem ducts only when an adequate diffusion-pressure deficit has been created in the water of the vocuoles or the walls of cells in organs in the upper parts of the plant. Since evaporation of water from the walls of the mesophyll cells in the most frequent cause of such diffusion-pressure deficits, the process of transpiration has been generally linked in discussions with the mechanism of the ascent of sap. It is only because of its effect in increasing the diffusion-pressure deficit of the water in the mesophyll cells that transpiration sets in motion the entire train of water through the plant. Any other process which results in an increase in the diffusion-pressure deficit in the cells at the terminus of any plant axis may also induce the translocation of water. Upward movement of water often continues during the night hours after transpiration has virtually ceased. This lag results from the residual high diffusion-pressure deficit of the leaf cells at the end of the daylight period. Water will continue to enter these cells until they reattain the maximum turgidity of which they are capable under the conditions prevailing. Similarly movement of water will occur into any rapidly growing stem tip or fruit because the binding up to water in certain phases of the growth process creates a diffusion-pressure deficit in the cells of such organs, thus inducing the migration of water towards such centers of growth activity.

At the present time most plant physiologists are agreed that the cohesion theory is a correct representation of the principal mechanism by which water is transported through plants, be they the tallest of trees or herbs only a few feet in height. It is also generally agreed that root pressure plays a minor role in this process at least under certain conditions and in some species. There is no doubt that living cells are involved in the phenomena generally described under the term root pressure, and there is at least a possibility that living cells of the xylem are in some way essential to the maintenance or operation of the cohesion mechanism.

Rates of movements of water through plants. The rate at which water ascends through the xylem ducts may vary from a movement so slow as to be almost imperceptible to speeds of at least 75 cm per minute and probably higher. In general, daily variations in the rate of ascent of water approximately parallel daily variations in transpiration rate ; the maximum velocity of the transpiration stream is often not attained, however, until several hours after the peak of transpiration.

Lateral movement of water. A cell-to cell lateral movement of water in a radial direction undoubtedly occurs along the vascular rays in the steams of most species of plants. In woody stems there is probably also a lateral movement of water around the stem in a tangential direction. Except in trees in which the "grain" of the wood is twisted, the conducting vessels on one side of the tree generally connect with branches on that side of the tree at their upper extremity, and with roots on the same side of the tree at their lower extremity. If no lateral movement of water occurs in woody stems, it would be expected that removal of the roots from one side of a tree would result in a dearth of water, or perhaps even death of the leaves or branches on that side of the tree.

Experiments have been performed on apple, peach, oak, and other woody species in which the roots on one side of the plant were removed in an attempt to ascertain whether or not lateral movement of the water occurs. Although the water content and growth of the plants treated in this manner diminished, there was no difference in the moisture content of the leaves on the two sides of the tree. Neither did the leaves on the side from which the roots had been removed show any greater tendency to wilt on clear, warm days than those on the other side. These results indicate very strongly that lateral movement of water occurs in woody stems, and that the water conductive system of plants acts as a unit system.

Downward movement of water. There are a number of experiments on record which indicate that downward movement of water can occur in stems. For example, Dixon (1924) found that, if the tip of the leaf of a potato plant be cut off under an eosin solution, the liquid will descend through the xylem tissue of the leaf, petiole, and stem, and eventually reach the underground organs. In general, however, such an effect is to be expected only when an internal water deficit exists in a plant.

The cohesion theory of the movement of water will account equally well for the conduction of water in either an upward or a downward direction through the plant. Whenever conditions are such that the diffusion-pressure deficit of the cells of an organ, which is basally situated relative to another, is greater than that of the cells of a more nearly apical organ, a reversal in the direction of the movement of the water will occur.

Sap flow in the maple. Sap flow from cuts or bore holes in maple trees occurs only during the late winter and early spring months when cold nights alternate with days upon which the temperature rises above freezing. The flow is largely or entirely confined to the daylight hours. A single sugar maple tree will generally yield from 25 to 75 liters of sap per season, the sucrose content of which is usually between 2 and 3 per cent. In other woody species that exhibit a prefoliation sap flow, such as birch and grape, exudation usually will not occur until later in the spring at a period when freezing temperatures have become uncommon. Xylem sap flow in such species is a root pressure phenomenon and continues during both day and night hours.

Sap flow in the red maple is not a result of root pressure because sap exuded as freely from borings into the xylem of cut trees with their bases immersed in water as from similar borings into rooted trees during the late winter and early spring whenever the necessary alternation of cold nights and warm days occured. These investigators suggest that the crystallization of water within the trunk and branches, especially in the outer layers of the xylem, indirectly causes upward movement of water through the tree. As a result of the conversion of some of the water present into ice, the osmotic pressure of the cells increases, which in turn induces upward movement of water through the xylem. It must also be assumed that at least some of the roots are in soil layers warm enough to permit the absorption of water. When, under the influence of daytime rising air temperatures or direct insolation, the temperature of the trunk and branch tissues reaches or exceeds the freezing point, the ice melts and the tissues become flooded with sap. This sap exudes through any sort of an incision made into the outer xylem of the tree.

We know that plants require water ; deprived of it, they will wilt and die. In fact, the typical land plant consumes prodigious quantities of water, vastly more than any of the other substances that enter it, and some 100 times more than its content of water at maturity. Stated differently, the rate of water loss is 100 times greater than the rate of entry of carbon dioxide. Most of this water therefore does not remain within the plant, but passes through it to the atmosphere. The process is called *transpiration.* The transpiration stream carries to the aerial parts of the plant nutrient ions absorbed by the roots, although the rate of transpiration is not necessarily a limiting factor in the absorption of nutrients.

The seemingly wasteful consumption of water through transpiration is nonetheless essential to the growth of plants. In nature and even is agriculture, it is a rare and fortunate plant that enjoys an optimum supply of water throughout its life. Without irrigation, most crops are inhibited, often seriously, from achieving full growth and development.

If we are concerned with the processes controlling the growth of plants in nature or if we are concerned with the world-wide problems of agriculture, then we must recognize that there is perhaps no single factor that is more crucial and none more amenable to our intervention (at least in principle) than that of control over transpiration in plants.

At one stage, the movement of water in plants involves its diffusion as a gas. We shall therefore want to consider some of the laws of gaseous diffusion, including the behaviour of CO_2 and oxygen as well as water.

We shall see that long-distance water movement occurs primarily through a system of capillaries called *xylem.* In contrast, the movement of organic solutes occurs in an independent set of extraordinary channels called *phloem*. We shall examins this process under the term *translocation*.

The Path of Transpiration

Let us first survey the phenomena of transpiration, looking especially at the anatomy of water movements. We shall here take as facts certain notions which we may wish to examine more critically in the sections that follow.

Water is absorbed from soil through young roots, especially by root hairs (Figs. 5.10 and 5.11). The water appears to move through the cortex of roots and through their cell walls. Conical cells are closely interconnected by strands of protoplasm called *plasmodesmata*. The collection of interconnecting protoplasm is often called the *symplast*.

The peculiar anatomy of the endodermis, composed of cells with radially thickened and suberized cell walls, has led many physiologists to infer that it is a barrier to water movement, that water at this point must pass through the cells rather than the cell walls. Although this notion is of only marginal relevance to the problems of water movement, it becomes of major importance for the leading hypothesis on ion transport.

Whatever are the more subtle aspects of radial movements of water across the root, once water enters the stele its distribution through the plant appears to be a matter of plumbing, rather elegant plumbing but nonetheless plumbing.

Pteridophytes, gymnosperms, and angiosperms evolved conductive cells called *tracheids* (Fig. 5.12). These cells are elongated, ranging typically from a few tenths of a millimeter to a few millimeters, and exceptionally to 10 or 20 mm in length, with strongly lignified walls. Since water must move across many more cross walls in tracheids than in vessels, the efficiency of water conduction is substantially less. The inefficiency is somewhat compensated by frequent lateral connections, known as bordered pits (Fig 5 13). These pits appear to be cleverly designed check valves, which may be able to seal off a tracheid which has suffered an embolism.

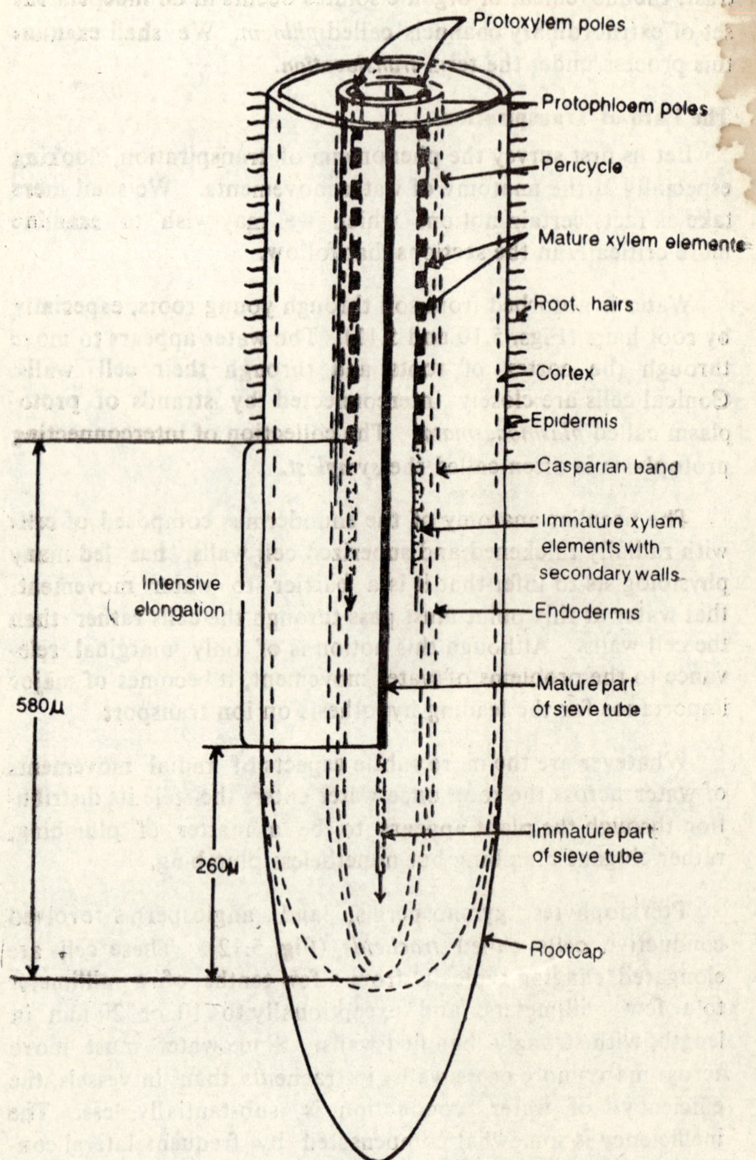

Fig. 5.10. Diagram of a longitudinal section of a tobacco root tip, showing the regions important for water absorption.

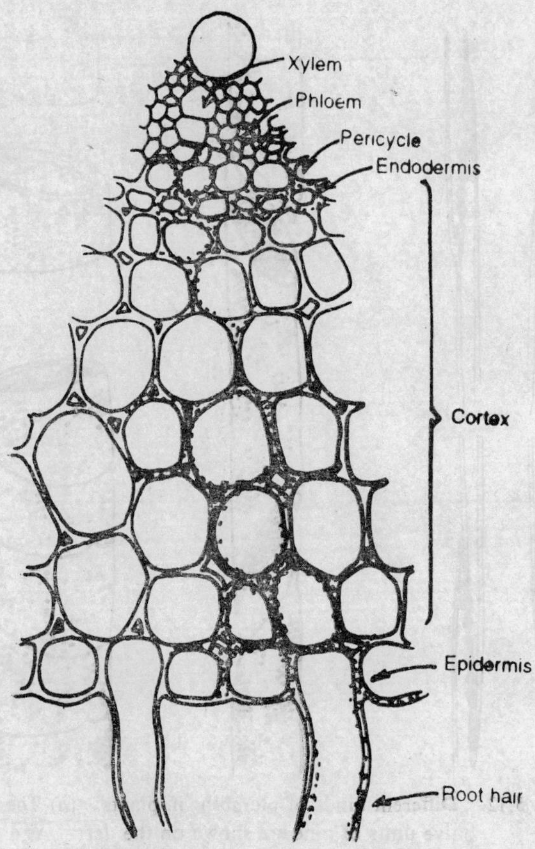

Fig. 5.11. The path of water across a root tip (Kramer, 1959). Water is thought to enter root hairs, follow cell walls, and traverse the cortex to the xylem.

Angiosperm wood, in addition to tracheids, also contains the hydrodynamically more efficient xylem *vessels*. These structures may be up to several meters long and a few tenths of a millimeter in diameter and can conduct impressive volumes of water. The vessels are the mortal remains of vertical chains of cells whose lateral walls are sealed with lignin and other amorphous polysaccharides and strengthened with helical thickenings of cellulose. The end plates have dissolved away to form long capillary channels (Fig. 5.12). The protoplasm has disappeared.

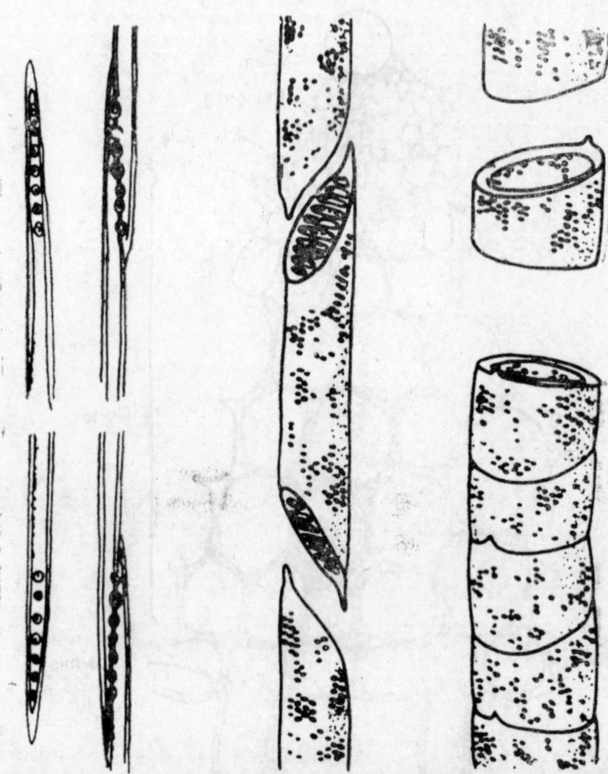

Fig. 5.12. Different kinds of plumbing in plants. (*a*) The conductive units of pine are shown on the left. As a gymnosperm, it has only tracheids, shown in the photomicrograph above and in longitudinal diagram below. Tracheids connect with one another through bordered pits. (*b*) The center units are vessels of birch ; the end walls are fully perforated and conduct freely. (*c*) The more efficient conducts on the right are vessels of oak. Here the end walls are completely eliminated with the result that water moves through long rigid tubes.

Water in vessels is typically under negative pressure. Since the introduction of bubbles into these capillaries will destroy the conductive system, the capillary column will often be broken in winter when the water freezes. In order to succeed in so-called temperate climates, dicots learned simply to

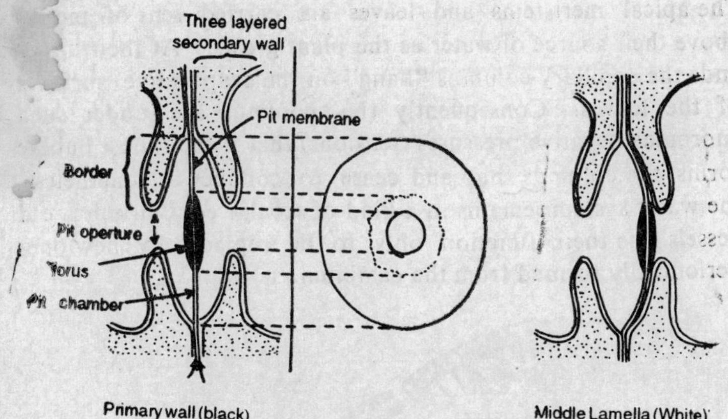

Fig. 5.13. The function of bordered pits. In their normal state (on the left) the hydrostatic pressure in cells connected by bordered pits is nearly equal. Water diffuses across the thin portion of the membrane. As shown on the right, if the pressure in one cell is lost, as through a cut, the membrane is forced to the side and the central portion of the membrane or torus, is pushed against the opposite opening, thus sealing off the embolism.

abandon air-filled vessels and cut off new vessels from the cambium each spring before new leaves expand.

All along the conductive system a modest amount of lateral transport supplies the living cells of the stem with inorganic nutrients and replenishes the relatively small amounts of water lost by evaporation. Most of the transpiration strean continues upward to the leaves. There, small amounts of water and the remainder of the nutrients enter the leaf parenchyma (Fig. 5.14). The bulk of the water passes along the parenchyma cell walls and evaporates into the intercellular spaces. Finally the water diffuses out into the atmosphere through the *stomata*.

The water system of a tall tree or liana (woody vine) is an astonishing device, especially when viewed in developmental sequence. The water columns are first filled by capillarity and osmosis as the living protoxylem cells expand. The water remains as the end walls dissolve and the protoplasm disappears.

The apical meristems and leaves are carried tens of meters above their source of water as the plant grows. At their upper ends the capillary columns "hang" on the evaporative surfaces of the leaves. Consequently the columns are under such enormous negative pressures (tensions) that if the tiniest bubble forms, the columns snap and cease to conduct. Nonetheless, the water system remains in a kind of secular equilibrium ; old vessels lose their function only to be replaced by new ones periodically formed from the cambium.

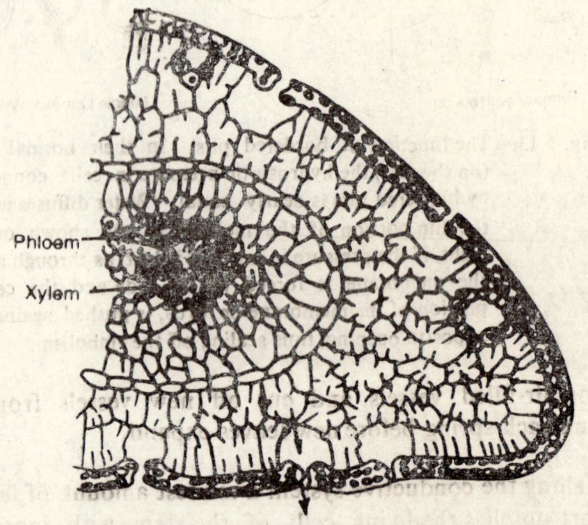

Fig. 5.14. Path of water across a pipe needle. Water is carried into the leaves through branches of xylem, moves through the cell walls of the leaf parenchyma, evaporates into the intercellular spaces, and thence diffuses through the stomata to the outside.

We have said that the transpiration stream brings inorganic nutrients to the aerial parts of the plant, that the volume of water transpired is larger than required for this purpose, and also that most of the water evaporates through the stomata. It is obvious that, if the rate of water loss is not subject to some kind of control, the plant's tissues will suffer from desiccation, as would those of any organism. Even though the leaves and stems are enveloped in a waxy cuticle which is almost impermeable to water, the stomata are so ingeniously designed that, when they are open, water can evaporate through them almost as rapidly as if the entire leaf surface were open to the atmosphere ! The seeming death wish on the part of plants becomes understandable when we recognize that the stomata are programmed to admit carbon dioxide for photosynthesis. Paradoxically in order to carry on maximal rates of photosynthesis, plants must accept a potentially dangerous loss of water. Land plants are thus forever poised between the Scylla of carbon starvation and the Charybdis of desiccation !

Considering the seemingly limitless capacity of plants for chemical synthesis, we can only conclude that the failure of plants to develop an epidermis that would admit carbon dioxide but retain water vapour is some cosmic oversight.

In the sections below we shall inquire more rigorously into the movement of water in plants. We shall then consider as case studies some of the limiting factors governing water movements in the several stages of transpiration.

Hydrostatics and Equilibrium Relations of Water in Plants

Exactly as in the case of chemical transformations net movements of water can occur only in the direction of a decrease in the free energy of the system. We must therefore inquire into the factors within the soil-plant-atmosphere system which may affect the free energy of water.

Although investigators concerned with soil physics have reported states of water in terms of its free energy, plant physiologists have usually used terms such as "diffusion pressure deficit," which are difficult to relate rigorously to thermodynamic quantities. Recently there has been general agreement to use the them *water potential*, ψ defined as the chemical potential per unit molal volume of water.

$$\psi \equiv \frac{\Delta \mu_w}{\bar{V}} \quad \cdots(1)$$

where $\bar{V}$ is the partial molal volume of water (18 cc mole^{-1}) and $\Delta \mu_w$ is the difference between the chemical potential of water in the solution and the chemical potential of pure water at 0° and 1 atm pressure. Chemical potential has its usual thermodynamic meaning of the change in free energy per mole at constant temperature and pressure (Equation 2).

$$\mu_w \equiv \text{chemical potential of water} \equiv \left(\frac{\partial G}{\partial n}\right)_{T,P} \qquad \ldots(2)$$

The term water potential at once keeps physical chemists content and has intuitive appeal to biologists ; as with other potentials, water will tend to flow from higher to lower water potentials. Similarly, the water of solutions with identical ψ values will be in equilibrium ; that is, the net flux will be zero.

Water potentials can be thought of as composed of three terms, attributable to osmotic pressure, hydrostatic pressure, and the so called matric potential or matric effect (Equation 3).

$$\psi = \psi_o + \psi_P + \psi_M \qquad \ldots(3)$$

ψ and the terms that compose it have the units of pressure. Under standard conditions (pure water at 0° and 1 atm), ψ=1 atm.

We shall give here a brief account of how each term is generated.

ψ_o and Osmotic Pressure

If a solution contains solutes, the water potential will be less than that of pure water. This happens quite simply because the presence of solutes decreases the concentration of water itself. More precisely, it is the *activity* of water which is decreased. (Activity is another of the useful fictions invented by physical chemists : it is the concentration of a solute or solvent in an ideal solution.)

Quantitatively, the decrease in water potential due to solutes, ψ_o, is related to the activity of water by Equation 4.

$$\psi_o = -\frac{RT}{\bar{V}} \ln \frac{a^{\circ}}{a} \qquad \ldots(4)$$

where R is the universal gas constant, T the absolute temperature, $\bar{V}$ the partial molal volume of water, and a° and a the activities of pure water and the water in the solution.

ψ_o is also the negative of the osmotic pressure π.

$$\psi_o = -\pi \qquad \ldots(5)$$

If the solution is "ideal,"

$$\psi_o = -\frac{RT}{\bar{V}} \ln \frac{1}{N_w} \qquad \ldots(6)$$

where N_w is the mole fraction of water.

For solutions which are both dilute and ideal,

$$\ln \frac{1}{N_w} \simeq \Sigma n_s$$

where n_s is the number of moles of solute. This leads to the van't Hoff relation

$$\pi = \frac{\Sigma n_s RT}{\bar{V}} \qquad \ldots(7)$$

Note that since the osmotic pressure π must always be positive, the ψ_o term is always negative.

ψ_P and Hydrostatic Pressure

The pressure term, ψ_P, is numerically equal to the hydrostatic pressure.

$$\psi_P = P \qquad \ldots(8)$$

Hydrostatic pressure may be either positive or negative. The inward pressure on a cell exerted by the cell wall is called the *turgor pressure* and is positive. Negative pressure or tension can exist in xylem elements.

ψ_M of Matric Potential

The peculiar properties of water are largely accountable for the tendency of water molecules to form hydrogen bonds with other water molecules, with solutes, and with hydrophilic colloids. Water has a large electric dipole moment, so that the electron-rich (oxygen) region of water molecules tends to align toward electron-poor centers (cations), while the hydrogen ends are oriented around anions. Ions typically attract a shell of water up to 2 Å deep, depending on the electric field strength. The shell is called water of hydration (Fig. 5.15).

The hydrogen bonding of water with oxygen-, nitrogen-, or sulphur-containing radicals of large molecules, such as proteins and polysaccharides, may be extremely extensive. The association of water with large molecules results in the phenomenon of *imbibition*, whereby seeds absorb large amounts of water even from relatively dry soils.

Finally, water may be bound in tissue capillaries by van der Walls' forces.

All of these forces tend to lower the potential of water and are lumped as ψ_M. The determination of ψ_M is largely empirical and is measured as the water potential after the hydrostatic pressure and solute effects have been eliminated or substracted out.

Calculations Involving Water Potentials

In order to predict the directions or rates of water movements in the soil-plant-atmosphere system, we should gain some experience in calculating water potentials of various plant compartments.

Example 1. What is the water potential of a 1 molal solution of sucrose at 25° and 1 atm pressure ?

Let us assume that a 1 molal solution is "ideal." Starting with Equation 3.

$$\psi = \psi_o + \psi_P + \psi_M$$

we may take as given that ψ_p and $\psi_M = 0$.

From Equation 6.

$$\psi_o = -\frac{RT}{\bar{V}} \ln \frac{1}{N_w}$$

$$R = \frac{18.06 \text{ cc atm}}{\text{deg mole}} \quad T = 298°K$$

$$\bar{V} = 18.07 \text{ cc}$$

$$N_w = \frac{55.5}{56.5}$$

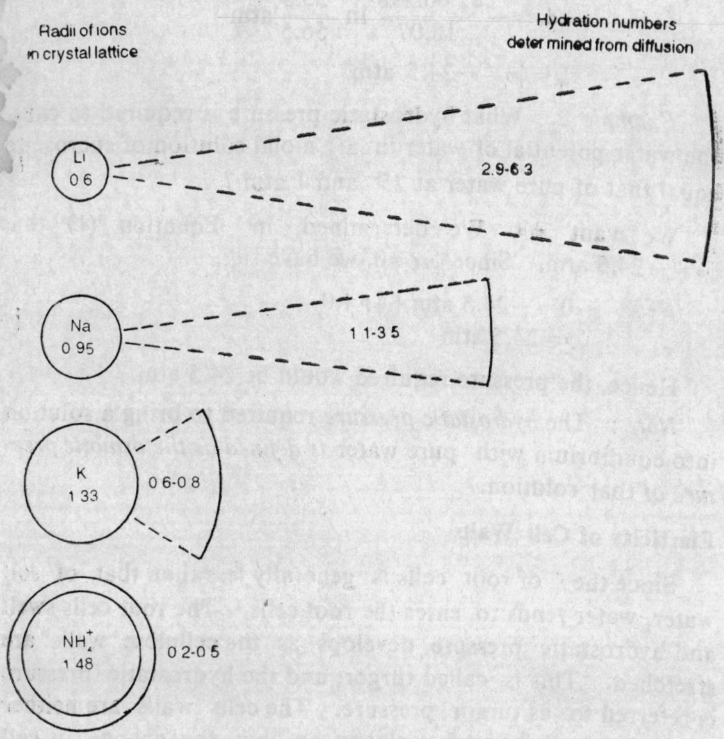

Fig. 5.15. **Relative size of some alkali ions. The radii of ions can be determined accurately from measurements of crystal lattices. But the effective size of an ion in aqueous solution is determined by its charge which attracts water molecules around it. The smaller ions have a more intense charge and are able to organize one or more layers of less oriented water molecules. In this way a sodium ion during difusion or ionophoresis behaves as a larger particle than does a potassium ion. The actual size of these hydration shells cannot be determined unambiguously or in absolute units. The numbers and the radii shown are only relative.**

(that is, the mole fraction of water in the solution is equal to the number of moles of water divided by the total number of moles present).

$$\phi_o = -\frac{82.06.298}{18.07} \ln \frac{55.5}{56.5} \text{ atm}$$

$$\psi = \psi_o = -24.5 \text{ atm}$$

Example 2. What hydrostatic pressure is required to cause the water potential of water in a 1 molal solution of sucrose to equal that of pure water at 25° and 1 atm ?

We want ψ_P. We determined in Equation (1) that $\psi_o = -24.5$ atm. Since $\psi_M = 0$, we have

$$0 = -24.5 \text{ atm} + \psi_P + 0$$

$$\psi_P = 24.5 \text{ atm}$$

Hence, the pressure required would be 24.5 atm.

Note : The *hydrostatic pressure* required to bring a solution into equilibrium with pure water *is defined as the osmotic pressure of* that solution.

Elasticity of Cell Walls

Since the ψ of root cells is generally less than that of soil water, water tends to enter the root cells. The root cells swell and hydrostatic pressure develops as the cellulose walls are stretched. This is called turgor, and the hydrostatic pressure is referred to as turgor pressure. The cells walls are neither perfectly nor infinitely elastic so that the change in cell volume is a complex function of the external water potential (Fig. 5.16).

Example 3. The volume of a cell is measured, and then the cell is placed in water at 25°, in which it is observed to swell. It is then placed in successively increasing concentrations of mannitol ; each time it shrinks some. When it is put into 0.2 M mannitol it is found to be the same size as it was initially. It continues to shrink in stronger solutions, until in 0.5 M mannitol very slight plasmolysis is seen, while in 0.6 M mannitol it is strongly plasmolyzed. In 0.5 M mannitol the volume of the cell is 5 per cent less than it was initially. Calculate, in atmosphere :

(*a*) The cell's ψ_o at zero turgor

(*b*) The cell's ψ_o in its initial state

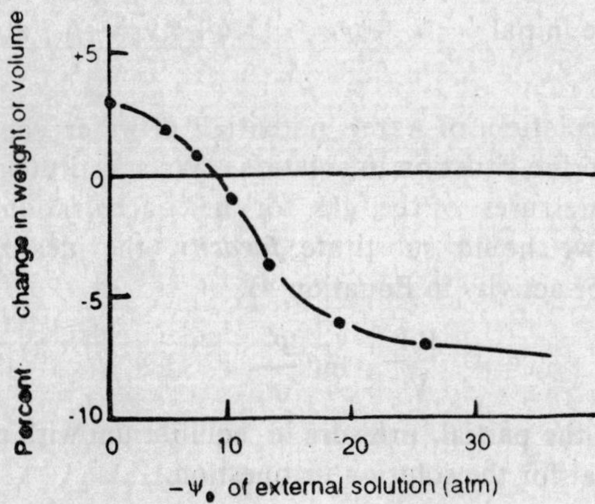

Fig. 5.16. Cell volume as a function of external osmotic pressure. As the osmotic pressure increases, water moves out of plant cells. Since the cell walls are somewhat elastic, the cell walls contract and the volume of the cell decreases.

(*c*) The cell's ψ_P in its initial state

(*d*) The cell's ψ in its initial state

As the cell is placed in progressively more concentrated mannitol, water will begin to leave the cell, causing a decrease in volume until the ψ_o of the external solution reaches the ψ_o of the cell. At this point the elastic cell walls begin to relax; the hydrostatic pressure ψ_P falls. At zero turgor where $\psi_P=0$ and (ignoring ψ_M) ψ_o (outside)$=\psi_o$ (inside), the cell contents will just begin to pull away from the cell walls. This is called *incipient plasmolysis.*

(*a*) In the example, incipient plasmolysis occurs in 0.5 M mannitol ; hence, from Equation (7).

$$\psi_o=-0.5\times0.08206\times298=-12.2 \text{ atm}$$

(*b*) Since the cell volume in 0.5 M mannitol is 95 per cent of the initial volume, it follows that the initial ψ_o must have been

$$\psi_o=-0.95\times0.5\times0.38206\times298=-11.6 \text{ atm.}$$

(*c*) The effect of 0.2 M mannitol is equivalent to the original hydrostatic pressure ; hence,

$$\psi_P = 0.2 \times 0.08206 \times 298 = 4.9 \text{ atm}$$

(*d*) The initial $\psi = \psi_O + \psi_P = -11.6 + 4.9 = -6.7$ atm

Air

The calculation of water potentials of water vapour in air is similar to the situation in solutions. We substitute the notion of partial pressures of the gas for the concentrations. (More precisely, we should substitute *fugacity*, the idealized partial pressure, for activity in Equation 4).

$$\psi_O = -\frac{RT}{\bar{V}} \ln \frac{p^\circ}{p} \qquad \text{...(9)}$$

where p° is the partial pressure in equilibrium with pure water and p is that for the solution in question.

Example 4. Calculate the water potential of air at 90 per cent relative humidity at 25°.

$$\psi = \psi_O + \psi_P + \psi_M$$

$$\psi_P = \psi_M = 0$$

$$\psi_O = \frac{RT}{\bar{V}} \ln \frac{p^\circ}{p}$$

$$\psi_O = -\frac{82.06 \times 298}{18.07} \ln \frac{100}{93} \text{ atm}$$

Hence,

$$\psi = \psi_O = -141 \text{ atm}$$

The astonishingly low water potential of even moist air means that equilibrium during the day between water in the leaf and in the air is virtually impossible.

Hydrodynamics : Laws Governing the Rate of Fluid Movement

We considered the factors governing water potentials in and around plants. Differences in water potential will determine the *direction* taken by water movements, but exactly as in the case of chemical transformations, the *rate* of water movement is determined by additional factors.

The movement of liquids in capillaries is well understood. From Poiseuille's law (Equation 10), we can predict rates of volume flow J_v

$$J_v = \frac{\pi r^4 \Delta P}{8\eta l} \qquad \text{...(10)}$$

given the radius of the capillary, r ; the viscosity η ; and $\Delta P/l$, the mean pressure gradient along the capillary. From the r^4 factor we may infer that in comparison to a tracheid of 10μ radius, a vessel with a radius of 100μ would have 10^4 greater specific conductivity !

We can similarly predict the flow of gases through stomata from Fick's law (Equation 11).

$$J = -DA\frac{\partial s}{\partial x} \text{ or } \frac{J}{A} = -D\frac{\partial s}{\partial x} \qquad \text{...(11)}$$

where D is a diffusion coefficient, $\partial s/\partial x$ is the concentration gradient of a solute s with respect to distance x, and A is the cross-sectional area.

Serious difficulties arise when we come to a membrane. The property of membranes to pass fluids is called *permeability*.

Nearly all substances move in or out of cells less rapidly than would be expected on the basis of free diffusion. Some substances (for example, proteins) appear to move across cell boundaries at essentially zero rates. We conclude that cells and tissues posses restricted permeability. We say that membranes are differentially permeable or, ambiguously, that they are "semipermeable."

Let us imagine an idealized membrane (Fig. 5.17) of thickness Δx separating an outer solution I from an inner solution II. The solutions contain water w and solute s. The chemical potentials μ_i of the water and solute are shown as μ_w^{I} μ_w^{II}, μ_s^{I}, μ_s^{II}, etc.

Intuitively (that is, viewing the system as simple plumbing), we can imagine characterizing the membranes in part by the rate of flow of fluid as a function of the pressure drop across the membrane.

Let $J_v \equiv$ total flow across the membrane. If the osmotic pressures of I and II are equal, we could expect that flow would be simply proportional to the pressure drop ΔP :

$$J_v = L_P \Delta P \qquad \ldots(12)$$

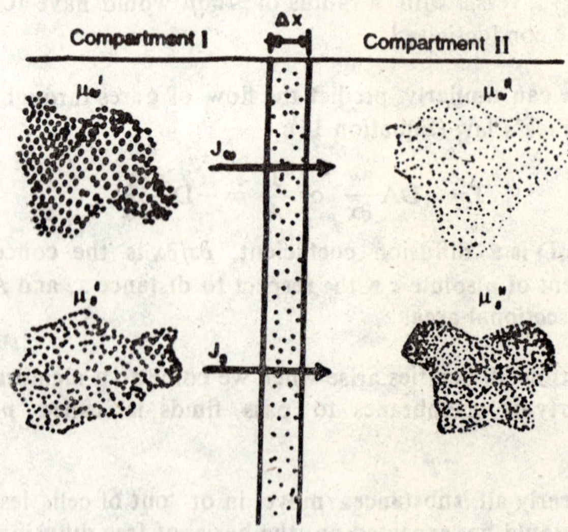

Fig. 5.17. Model for membrane transport. Two homogeneous (that is, stirred) compartments I and II are separated by a membrane of thickness ΔX.

μ_wI = chemical potential of water in solution I
μ_sI = chemical potential of solute in solution I
μ_wII = chemical potential of water in solution II
μ_sII = chemical potential of solute in solution II
J_w = flow of water
J_s = flow of solute

In the diagram the μ_wI is shown as being greater than μ_wII so water would tend to flow from left to right. On the other hand, $\mu_s\text{II} > \mu_s\text{I}$ and the solute would tend to flow from right to left, so that the value of J_s would be negative.

where we define L_p as the hydraulic or filtration coefficient for unit pressure difference. We can also imagine characterizing the membrane by the rate of diffusion of solute across the membrane. With zero electrical potential difference, we expect according to Fick's law (Eq. 11).

$$J_s = -\omega \frac{ds}{dx} \quad ...(13)$$

where ω is the diffusion coefficient for the solute and ds/dx is the concentration gradient across the membrane. Assuming a homogeneous membrane, ds/dx should be (intuitively) proportional to the concentration difference, $s_I - s_{II}$. Hence Equation (13) could be rewritten

$$J_s = -k(\pi_I - \pi_{II}) \quad ...(13)$$

where k is an arbitrary constant.

Both of these kinds of measurements are of limited application since flows arise in the intact plant from differences of both osmotic pressure and hydrostatic pressure acting simultaneously. This problem was attacked intuitively by Ursprung and Blum, who hypothsized that the rates of flow across membranes are proportional to differences in water potential (Equation 14)

$$J_v = k(\psi_I - \psi_{II}) \quad ...(14)$$

or, ignoring matric potentials

$$J_v = k[(P_I - \pi_I) - (P_{II} - \pi_{II})]$$

This same relation is known in animal physiology as Starling's hypothesis.

Unqualified acceptance of Equation 14 resulted in certain distressing contradictions, which will be discussed below.

Since Equation 14 was not derived rigorously, biophysicists have searched for a theoretical approach to permeability based on thermodynamics. Katchalsky (1961) assayed the problem as follows :

"Modern research on biological permeability is mainly concerned with the molecular organization of the membranes. The ultimate goal of the investigation is the development of structural models which can be used for prediction of the physico-chemical properties of the membranes.

"During the last few years, another approach is emerging whose main concern is not the construction of adequate models, but the establishment of equations capable of organizing the experimental results in a consistent formal framework. This approach is based on the thermodynamics of irreversible processes. The thermodynamic approach admittedly cannot yield quantitative numerical predictions as derived from the treatment of a tangible model. However, it has the advantage of being general, free from contradiction and able to predict new correlations. There is, moreover, a possibility of bridging the two approaches by expressing the thermodynamic flow parameters in terms of physical coefficients, such as frictional coefficients, amendable to kinetic interpretation."

The following is an account of the basic transport equation greatly abbreviated from Katchalsky.

We shall use the same model as that in Fig. 5.17.

According to the thermodynamics of irreversible processes, the rate of flow of water is

$$J_w = L_{ww}X_w + L_{ws}X_s \qquad \ldots(15)$$

The symbols deserve some explanation : X_w and X_s are the gradients of chemical potential (cf. Equation 2 across the membrane).

$$X_w \equiv \frac{-d\mu_w}{dx}$$

$$X_s \equiv \frac{-d\mu_s}{dx}$$

L_{ww} and L_{ws} are called Onsager coefficients. It seems intuitively reasonable in Equation 15 that the flow of water should be proportional to the gradient of chemical potential of water. What we are unprepared for in the dogma of irreversible thermodynamics is that the flow of water *may* also be affected by the gradient of solute potential. I say *may* because the L_{ws} factor may be positive but it may also be zero.

By the same token, the flow of solute across the membrane, J_s, is also the resultant of two terms :

$$J_s = L_{ss}X_s + L_{sw}X_w \qquad ...(16)$$

From Equation 16 we see that the movement of solutes is a function of the solute potential gradient X_s but may also be affected by the gradient of water potential X_w.

According to the disciples of irreversible thermodynamics, the necessary and sufficient characterization of any membrane would consist in a specification of the coefficients L_{ww}, L_{ss}, and L_{ws} ($=L_{sw}$) for each solute.

In dealing with water movements we should like to use the same terms P and π, which proved useful for water equilibria.

We can write a pair of equations analogous to Equations 15 and 16, but with P and π substituting for X_s and X_w as the driving forces. J_w and J_s must also be replaced with J_v and J_D.

$$J_v \equiv \text{total flow across the membrane} = L_p\Delta P + L_{pD}\Delta\pi \qquad ...(17)$$

$$J_D \equiv \text{relative flow of solute } vs. \text{ water} = E_{pD}\Delta P + L_D\Delta\pi \qquad ...(18)$$

where ΔP is the pressure difference and $\Delta\pi$ is the difference in osmotic potential across the membrane. L_p, L_{pD}, and L_D are a corresponding set of Onsager coefficients.

We can appreciate the physical significance of the new Onsager coefficients by letting the two forces become zero separately. When $\Delta\pi$ is zero, J_v is the flow resulting from only a pressure difference, and L_p is seen to be the same coefficient of filtration that we encountered earlier.

$$J_v = L_p\Delta P \qquad ...(12)$$

When $\Delta P=0$, J_D is the exchange flow, so that L_D is the coefficient of diffusion across the membrane,

$$J_D = L_D\Delta\pi \qquad ...(19)$$

The significance of the cross coefficient L_{pD} can be seen in

Equation 17 when $\Delta P=0$. Then $J_v=L_{pD}\Delta\pi$ which means that L_{pD} is the coefficient for volume flow with unit osmotic gradient.

Another way of seeing the significance of L_{pD} is simply to adjust the pressure so that there is no volume flow ; that is, $J_v=0$. Then

$$\frac{\Delta P}{\pi}=-\frac{L_pD}{L_p}=\sigma \qquad ...(20)$$

But for an ideal membrane $P=\pi$ under equilibrium conditions ; therefore, σ is the ratio of the *apparent* osmotic pressure to the *theoretical* osmotic pressure. The ideality of a membrane is thus expressed by σ, the reflection coefficient.

$$\sigma\equiv-\frac{L_pD}{L_p}$$

The traditional (that is, ideal) differentially permeable membrane will have unit volume flow for unit osmotic gradient, so that $\sigma=1$. But for a completely nonselective membrane under these conditions, osmotic flow is zero, and hence $L_{pD}=0$. For real membranes, therefore, σ varies between 0 and 1. Equation 17 may now be rewritten

$$J_v=L_p(\Delta P-\sigma\Delta\pi) \qquad ...(21)$$

This leaves L_D as the only coefficient which cannot be conveniently measured. It turns out that the solute permeability coefficient ω is related to the Onsager coefficient by Equation 22.

$$\omega=\frac{L_pL_D-L_{pD}^2}{L_p}\bar{C}_s \qquad ...(22)$$

where $\bar{C}_s$ is the mean concentration of solute in the membrane. Thus a membrane can (in principle) be completely characterized by two conventional coefficients and a new one arising from the Onsager equations :

L_p=hydraulic or filtration coefficient

ω=solute permeability coefficient

σ=reflection coefficient

Evaluation of membrane properties and the idea) semipermeable membrane

The hydraulic coefficient can be evaluated fairly directly for the relatively spacious membranes of giant algae, such as the *Characea*, but with much greater difficulty and uncertainty in cellular organisms.

Measurements of the solute permeability coefficient ω have also been made, again most directly with giant algal cells. The permeability of membranes toward nonelectrolytes follows a very curious rule, first identified by Overton, an English physiologist working in Switzerland in the last century. According to Overton's rule, natural membranes behave as if they contained a layer of lipid (Fig. 5.18). Since a substance could

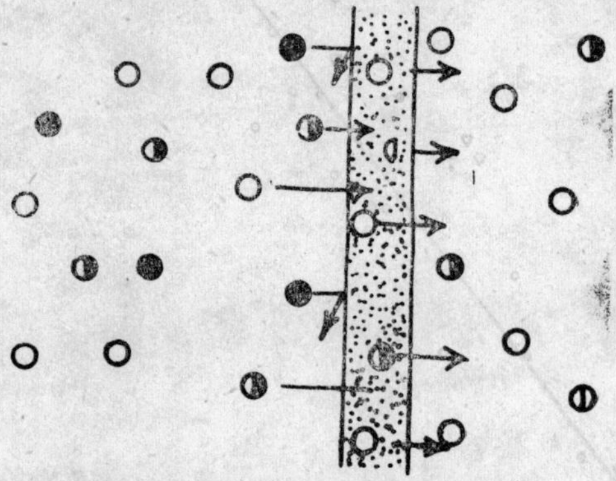

Fig. 5.18. Diagram of natural membranes according to the lipid hypothesis of Overton. According to Overton's hypothesis, solutes traverse biological membranes only by passing through a lipid phase in the membrane. Solutes that are insoluble in lipids (●) cannot pass. Solutes that are soluble (○) or partially soluble (⊖) in lipid penatrate the membrane in proportion to their distribution coefficients between the lipids of the membrane and water.

traverse the membrane only by traversing the lipid layer, the permeability is related to the lipid solubility of the substance, more specifically the distribution coefficient between lipid and water.

Lipid solubility turns out not to be the only factor in permeability. Molecular weight also influences permeability to about the 1.5 power. Collander has determined the permeability of the cell membrane of *Nitella*, one of the *Characea*, toward a large number of substances (Fig. 5.19).

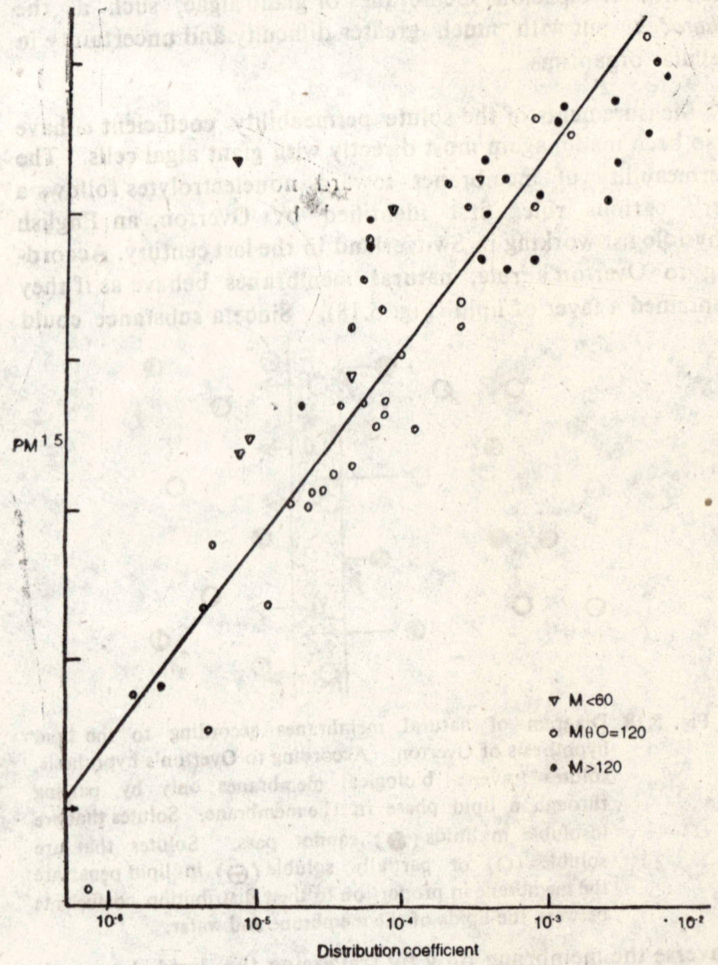

Fig. 5.19. Permeability of *Nitella macronata* cells as a function of oil-water distribution coefficients and molecular weights of non-electrolytes. The oblinate is $PM^{1.3}$, where P is the permeability coefficient in cm sec^{-1} and M is the molecular weight. The selection of the exponent is empirical. The abscissa is the distribution coefficient of the substances between olive oil and water. The empirical relation is P a $K^{1.3}$ $M^{-1.5}$. Small molecules, natably water deviate toward higher permeability coefficients.

One of the striking exceptions to the Overton rule is the permeability of all membranes toward water (see Fig. 5.19). The observed values are as much as 1,000 times greater than would be predicted from he Overton rule.

A determination of the reflection coefficient σ will indicate the ideality of a membrane. In an ideal semipermeable membrane,

$$\sigma = 1$$

$$L_{pD} = -L_p$$

In such a case the volume flow J_v in Equation 17 becomes

$$J_s = L_p(\Delta P - \Delta \pi) \qquad \ldots(23)$$

or from Equations 1, 5, and 8

$$J_s = L_p \Delta \psi$$

This states that volume flow is simply proportional to the difference between the pressure and osmotic gradients across the membrane. This of course is our old friend, Equation 4, from Ursprung, Blum, and Starling.

It was only reasonable that a gradient of osmotic pressure should cause difference in the activities of water on the two sides of a membrane and hence a *net diffusion* of water from higher to lower activity. But it was *not* obvious how osmotic pressure would generate a flow *equivalent* to that produced by hydrostatic pressure.

One can define an osmotic permeability coefficient P_o for water as

$$P_o = \frac{L_p RT}{\bar{V}_o}$$

with the units of cm sec^{-1}.

If osmotic flow occurs only by diffusion, the coefficient for the movement of water due to diffusion can be related to hydraulic flow by the equation

$$\frac{L_p RT}{\bar{V}_w} = P_d = P_o \qquad \ldots(24)$$

where P_d is the diffusion permeability coefficient of the membrane for water. Attempts were made, therefore, to test Equation 23 through the prediction of Equation 24. P_o was evaluated by measuring the bulk flow of water induced by differences in osmotic pressure ; P_d was evaluated by measuring the movement of labeled water, DHO or THO, into or out of cells that were at hydrostatic equlibrium. The results (Table 5.3 showed that P_o exceeded P_d by a factor of 10.

Solomon, Ussing, and others argued that these results could be rationalized if the transport of water (generated by $\triangle P$ or $\Delta\pi$) occurred not by diffusion alone but also by bulk flow through microcapillary pores in the membrane.

It is possible that pores may be detected by electron microscopists armed with greater resolution, but there is a simpler solution : it may be that membranes tested are not ideal that $\sigma<1$. Doubts of the existence of these pores are based on

Table 5.3. Radio of osmotic and diffusion permeability coefficients of water of some natural membranes. If osmotic flow of rates occurs by diffusion across memberanes P_o, the osmotic permeability coefficient, should equal P_d, the diffusion permeability coefficient. Most of the P_d values reported are in fact much lower than P_c.

Tissue	P_o *cm sec*$^{-1}$ *10*$^{-4}$	P_d *cm sec*$^{-1}$ *10*$^{-4}$	P_o/P_d
Arena coleoptiles			9.3
Telypellopsis stelligera single cells			5.3
Nitella mucronate	370	6.30	59
Potato tuber discs			6.7
Vicia faba root segments			35
Valonio veatricosa single cells	2.36 ± 0.17	2.4 ± 0.29	0.98

two lines of reasoning : (1) Equation 21 stands on thermodynamic not empirical or initiative, foundations, so that no special mechanism is needed to "explain at" ; (2) the existence of unstirred layers of solution on the order of 10μ thick is sufficient to make calculation of P_d extremely doubtful. Such unstirred layers are far less likely to exist under conditions of bulk flow.

In fact, the one known case where P_o equals P_d is that of *Valonia* (Table 5.3), a gaint unicellular alga. The dimensions of the alga provide two principal advantages ; (1) the effects of the unstirred layer will be small compared with the dimensions of the cell, and (2) the cells can be perfused internally, which facilitates the measurement of the permeability of rapidly permeating substances. Thus, where Equations 23 and 24 can be evaluated with confidence, the results are consistent with a simple, non-perforate membrane model.

The same problem of unstirred layers prevents the accurate estimation if σ-values. Thus until improved techniques are at hand, *e.g.*, until we can object isolated membranes to direct physical chemical analysis, the full characterization of natural membranes remains incomplete.

Membrane structure

The molecular structure of membranes is also disputed. There are two extreme models—the lipieds andwich or lipid bilayer of Davson, Danielli, and Robertson and the lipoprotein subunit hypothesis of Benson. These alternate models and some variants are represented in Fig. 5.20.

The lipid bilayer model finds support in electron microscopy where membranes can sometimes be visualized as a triple layer of two 20-Å electron-dense layers, thought to be protein, enclosing a 35-Å electron-transparent layer, thought to be lipid. But the optical rotatory dispersion spectra of a number of natural membranes show substantial helical structures among the proteins, which is inconsistent with the spread form or monolayer predicted from the bilayer model.

Benson finds support for his alternate view of lipoprotein subunits in the pock-marked appearance of freeze-etched

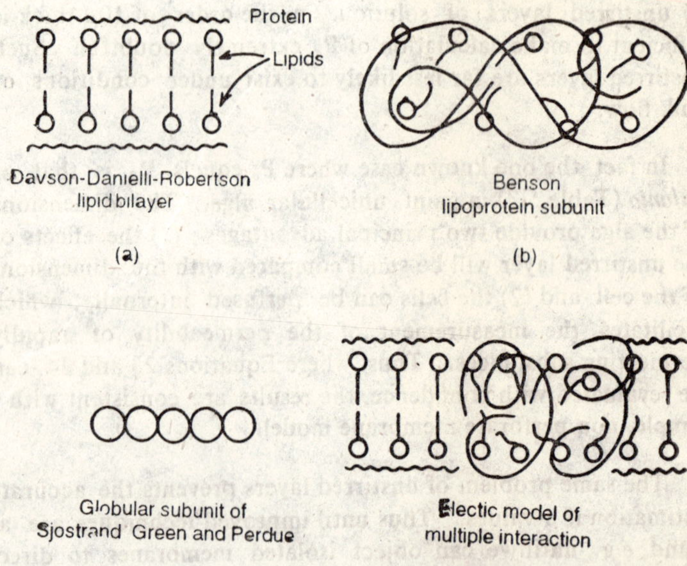

Fig. 5.20. Molecular models of membranes after Rothfield and Finkelstein, 1968). (*a*) A sheet of protein forms a monolayer over a double layer of lipids. The polar ends of the lipids face the protein, and the nonpolar ends bond hydrophobically to one another. (*b*) A lipoprotein in a random coil bonds lipids in various orientations. (*c*) Globular lipoprotein subunits form a two-dimensional sheet—an extension of the lipoprotein model. (*d*) Some regions of the membrane are oriented as lipid bilayers and others as randomly coiled lipoproteins.

chloroplast membranes. Sjostrand, Green, and Perdue, in an extension of the lipoprotein subunit model, visualize a two-dimensional sheet of lipoproteins in which different proteins are imbedded in the same sets of lipids, forming a functional mosaic.

It is safe to conclude that the several-models of membrane structure remain unproven.

Kinds of evidence required for the characterization of fluid movements in plants

Since we cannot at this time expect to achieve the kind of

characterization of membrane properties required by thermodynamics, the best we can do is a semiquantitative treatment as follows :

1. Information on the variables affecting water potential in a given plant compartment.
2. Data on the anatomical pathways of fluid movement, including the ultrastructure of the membranes, plysmodesmata, and cell walls involved.
3. Measurement of parameters, *e.g.*, vessel radii affecting fluid movements through nonmembranous paths, and factors affecting these parameters.

Case study : water movement in soil and roots

The water potential of soils depends in a characteristic way on the water contents of the soil. The relation is approximately hyperbolic (Fig 5.21) (Veihmeyer, 1956; Kramer, 1959),

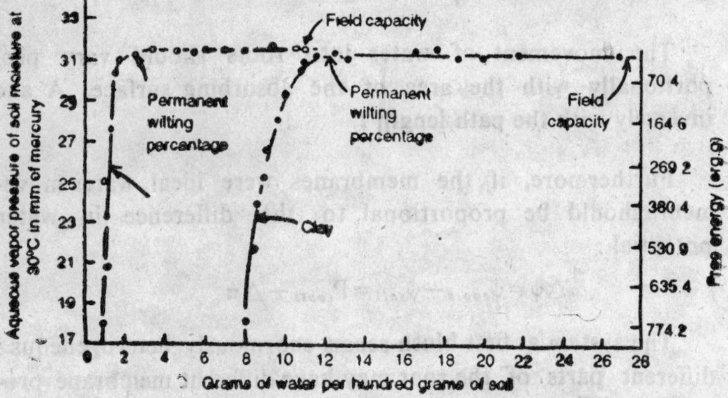

Fig. 5.21. Osmotic pressure and related values of soil water, as functions of water contents. The water potential of a soil, normally equal to the osmotic pressure, is a complete function of water content and is dependent on hydrogen bonding and stronger forces between water and the soil constituents. In the figure, it is clear that clay particles bind water relatively more tightly than sand does.

We may note, for example, that the water content of sand must be decreased to one-tenth of *field capacity* before the potential falls appreciably, whereas the water potential in the clay soil begins to fall at about one-half of field capacity. Clearly, substantial amounts of water are bound by clay. The "permanent wilting percentages" of the soils are also noted in Fig. 5.21. At first blush we might imagine that at these values the water potentials of water in the soil and in the roots are equal. Roots of mesophytic plants in fact have water potentials down to about —7 atm, which is much *less* than the ψ of soils at the "permanent wilting percentage." Clearly some factor or factors other than water potential determine the point of irreversible wilting of plants.

We presume that the rate of water movement into roots follows Equation 23 (or Equation 21 if the membranes are not truly differentially permeable). From this equation we can deduce some of the physiological variables that should govern the movement of water into roots and test if roots in fact behave in this way.

The movement of water into roots should vary proportionally with the area of the absorbing surface, A and inversely with the path length *l*.

Furthermore, if the membranes were ideal, water movement should be proportional to the difference in water potential :

$$\Delta\psi=\psi_{roots}-\psi_{soil}=P_{roots}-\Delta\pi$$

The system at first blush seems enormously heterogeneous : different parts of the root may have different membrane properties, different π, and different P. Moreover the external surface is not the only, and perhaps not even the, limiting barrier to water movement. From any one point on the root surface, water on its journey to the xylem may have to pass a dozen or more membranes with different parameters (Fig. 5.10).

However, certain generalizations are possible : root hairs comprise a major fraction of the absorption surface. These structures seem to be most abundant under moist conditions, when they are needed least. But in plants that bear root hairs, they account even at the minimum for over 60 percent of the total root surface. It seems agreed that maximum water absorption takes place on the parts of the root where root hairs are most abundant. This region also concides with that of xylem differentiation and lack of suberization.

Rates of water movement for several plant roots are shown in Table 5.4. The prodigious extent of roots is almost legendary. Annual crop plants in good soil typically penetrate to 2 m depth ; trees and some xerophytes may penetrate to 10 m more.

The general applicability of Equation 17 can be tested with the system shown in Fig. 5.22. Brouwer (1953) tested

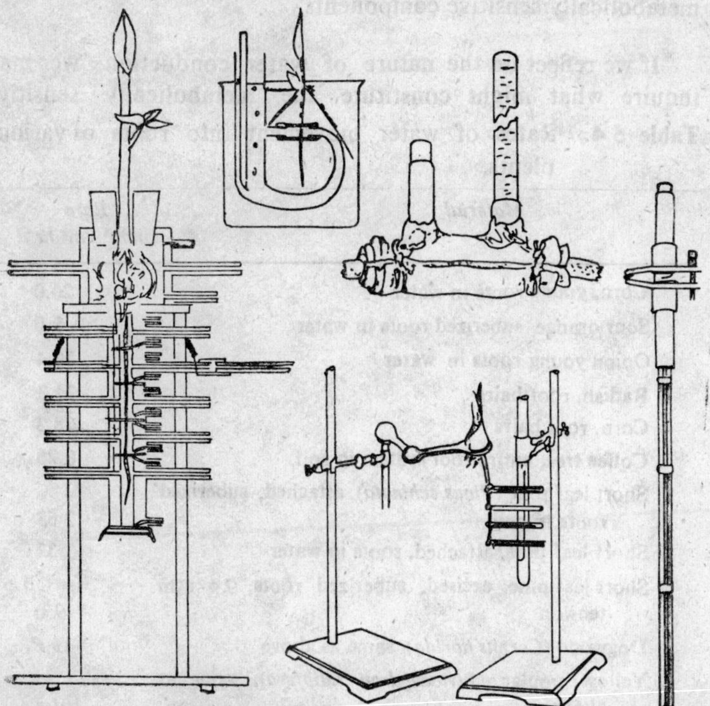

Fig. 5.22. Apparatus for measuring water conductivities of roots (collected by Kromer, 1956).

Vicia faba roots with one such system in which the water deficits of the medium and the tissue could be varied independently. He found that the conductivity was constant with varying water potential of the medium but was not constant with changing internal water potential (Fig. 5.23).

This is remarkably general finding. Increased water deficits, normally resulting from increased transpiration, depress somewhat the conductivity of tissue near the tip but greatly increase the conductivity of order tissue further from the tip. The result (Fig. 5.23) is that with increased transpiration, there is a shift in the region of maximal water absorption toward order tissues.

Water conductivity is also strongly depressed by anaerobiosis and cyanide poisoning. Hence conductivity contains metabolically sensitive components.

If we reflect on the nature of water conduction, we may inquire what might constitute the metabolically sensitive

Table 5.4. Rates of water movement into roots of various plents.

Material	*Rate* $mm^3\ cm^2\ hr^{-1}$
Corn, young roots in water	20.0
Sour orange, suberized roots in water	5 0
Onion young roots in water	50.4
Radish, root hairs	33.3
Corn, root hairs	28.3
Coffee tree, entire root system in soil	0.25
Short leaf pine (*Pinus echinata*), attached, suberized roots in soil	2·63
Short leaf pine, attached, roots in water	3.37
Short leaf pine, excised, suberized roots, 0.4 atm tension	9.0
Dogwood (*Cornus florida*), same as above	15.6
Yellow poplar (*Liviodendron tulipfera*), same as above	101.4

elements in this expression. There are several possibilities: metabolism may control the path length (thickness of the membrane) or the membrne characteristics.

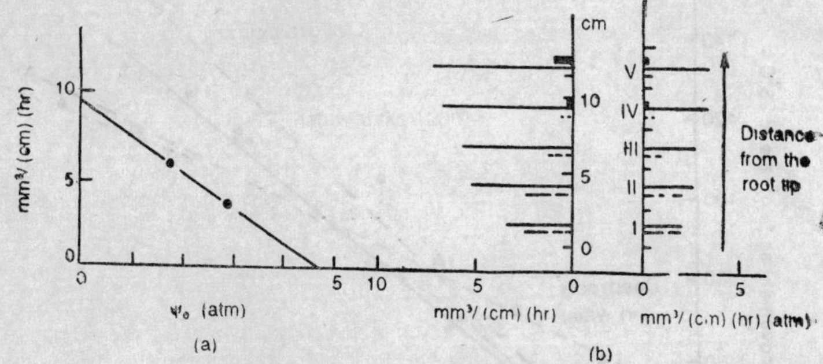

Fig. 5.23. Water conductivities of *Vicia faba* roots under conditions of varying water potential. The data were obtained using the apparatus shown in Fig. 5.22. (*a*) Rates of water movement as a function of applied osmotic pressure. (*b*) Rates of water flow (left) and conductivities (right) in different root zones. Water potentials were ...1.3 atm (dotted lines) and -2.5 atm (light solid lines). The heavy solid lines are for roots poisoned with 10^{-3} M KCN and at $\psi = -2.5$ atm.

Table 5.5. Rate of water absorption by intact and de-topped plants.*

Species	*Transpitation***		*Exudation****		*Exudation is Percentage of Transpiration*
	First Hour	*Second Hour*	*First Hour*	*Second Hour*	
Coleus blumei	8.6	8.7	0.30	0.28	3.2
Helianthus annous	4.3	5.0	0.02	0.02	0.4
Hibiscus moschentos	5.8	6.7	—0.01	0.05	0.7
Impatiens sultanii	2.1	1.5	—0.22	—0.06	
Tomato (*Lycopersion esculentum*) :					
Ser. 1	10.0	11.0	—0.62	0.07	0.6
Ser. 2	7.5	8.7	0.14	0.27	3.1

*Kramer, 1959.

**Rates are in milliliters of water per half hour and are average 5-8 plants.

***Tops removed.

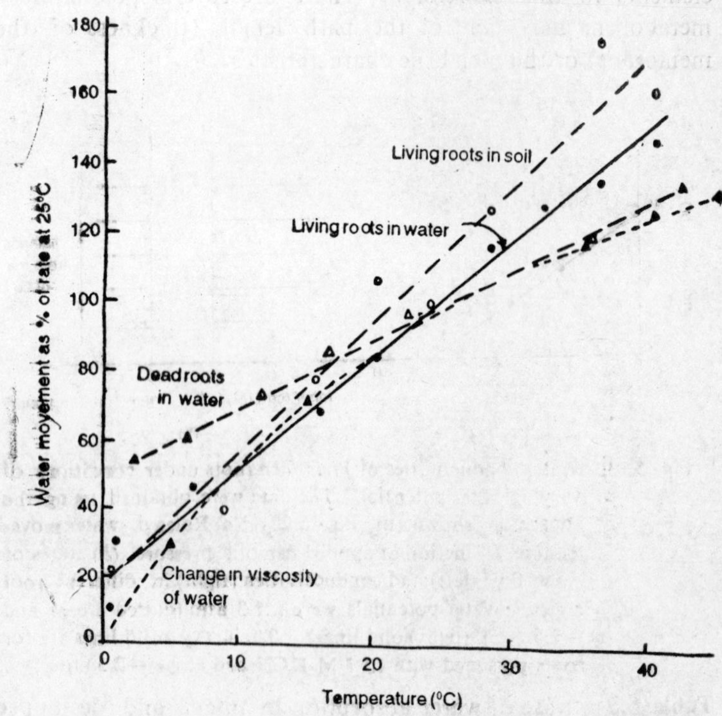

Fig. 5.24. Water movements in living and head *Heliantbus* roots as a function of temperature. The corresponding change in the viscosity of water is shown for comparison.

Of the terms that compose the water potentials of root tissues, the hydrostatic pressure P is typically more important than the osmotic pressure $\Delta\pi$. Because of evaporation from the leaf surfaces and the development of large water tensions in the xylem, P is normally negative. In contrast, $\Delta\pi$ (and hence $-\psi_o$) is rarely larger than 6 atm. The importance of P can be seen by comparing the rates of water absorption in intact with the 100-fold lower rates of de-topped plants (Table 5.5).

The conductivity of water in roots is very sensitive to temperature. In fact the water movement through killed *Helianthus* roots at different temperatures almost exactly parallels the viscosity of water (Fig. 5.24). Living roots show a steeper response to temperature (Fig. 5.25), especially in the case of warm-weather plants. That there should be species differences in the response to temperature strongly supports the notion of a metabolically cotrolled component in conductivity. The correlation with viscosity reminds us of the microcapillary model for bulk flow through membranes.

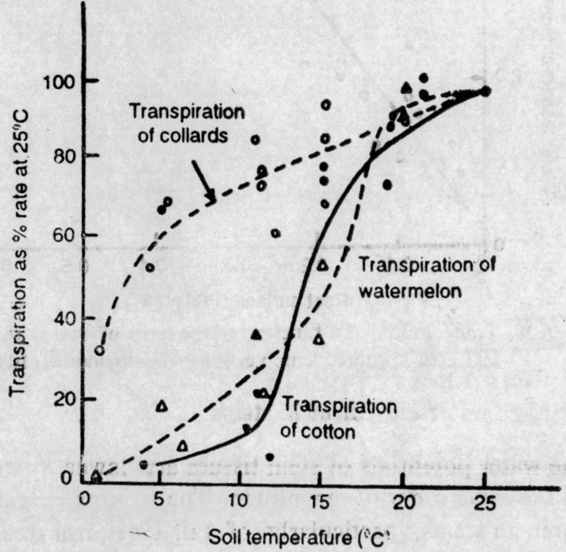

Fig. 5.25. Transpiration as a function of soil temperature in different species. Note that the transpiration of collards (*Brassica oleracea* var. *acephela*), which is a cool-weather crop, is much less affected by temperature than transpiration is in the two plants accustomed to warm soils.

Paradoxically, we may think of the roots as organs of water absorption but equally well as barriers to absorption. Indeed, below some critical ratio of root to leaf area, transpiration decreases linearly with root area (Fig. 5.26) but if the roots are cut off and the stems placed in water, transpiration is more rapid than in the intact plant.

Thus, in sum, it seems that roots are more than merely pieces of plumbing, but along with the stomata, roots are carefully controlled devices designed to respond to adverse conditions of water supply.

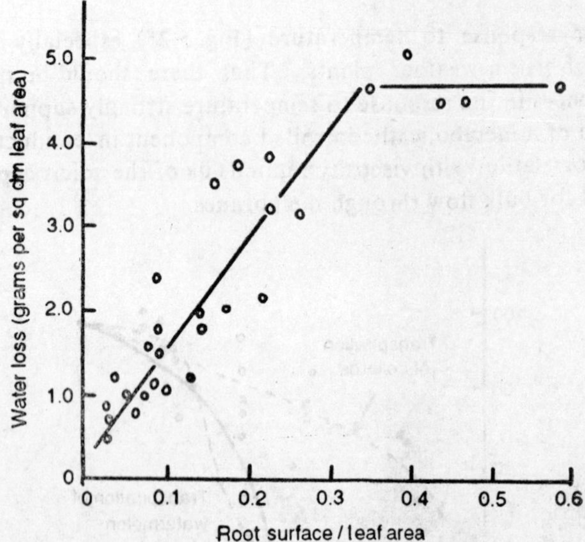

Fig. 5.26. Transpiration as a function of the ratio of root, surface to leaf area in rooted lemon cuttings (Bialoglowski, 1936).

Case study : water movement in stems

The water potentials of stem tissues are lower than roots, but of the same order of magnitude. The striking characteristic of water in stems, particularly of tall vines arm trees, is the existence of *negative pressure* (tension) in the xylem. Since xylem sap is fairly dilute (that is, has a small ψ_o) and since it must be in near equilibrium with surrounding tissues, the hydrostatic pressure must become negative as the water potentials of these tissues fall.

The conductive tissue of xylem consists of tracheids and vessels. These structures resemble capillaries, and we may consider the physics of water movement through them accordingly.

The factors governing the rate of movement of water through capillaries are expressed in Poiseuille's law (the same as Equation 10).

$$J_v = \frac{\pi r^4 \triangle P}{8\eta l} \quad \cdots (25)$$

where η is the viscosity, $\Delta P/l$ is the mean pressure gradient along the capillary, r is the radius of the capillary, and l is the length.

As we noted above, a vessel with a radius of 100μ would have a 10^4 greater specific conductivity than a tracheid of 10μ radius would have.

But does the water in xylem obey Poiseuille's law ? We restate Equation 25 as the sum of all conducting elements in a stem :

$$J_v = \frac{\pi}{8\eta} \frac{\Delta P}{l} \sum_{i=1}^{i=n} r_i^4 \quad \ldots(26)$$

A comparison of actual *vs.* expected conductivities is shown for several species in Table 5.6. The water conduction of the wood of lianas (first three species) is precisely that expected.

Table 5.6. Efficiency of water conductivity in various plants. Comparison of the observed rates of water movement (v_o) compaired with rates predicted (v_i) from Poiseuille's law.

Plant	*Conductivity Efficiency* $v_o/v_i \times 100$
Vitis vinifera	100
Aristolochia sipho	100
Atragene alpina	100
Root wood of oak	84
Root wood of beech	37 5
Helianthus annuus	32
Rhododendron ferrugoineum	20
Rhododendron hirsutum	13

The conductivities of oak wood are nearly equal to theoretical, whereas those of the other species range down to a small fraction of the expected values. Generally close agreement between predicted and observed rates of flow through the xylem of tomato was also observed by Dimond (1966).

Deviations from Poiseuille's law should be anticipated tracheids and vessels have end walls. Short vessels and especially tracheids must show conductivities less than that calculated for an open-ended tube.

The lengths of tracheids are fairly easy to determine in longitudinal sections (Table 5.7). Vessels present more of a problem. Huber's method (1956) is to macerate a given length of woody tissue and count the percentage of end walls. This procedure is directly analogous to end-group analysis in carbohydrates and proteins. Another is to measure the rise of mercury into the capillaries of the wood. Scholander *et al.* (1957) employed still another method with lianas, when segments of the vine were cut at progressively lower levels, water was sometimes released from the lower end. The release of water occurs when a vessel is cut for the first time. From the volume released and the diameter of the emptied vessels, the lengths of the vessels can be calculated. A summary of some typical vessel lengths is shown in Table 5.8. It has even been proposed that in some plants, single vessels may extend from the roots to the leaves.

Huber (1956) calculated an empirical quantity called the "relative conduction surface". It is the ratio of the area of the conducting elements to the fresh weight of the plant. The

Table 5.7. Typical lengths of tracheids.*

Plant Type	*Length, mm*
Medullosae	24
Araucaria	5(1.5—9)
Conifers	2—3
Populus	1
Fraxinus	0.12—0.29

*Collected by Huber, 1956.

distribution of values for this ratio is shown in Fig. 5.27. A very large fraction of the values fall between 0.2 and 0.8. Since the conductivity is related not to the area but to the square of the area of the xylem elements, the degree of cons-

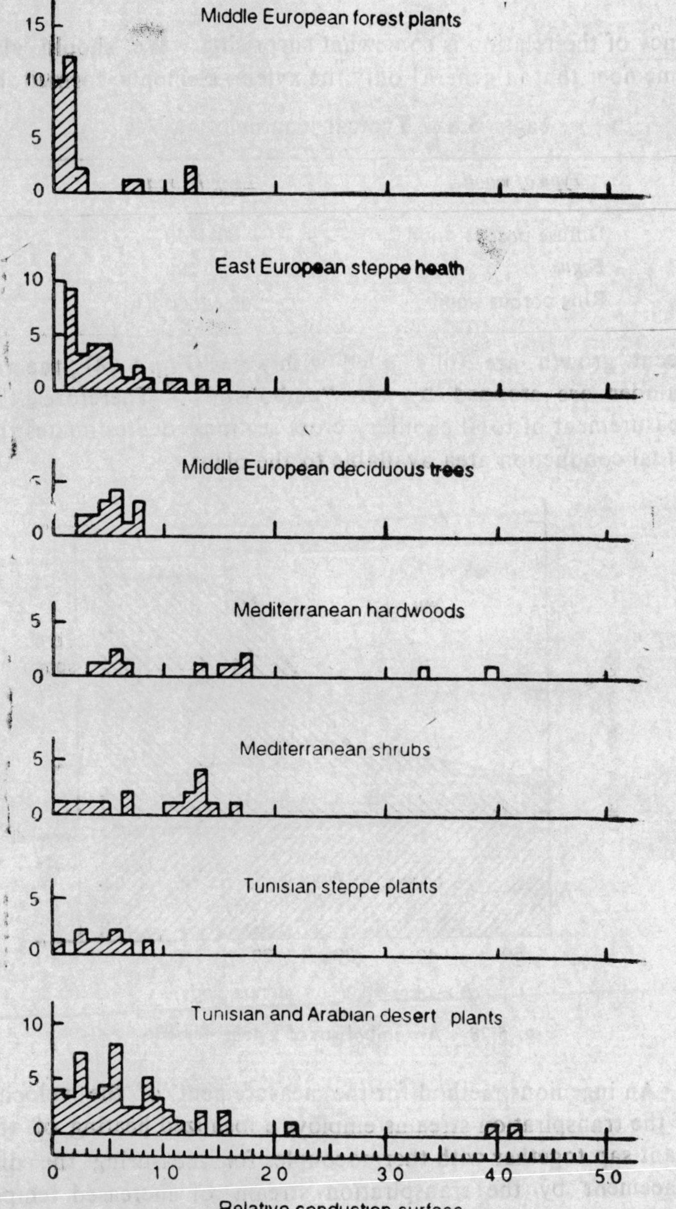

Fig. 5.27. Distribution of relative conduction-surface values for a variety of plants.

tancy of the relation is somewhat surprising. We should also remember that in general only the xylem elements formed by

Table 5.8. Typical lengths of vessels

Type of wood	*Length, mm*
Diffuse-porous wood	80—150
Fagus	200
Ring-porous wood	500—1800

recent growth are fully filled with water (Fig. 5.28) ; the remainder are stopped by air "embolism". Therefore the measurement of total capillary cross sections overestimates the actual conduction area available to the plant.

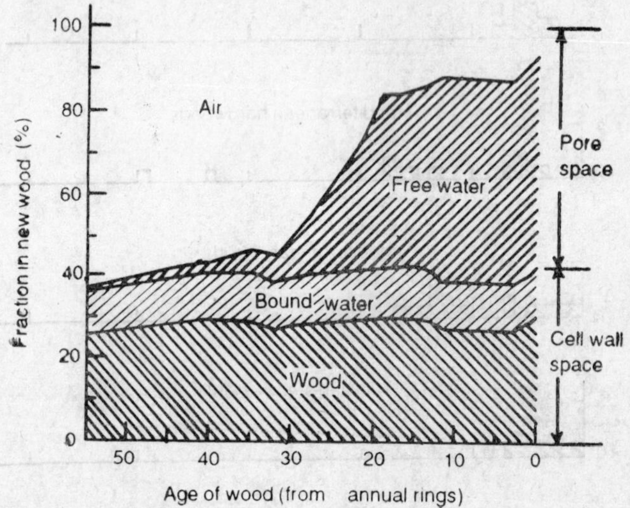

Fig. 5.28. Air embolism of xylem elements.

An ingenious method for the measurement of the velocity of the transpiration streams employs a localized heating of the plant sap together with thermocouples for measuring the displacement by the transpiration stream of increased temperature along the stem (Fig. 5.29). Zimmermann (1964*a*) employed a variant of this scheme to prove that there was no

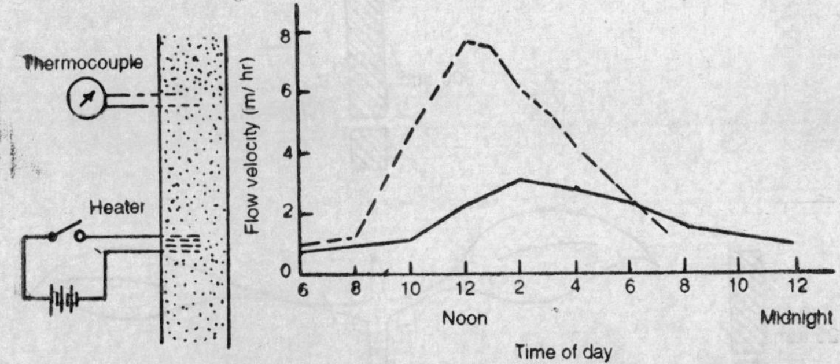

Fig. 5.29. The measurement of transpiration through heat conductivity. (*Left*) A heating element is inserted into the xylem and a thermo-couple some distance above it measures the steady-state difference in temperature. Faster transpiration is recorded as increased temperature. (*Right*) Transpiration rates in twigs (dotted line) and trunk (solid line) of a tree over the course of a day.

interruption in water conduction in wood at temperatures in the vicinity of 0°, provided that no actual freezing occurred.

Movement of Gases—Water Vapour and Co_2—Across Leaf Surfaces

During the day, when stomata are open, leaf water potentials may become very low (Table 5.9). Values of −15 to −20 atm and lower are not unusual.

Table 5.9. Osmotic Pressure and Water Potentials of Leaves*

Time of Day	*Andropogan trifide Osmotic Pressure, atm.*	*Ambrosia scoparius Water Potentials, atm.*
Early morning	20-22	—7-10
Noon	22-24	—12-15
Mid-afternoon	25-27	—15-17

*Water potentials of *Ambrosia* leaves ; osmotic pressure of *Andropogon* leaves.

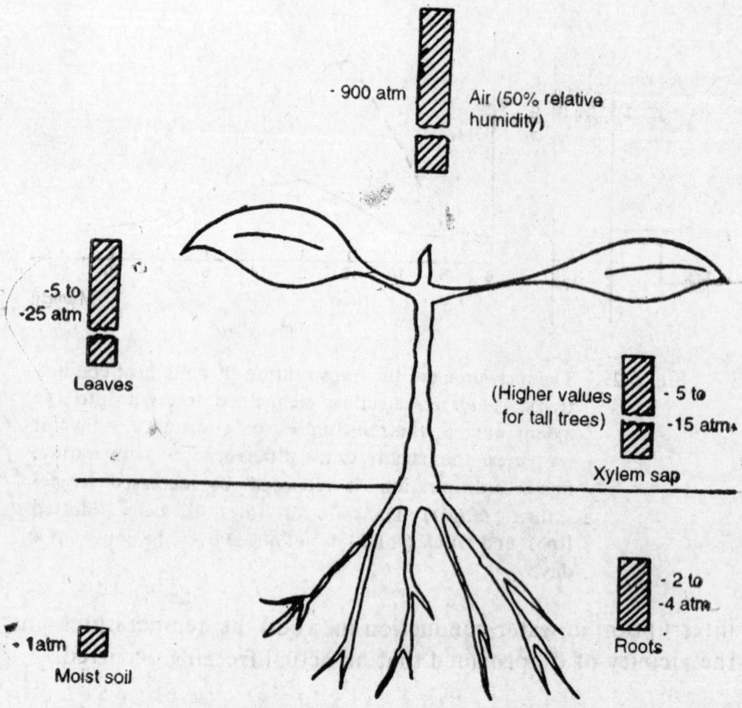

Fig. 5.30. Some possible values of water potentials in soil-plant-air systems. *Note* ; The actual values of water potentials will obviously vary with the availability of water, transpiration rate, etc.

Possible values of water potentials from soil to air are shown in Fig. 5.30. Clearly the largest free-energy changes occur between leaf and atmosphere.

Example 5. In a saturated atmosphere, is the decrease in water potential across the surface of an illuminated leaf sufficient to drive transpiration.

Let us grant that the sun raised the temperature of the leaf by only 1°, say from 25 to 26°. This would increase the vapour pressure of water in the leaf from 23.69 to 25.13 mm Hg. The $\Delta\psi$ for the movement of water from the heated interior of the leaf

$$\Delta\psi = -\frac{RT}{\bar{V}} \ln \frac{25.13}{23.69} = -80 \text{ atm}$$

is clearly greater than the $\Delta\psi$ for the movement of water into soil at the permanent wilting percentage (approx. −15 atm).

We calculated earlier that the principal free-energy decrease in the transpiration stream occurs across the leaf surface. The opening and closing of the stomata are therefore likely to be major factors in the control of water movement. The part the stomata play is determined, of course, not by free-energy changes alone (that is, hydrostatics) but by the factors governing the rates of water movement (hydrodynamics). We can construct a hydrodynamic model of the leaf-stomata-atmosphere system from a consideration of the path of water movement (Fig. 5.31). The intercellular air spaces of leaves are in effective contact with most of the leaf parenchyma, including the palisade as well as the spongy parenchyma. The leaf parenchyma cell walls therefore present an evaporation surface to there numerous intercellular air spaces of the leaf. Water vapour can also diffuse through the epidermis and cuticle.

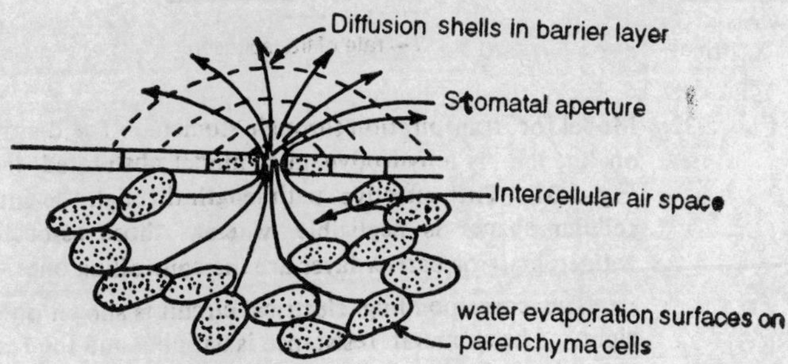

Fig. 5.31. Path of water from parenchyma cells to the atmosphere. Water evaporates from the cell walls of leaf parenchyma, diffuses through the intercellular air spaces and through the stomatal aperture to the atmosphere. In still air, a "barrier layer" or "diffusion shells" lie between the leaf surface and the atmosphere.

In this segment of the transpiration steam, water movement is by diffusion, at least within and through the leaf. Fick's law (Equation 11) in a time-varying system is terribly messy. Let us make the physiologically reasonable assumption of a steady state, with one concentration (activity or fugacity) of water in the atmosphere outside the leaf (p_a) and another at the evaporating surface of the parenchyma cells (p_p). Now, instead of trying to calculate the tortuous path of diffusion of water vapour through the intercellular spaces, let us lump all the stomata and all the intercellular air spaces into two "apparent path lengths" of a diffusion system of unit are (Fig. 5.32).

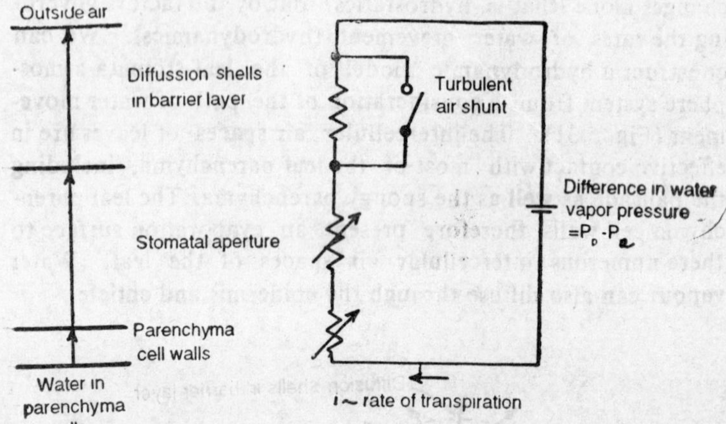

Fig. 5.32. Model for transpiration through stomata. The diagram on the left is a hydrodynamic model abstracted from Fig. 5.31. The effective path length through the intercellular spaces is negligible, whereas those associated with stomata or barrier layer are the controlling ones.

The corresponding electrical circuit is shown on the right. The stomatal resistance is variable and the resistance through the barrier layer may be effectively shunted in turbulent air.

Fick's equation (11) at steady state then reduces to Equation 27.

$$T = \frac{DA(p_a - p_p)}{L+S} \quad ...(27)$$

where T is the rate of transpiration, p_a and p_p are the partial pressures (or fugacities) of water in the atmosphere and at the evaporating surface, respectively, D is the diffusion constant for water in air, A is the cross-sectional area, and L and S are the apparent path lengths for diffusion in intercellular air spaces and through the stomata, respectively Equation 27 gives transpiration as a function of stomatal opening, but we must carry the analysis further before we can evaluate it. The apparent path length through the stomata, S, can be evaluated by a consideration of stomatal geometry (Fig. 5.33).

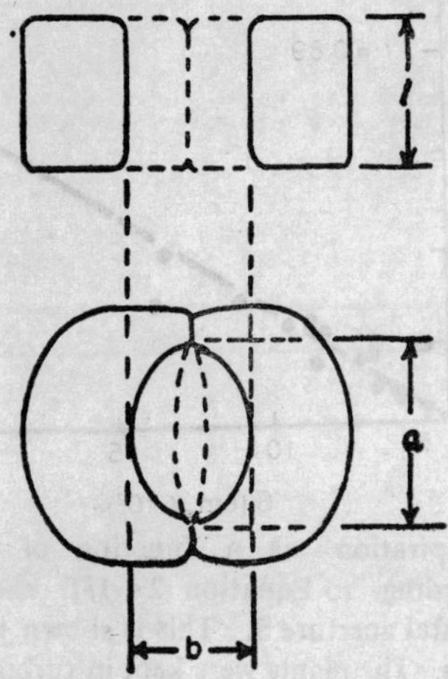

Fig. 5.33. Geometry of stomatal diffusion path. The outline of an open stomate appears in the solid line and the closed stomate in dotted line. To calculate the effective diffusion path, the distance from inside to outside is designated *l*, and the cross section is taken to be an ellipse with major and minor axes of *a* and *b*.

The fluid path length S can be approximated by the relation

$$S=\frac{l/\pi ab+l/2\sqrt{ab}}{n} \quad \ldots(28)$$

where *l*, *a*, and *b* are the depth, width, and length of the stomatal aperture, and *n* is the number of stomata per square centimeter.

Equation 27 can be inverted

$$\frac{1}{T} = \frac{L}{DA(p_a - p_p)} + \frac{S}{DA(p_a - p_p)} \quad ...(29)$$

which, for steady-state conditions, is a linear equation. L may now be evaluated by extrapolating 1/T for S=0. Zelitch and Waggoner have shown that a linear relation obtains for tobacco leaves (Fig, 5.34).

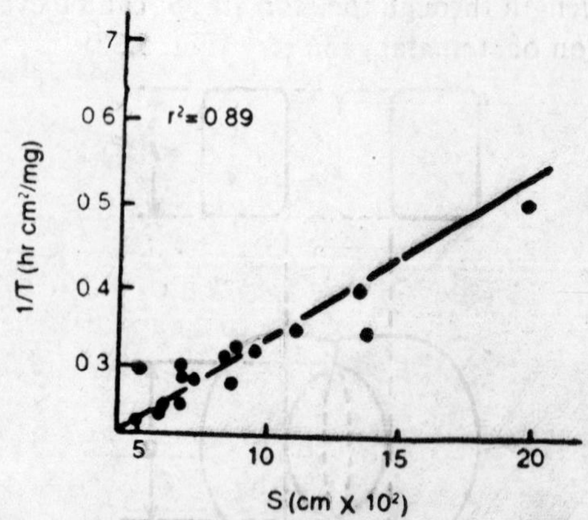

Fig. 5.34. Transpiration as a function of stomatal opening. According to Equation 29, 1/T should be linear with stomatal aperture S. This is shown to be so for tobacco leaves. The plants were kept in turbulent air, so that the value of L (*y* intercept) is very small.

We see in Fig. 5.34 that the reciprocal of transpiration 1/T, can be a linear function of S. We took L, the apparent path length for diffusion in air, to be constant. Just outside the stomatal pore there will be a layer of quiet air, the barrier layer which represents part of the apparent path length L. The diffusion of water vapour from the pore through the barrier layer is commonly identified by the term "diffusion shells" in reference to the surfaces representing equal concentration of water vapour (Fig. 5.31). Brown and Escomb (1900) long ago demonstrated that the diffusion of gases through a system of pores through a membrane is remarkably efficient, up to 40 per cent as great as diffusion from an open vessel of the smae area :

The thickness of the barrier layer is determined by the rate of movement of air outside the leaf. We can expect that at a sufficiently rapid rate of air movement, the component of L due to the barrier layer becomes neglible and L becomes equal to the diffusion path within the leaf. In fact with high air turbulence, L virtually disappears (Fig. 5.34). The apparent diffusion path of water vapour in air is therefore due almost exclusively to diffusion through the barrier layer.

Up to now we have treated the stomata kinetically, but what controls the opening and closing of these pores ? There is a general agreement that the stomata are controlled by the turgor of the guard cells. This can be demonstrated by placing a section of leaf in a hypertonic sucrose solution medium. But in nature these turgor changes are triggered by a bewildering variety of factors. If a generalization is possible, we may say that the guard cells are programmed to close in response to conditions that either are unfavourable to photosynthesis or might be indicative of impaired photosynthesis : low light intensity, high internal CO_2 concentration, low water supply, and poor nutrition. The detailed effects of a number of these factors are reviewed by Heath (1959).

The mechanism of the control of turgor changes is a subject of intense dispute. Attention has focused on the chloroplasts of the guard cells: functional guard cells always contain chloroplasts ; the action spectrum for stomatal opening corresponds to typical chloroplast absorption spectra (Fig. 5.35); glycollate, a typical product of photosynthesis, is unusually effective in promoting opening, and inhibitors of glycollate catabolism are effective in inhibiting stomatal opening. If, as seem likely, the closing of stomata is caused by a passive loss of water from cells, and opening is active (that is, metabolically dependent), we should expect that low temperature and inhibitors should inhibit stomatal opening but not closing. Zelitch has shown that this is indeed so.

The immediate mechanism of opening is almost certainly osmotic, and the most likely immediate cause is a large increase

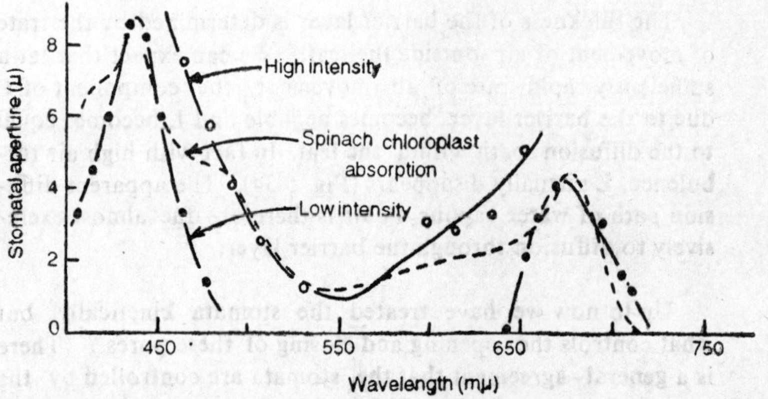

Fig. 5.35. Action spectrum for maintenance of stomatal openings. Strips we copeeled from the lower epidermis of *Senecio odoris* (Defl.) leaves in the light and floated on water. The stomata remained open in white light and closed in the dark. The action spectrum at two leaves of intensity (○, 1.9×10^{-9} and ●, 1.14×10^{-9} einstein cm^{-2} sec^{-1}) is compared with the absorption spectrum of spinach chloroplasts (dotted line).

in the potassium concentration in the guard cells. For the present one can only guess at the identity of the putative potassium pump and the means by which the various environmental factors control it.

Substances that combine with sulphydryl groups are strikingly effective in inhibiting opening. Metal-complexing agents also inhibit opening, but at certain concentrations, they inhibit closing as well (Fig. 5.36).

The fact that the phenylmercuri compounds when sprayed onto leaves penetrate only the epidermal layer provides an opportunity for a closer examination of stomatal physiology. After treatment with these substances, the stomata are "fixed" in various positions of closing for long periods of time with no direct effect on the parenchyma. Since different values of S can be obtained in this way, it becomes possible to test several deductions from the hydrodynamic model of gas movements through the stomata.

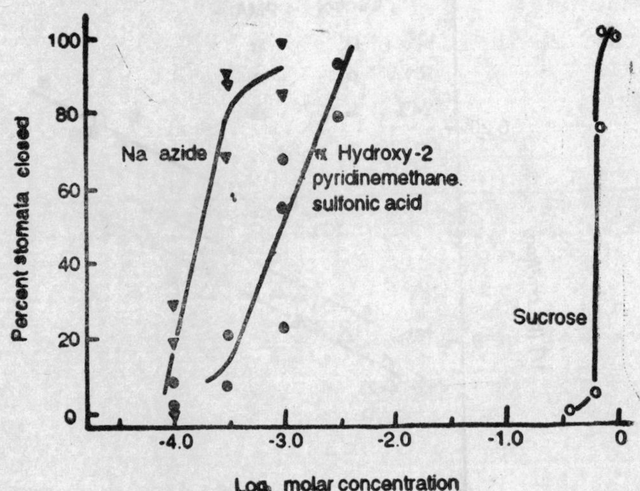

Fig. 5.36. Stomatal closure in different concentrations of metabolic inhibitors and sucrose. Stomata of tobacco leaf discs normally remain open when floating on water in the light. Additions of azide and inhibitors of glycollate oxidase, such as α-hydroxy-2-pyridinemethane sulphonic acid, cause rapid closure. That the effect of the inhibitors is not osmotic is shown by the much higher concentrations of sucrose required to cause closure.

One inference is that, since the principal ΔG change is across the leaf surface, resistance to water movement is through the stomata. If this were so, the rate of transpiration would be dependent only on S (and L) and would be independent of soil water potentials. (In the absence of the inhibitor, water stress will influence stomatal opening.) Shimshi (1963) has shown that, contrary to expectation (Fig. 5.37), soil moisture alters the 1/T intercept (but not the slope) of 1/T *vs.* S plots. Thus there is a significant resistance to water movement within the plant itself. This additional resistance, which we can calculate from the X intercepts at different water potentials, may be in the parenchyma cell walls.

CO_2 Movements through the Stomata

We have seen how the movements of water vapour through the leaf can be dealt with as a simple kinetic system. The

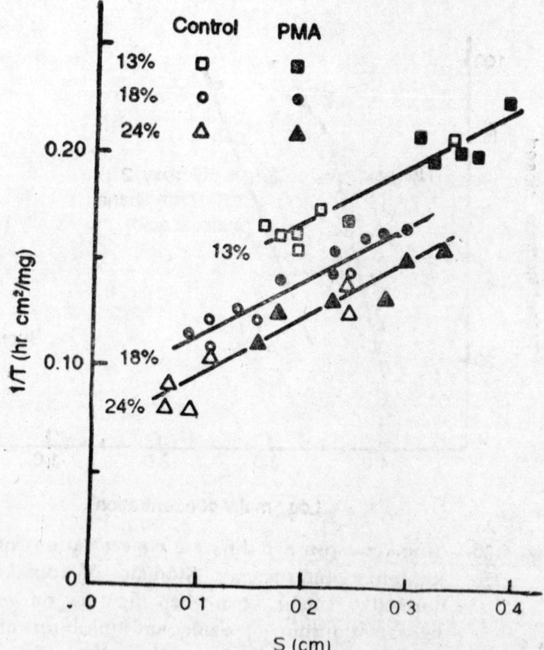

Fig 5 37. Transpiration as a function of S and soil moisture. Maize plants were grown in soil of different water contents. The value 1/T was in each case linear with S, but the 1/T intercepts were different for the different ψ-values. Shimshi infers that with increasing moisture stress, there is a corresponding increase in the resistance to evaporation from the parenchyma cell walls.

Phenylmercuric acetate (PMA) was employed to obtain large values of S.

movement of another gas, CO_2, also crucial to the life of the plant, should follow virtually the same scientific model.

We can modify Equation 29 in two ways and obtain a mathematical model which should predict the movement of CO_2. The diffusion constant for CO_2 in air is different from that of water vapour ; so that D becomss D′. Also CO_2 must pass through the water layer surrounding the cell and diffuse to the chloroplasts, the presumed centers of CO_2 absorption. This requires another apparent diffusion path length, M. The partial pressures p_a and p_p must be changed to correspond to the partial pressures of CO_2 in the air and at the chloroplasts, respectively. Even though this distance M may be very small, the diffusion constant through water is enormously less (1/2,500) than that through air ; consequently the M term may well be significant.

The resulting equation (30) now expresses photosynthesis p as a function of M, L, and S. It is of course essential in using this equation that we employ conditions under which the rate of photosynthesis is limited by the CO_2 concentration and moreover that photosynthesis is a linear function of the CO_2 concentration.

$$\frac{1}{p} = \frac{M}{D'_w A(p_a - p_c)} + \frac{L+S}{D'_a A(p_a - p_c)} \qquad \ldots(30)$$

If we eliminate L by the use of turbulent air, we can evaluate the M term. It is in fact a significant component of the resistance to CO_2 movement (Fig. 5.38).

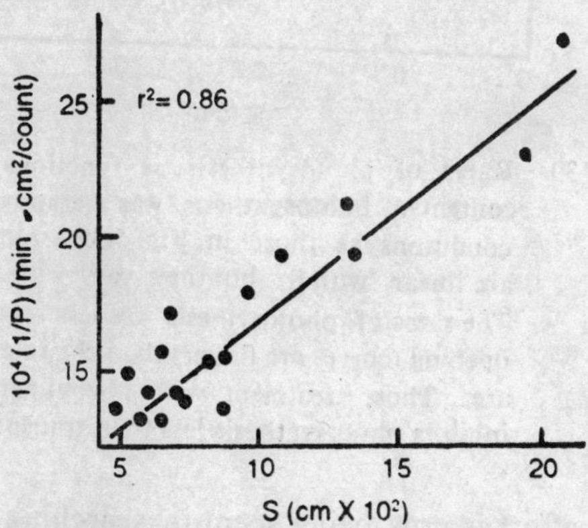

Fig. 5.38. Photosynthesis as a function of stomatal aperture. Photosynthesis in tobacco measured under the same conditions as Fig. 5.34.

One of the interesting conclusions from a comparison of the effects of stomatal aperture on transpiration and photosynthesis is that with moderate decreases in stomatal apertures, there is a disproportionately large decrease in transpiration (Fig. 5.39).

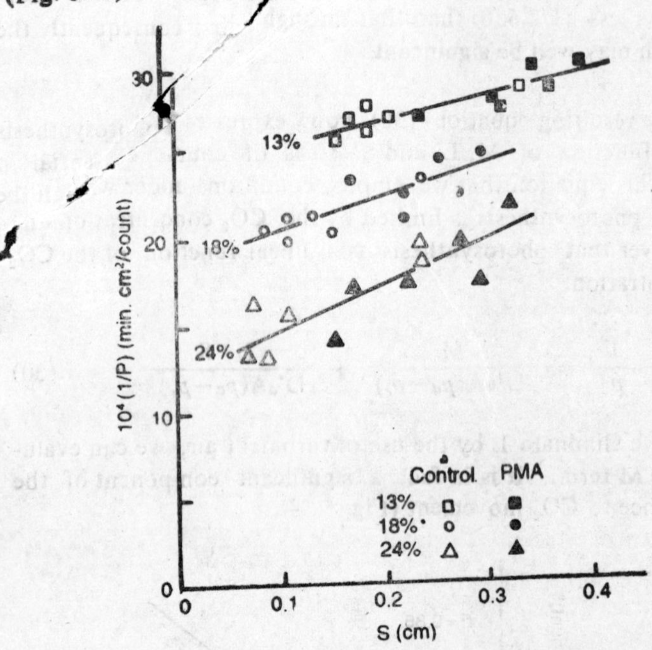

Fig. 5 39. **Rates of photosynthesis as functions of S and water contents. Photosynthesis was measured under the same conditions as those in Fig. 5.39. Again the 1/P values are linear with S, but they vary with water potential. The rates of photosythesis are less sensitive to stomatal opening (curves are flatter) than the rates of transpiration are. Thus, treatment with phenylmercuric compounds inhibits photosynthesis less than transpiration.**

Much effort is now being spent in searching out the nature of stomatal control. Attempts are also being made to develop artificial cuticles with an eye toward rational measures of water conservation in arid land agriculture.

●●